Lecture Notes in Computer Science

Lecture Notes in Artificial Intelligence 14648

Founding Editor

Jörg Siekmann

The series Lecture Notes in Artificial Intelligence (LNAI) was established in 1988 as a topical subseries of LNCS devoted to artificial intelligence.

The series publishes state-of-the-art research results at a high level. As with the LNCS mother series, the mission of the series is to serve the international R & D community by providing an invaluable service, mainly focused on the publication of conference and workshop proceedings and postproceedings.

De-Nian Yang · Xing Xie · Vincent S. Tseng ·
Jian Pei · Jen-Wei Huang · Jerry Chun-Wei Lin
Editors

Advances in Knowledge Discovery and Data Mining

28th Pacific-Asia Conference
on Knowledge Discovery and Data Mining, PAKDD 2024
Taipei, Taiwan, May 7–10, 2024
Proceedings, Part IV

Editors
De-Nian Yang
Academia Sinica
Taipei, Taiwan

Vincent S. Tseng
National Yang Ming Chiao Tung University
Hsinchu, Taiwan

Jen-Wei Huang
National Cheng Kung University
Tainan, Taiwan

Xing Xie
Microsoft Research Asia
Beijing, China

Jian Pei
Duke University
Durham, NC, USA

Jerry Chun-Wei Lin
Silesian University of Technology
Gliwice, Poland

ISSN 0302-9743 ISSN 1611-3349 (electronic)
Lecture Notes in Artificial Intelligence
ISBN 978-981-97-2240-2 ISBN 978-981-97-2238-9 (eBook)
https://doi.org/10.1007/978-981-97-2238-9

LNCS Sublibrary: SL7 – Artificial Intelligence

This Springer imprint is published by the registered company Springer Nature Singapore Pte Ltd.
The registered company address is: 152 Beach Road, #21-01/04 Gateway East, Singapore 189721, Singapore

Paper in this product is recyclable.

General Chairs' Preface

On behalf of the Organizing Committee, we were delighted to welcome attendees to the 28th Pacific-Asia Conference on Knowledge Discovery and Data Mining (PAKDD 2024). Since its inception in 1997, PAKDD has long established itself as one of the leading international conferences on data mining and knowledge discovery. PAKDD provides an international forum for researchers and industry practitioners to share their new ideas, original research results, and practical development experiences across all areas of Knowledge Discovery and Data Mining (KDD). This year, after its two previous editions in Taipei (2002) and Tainan (2014), PAKDD was held in Taiwan for the third time in the fascinating city of Taipei, during May 7–10, 2024. Moreover, PAKDD 2024 was held as a fully physical conference since the COVID-19 pandemic was contained.

We extend our sincere gratitude to the researchers who submitted their work to the PAKDD 2024 main conference, high-quality tutorials, and workshops on cutting-edge topics. The conference program was further enriched with seven high-quality tutorials and five workshops on cutting-edge topics. We would like to deliver our sincere thanks for their efforts in research, as well as in preparing high-quality presentations. We also express our appreciation to all the collaborators and sponsors for their trust and cooperation. We were honored to have three distinguished keynote speakers joining the conference: Ed H. Chi (Google DeepMind), Vipin Kumar (University of Minnesota), and Huan Liu (Arizona State University), each with high reputations in their respective areas. We enjoyed their participation and talks, which made the conference one of the best academic platforms for knowledge discovery and data mining. We would like to express our sincere gratitude for the contributions of the Steering Committee members, Organizing Committee members, Program Committee members, and anonymous reviewers, led by Program Committee Chairs De-Nian Yang and Xing Xie. It is through their untiring efforts that the conference had an excellent technical program. We are also thankful to the other Organizing Committee members: Workshop Chairs, Chuan-Kang Ting and Xiaoli Li; Tutorial Chairs, Jiun-Long Huang and Philippe Fournier-Viger; Publicity Chairs, Mi-Yen Yeh and Rage Uday Kiran; Industrial Chairs, Kun-Ta Chuang, Wei-Chao Chen and Richie Tsai; Proceedings Chairs, Jen-Wei Huang and Jerry Chun-Wei Lin; Registration Chairs, Chih-Ya Shen and Hong-Han Shuai; Web and Content Chairs, Cheng-Te Li and Shan-Hung Wu; Local Arrangement Chairs, Yi-Ling Chen, Kuan-Ting Lai, Yi-Ting Chen, and Ya-Wen Teng. We feel indebted to the PAKDD Steering Committee for their constant guidance and sponsorship of manuscripts. We are also grateful to the hosting organizations, National Yang Ming Chiao Tung University and Academia Sinica, and all our sponsors for continuously providing institutional and financial support to PAKDD 2024.

May 2024

Vincent S. Tseng
Jian Pei

PC Chairs' Preface

It is our great pleasure to present the 28th Pacific-Asia Conference on Knowledge Discovery and Data Mining (PAKDD 2024) as Program Committee Chairs. PAKDD is one of the longest-established and leading international conferences in the areas of data mining and knowledge discovery. It provides an international forum for researchers and industry practitioners to share their new ideas, original research results, and practical development experiences in all KDD-related areas, including data mining, data warehousing, machine learning, artificial intelligence, databases, statistics, knowledge engineering, big data technologies, and foundations.

This year, PAKDD received a record number of 720 submissions, among which 86 submissions were rejected at a preliminary stage due to policy violations. There were 595 Program Committee members and 101 Senior Program Committee members involved in the double-blind reviewing process. For submissions entering the double-blind review process, each one received at least three quality reviews from PC members. Furthermore, each valid submission received one meta-review from the assigned SPC member, who also led the discussion with the PC members. The PC Co-chairs then considered the recommendations and meta-reviews from SPC members and looked into each submission as well as its reviews and PC discussions to make the final decision.

As a result of the highly competitive selection process, 175 submissions were accepted and recommended to be published, with 133 oral-presentation papers and 42 poster-presentation papers. We would like to thank all SPC and PC members whose diligence produced a high-quality program for PAKDD 2024. The conference program also featured three keynote speeches from distinguished data mining researchers, eight invited industrial talks, five cutting-edge workshops, and seven comprehensive tutorials.

We wish to sincerely thank all SPC members, PC members, and external reviewers for their invaluable efforts in ensuring a timely, fair, and highly effective paper review and selection procedure. We hope that readers of the proceedings will find the PAKDD 2024 technical program both interesting and rewarding.

May 2024

De-Nian Yang
Xing Xie

Organization

Organizing Committee

Honorary Chairs

Philip S. Yu	University of Illinois at Chicago, USA
Ming-Syan Chen	National Taiwan University, Taiwan

General Chairs

Vincent S. Tseng	National Yang Ming Chiao Tung University, Taiwan
Jian Pei	Duke University, USA

Program Committee Chairs

De-Nian Yang	Academia Sinica, Taiwan
Xing Xie	Microsoft Research Asia, China

Workshop Chairs

Chuan-Kang Ting	National Tsing Hua University, Taiwan
Xiaoli Li	A*STAR, Singapore

Tutorial Chairs

Jiun-Long Huang	National Yang Ming Chiao Tung University, Taiwan
Philippe Fournier-Viger	Shenzhen University, China

Publicity Chairs

Mi-Yen Yeh	Academia Sinica, Taiwan
Rage Uday Kiran	University of Aizu, Japan

Industrial Chairs

Kun-Ta Chuang	National Cheng Kung University, Taiwan
Wei-Chao Chen	Inventec Corp./Skywatch Innovation, Taiwan
Richie Tsai	Taiwan AI Academy, Taiwan

Proceedings Chairs

Jen-Wei Huang	National Cheng Kung University, Taiwan
Jerry Chun-Wei Lin	Silesian University of Technology, Poland

Registration Chairs

Chih-Ya Shen	National Tsing Hua University, Taiwan
Hong-Han Shuai	National Yang Ming Chiao Tung University, Taiwan

Web and Content Chairs

Shan-Hung Wu	National Tsing Hua University, Taiwan
Cheng-Te Li	National Cheng Kung University, Taiwan

Local Arrangement Chairs

Yi-Ling Chen	National Taiwan University of Science and Technology, Taiwan
Kuan-Ting Lai	National Taipei University of Technology, Taiwan
Yi-Ting Chen	National Yang Ming Chiao Tung University, Taiwan
Ya-Wen Teng	Academia Sinica, Taiwan

Steering Committee

Chair

Longbing Cao	Macquarie University, Australia

Vice Chair

Gill Dobbie	University of Auckland, New Zealand

Treasurer

Longbing Cao	Macquarie University, Australia

Members

Ramesh Agrawal	Jawaharlal Nehru University, India
Gill Dobbie	University of Auckland, New Zealand
João Gama	University of Porto, Portugal
Zhiguo Gong	University of Macau, Macau SAR
Hisashi Kashima	Kyoto University, Japan
Hady W. Lauw	Singapore Management University, Singapore
Jae-Gil Lee	KAIST, Korea
Dinh Phung	Monash University, Australia
Kyuseok Shim	Seoul National University, Korea
Geoff Webb	Monash University, Australia
Raymond Chi-Wing Wong	Hong Kong University of Science and Technology, Hong Kong SAR
Min-Ling Zhang	Southeast University, China

Life Members

Longbing Cao	Macquarie University, Australia
Ming-Syan Chen	National Taiwan University, Taiwan
David Cheung	University of Hong Kong, China
Joshua Z. Huang	Chinese Academy of Sciences, China
Masaru Kitsuregawa	Tokyo University, Japan
Rao Kotagiri	University of Melbourne, Australia
Ee-Peng Lim	Singapore Management University, Singapore
Huan Liu	Arizona State University, USA
Hiroshi Motoda	AFOSR/AOARD and Osaka University, Japan
Jian Pei	Duke University, USA
P. Krishna Reddy	IIIT Hyderabad, India
Jaideep Srivastava	University of Minnesota, USA
Thanaruk Theeramunkong	Thammasat University, Thailand
Tu-Bao Ho	JAIST, Japan
Vincent S. Tseng	National Yang Ming Chiao Tung University, Taiwan
Takashi Washio	Osaka University, Japan
Kyu-Young Whang	KAIST, Korea
Graham Williams	Australian National University, Australia
Chengqi Zhang	University of Technology Sydney, Australia

Ning Zhong	Maebashi Institute of Technology, Japan
Zhi-Hua Zhou	Nanjing University, China

Past Members

Arbee L. P. Chen	Asia University, Taiwan
Hongjun Lu	Hong Kong University of Science and Technology, Hong Kong SAR
Takao Terano	Tokyo Institute of Technology, Japan

Senior Program Committee

Aijun An	York University, Canada
Aris Anagnostopoulos	Sapienza Università di Roma, Italy
Ting Bai	Beijing University of Posts and Telecommunications, China
Elisa Bertino	Purdue University, USA
Arnab Bhattacharya	IIT Kanpur, India
Albert Bifet	Université Paris-Saclay, France
Ludovico Boratto	Università degli Studi di Cagliari, Italy
Ricardo Campello	University of Southern Denmark, Denmark
Longbing Cao	University of Technology Sydney, Australia
Tru Cao	UTHealth, USA
Tanmoy Chakraborty	IIT Delhi, India
Jeffrey Chan	RMIT University, Australia
Pin-Yu Chen	IBM T. J. Watson Research Center, USA
Bin Cui	Peking University, China
Anirban Dasgupta	IIT Gandhinagar, India
Wei Ding	University of Massachusetts Boston, USA
Eibe Frank	University of Waikato, New Zealand
Chen Gong	Nanjing University of Science and Technology, China
Jingrui He	UIUC, USA
Tzung-Pei Hong	National University of Kaohsiung, Taiwan
Qinghua Hu	Tianjin University, China
Hong Huang	Huazhong University of Science and Technology, China
Jen-Wei Huang	National Cheng Kung University, Taiwan
Tsuyoshi Ide	IBM T. J. Watson Research Center, USA
Xiaowei Jia	University of Pittsburgh, USA
Zhe Jiang	University of Florida, USA

Toshihiro Kamishima	National Institute of Advanced Industrial Science and Technology, Japan
Murat Kantarcioglu	University of Texas at Dallas, USA
Hung-Yu Kao	National Cheng Kung University, Taiwan
Kamalakar Karlapalem	IIIT Hyderabad, India
Anuj Karpatne	Virginia Tech, USA
Hisashi Kashima	Kyoto University, Japan
Sang-Wook Kim	Hanyang University, Korea
Yun Sing Koh	University of Auckland, New Zealand
Hady Lauw	Singapore Management University, Singapore
Byung Suk Lee	University of Vermont, USA
Jae-Gil Lee	KAIST, Korea
Wang-Chien Lee	Pennsylvania State University, USA
Chaozhuo Li	Microsoft Research Asia, China
Gang Li	Deakin University, Australia
Jiuyong Li	University of South Australia, Australia
Jundong Li	University of Virginia, USA
Ming Li	Nanjing University, China
Sheng Li	University of Virginia, USA
Ying Li	AwanTunai, Singapore
Yu-Feng Li	Nanjing University, China
Hao Liao	Shenzhen University, China
Ee-peng Lim	Singapore Management University, Singapore
Jerry Chun-Wei Lin	Silesian University of Technology, Poland
Shou-De Lin	National Taiwan University, Taiwan
Hongyan Liu	Tsinghua University, China
Wei Liu	University of Technology Sydney, Australia
Chang-Tien Lu	Virginia Tech, USA
Yuan Luo	Northwestern University, USA
Wagner Meira Jr.	UFMG, Brazil
Alexandros Ntoulas	University of Athens, Greece
Satoshi Oyama	Nagoya City University, Japan
Guansong Pang	Singapore Management University, Singapore
Panagiotis Papapetrou	Stockholm University, Sweden
Wen-Chih Peng	National Yang Ming Chiao Tung University, Taiwan
Dzung Phan	IBM T. J. Watson Research Center, USA
Uday Rage	University of Aizu, Japan
Rajeev Raman	University of Leicester, UK
P. Krishna Reddy	IIIT Hyderabad, India
Thomas Seidl	LMU München, Germany
Neil Shah	Snap Inc., USA

Yingxia Shao	Beijing University of Posts and Telecommunications, China
Victor S. Sheng	Texas Tech University, USA
Kyuseok Shim	Seoul National University, Korea
Arlei Silva	Rice University, USA
Jaideep Srivastava	University of Minnesota, USA
Masashi Sugiyama	RIKEN/University of Tokyo, Japan
Ju Sun	University of Minnesota, USA
Jiliang Tang	Michigan State University, USA
Hanghang Tong	UIUC, USA
Ranga Raju Vatsavai	North Carolina State University, USA
Hao Wang	Nanyang Technological University, Singapore
Hao Wang	Xidian University, China
Jianyong Wang	Tsinghua University, China
Tim Weninger	University of Notre Dame, USA
Raymond Chi-Wing Wong	Hong Kong University of Science and Technology, Hong Kong SAR
Jia Wu	Macquarie University, Australia
Xindong Wu	Hefei University of Technology, China
Xintao Wu	University of Arkansas, USA
Yiqun Xie	University of Maryland, USA
Yue Xu	Queensland University of Technology, Australia
Lina Yao	University of New South Wales, Australia
Han-Jia Ye	Nanjing University, China
Mi-Yen Yeh	Academia Sinica, Taiwan
Hongzhi Yin	University of Queensland, Australia
Min-Ling Zhang	Southeast University, China
Ping Zhang	Ohio State University, USA
Zhao Zhang	Hefei University of Technology, China
Zhongfei Zhang	Binghamton University, USA
Xiangyu Zhao	City University of Hong Kong, Hong Kong SAR
Yanchang Zhao	CSIRO, Australia
Jiayu Zhou	Michigan State University, USA
Xiao Zhou	Renmin University of China, China
Xiaofang Zhou	Hong Kong University of Science and Technology, Hong Kong SAR
Feida Zhu	Singapore Management University, Singapore
Fuzhen Zhuang	Beihang University, China

Program Committee

Zubin Abraham	Robert Bosch, USA
Pedro Henriques Abreu	CISUC, Portugal
Muhammad Abulaish	South Asian University, India
Bijaya Adhikari	University of Iowa, USA
Karan Aggarwal	Amazon, USA
Chowdhury Farhan Ahmed	University of Dhaka, Bangladesh
Ulrich Aïvodji	ÉTS Montréal, Canada
Esra Akbas	Georgia State University, USA
Shafiq Alam	Massey University Auckland, New Zealand
Giuseppe Albi	Università degli Studi di Pavia, Italy
David Anastasiu	Santa Clara University, USA
Xiang Ao	Chinese Academy of Sciences, China
Elena-Simona Apostol	Uppsala University, Sweden
Sunil Aryal	Deakin University, Australia
Jees Augustine	Microsoft, USA
Konstantin Avrachenkov	Inria, France
Goonmeet Bajaj	Ohio State University, USA
Jean Paul Barddal	PUCPR, Brazil
Srikanta Bedathur	IIT Delhi, India
Sadok Ben Yahia	University of Southern Denmark, Denmark
Alessandro Berti	Università di Pisa, Italy
Siddhartha Bhattacharyya	University of Illinois at Chicago, USA
Ranran Bian	University of Sydney, Australia
Song Bian	Chinese University of Hong Kong, Hong Kong SAR
Giovanni Maria Biancofiore	Politecnico di Bari, Italy
Fernando Bobillo	University of Zaragoza, Spain
Adrian M. P. Brasoveanu	Modul Technology GmbH, Austria
Krisztian Buza	Budapest University of Technology and Economics, Hungary
Luca Cagliero	Politecnico di Torino, Italy
Jean-Paul Calbimonte	University of Applied Sciences and Arts Western Switzerland, Switzerland
K. Selçuk Candan	Arizona State University, USA
Fuyuan Cao	Shanxi University, China
Huiping Cao	New Mexico State University, USA
Jian Cao	Shanghai Jiao Tong University, China
Yan Cao	University of Texas at Dallas, USA
Yang Cao	Hokkaido University, Japan
Yuanjiang Cao	Macquarie University, Australia

Sharma Chakravarthy	University of Texas at Arlington, USA
Harry Kai-Ho Chan	University of Sheffield, UK
Zhangming Chan	Alibaba Group, China
Snigdhansu Chatterjee	University of Minnesota, USA
Mandar Chaudhary	eBay, USA
Chen Chen	University of Virginia, USA
Chun-Hao Chen	National Kaohsiung University of Science and Technology, Taiwan
Enhong Chen	University of Science and Technology of China, China
Fanglan Chen	Virginia Tech, USA
Feng Chen	University of Texas at Dallas, USA
Hongyang Chen	Zhejiang Lab, China
Jia Chen	University of California Riverside, USA
Jinjun Chen	Swinburne University of Technology, Australia
Lingwei Chen	Wright State University, USA
Ping Chen	University of Massachusetts Boston, USA
Shang-Tse Chen	National Taiwan University, Taiwan
Shengyu Chen	University of Pittsburgh, USA
Songcan Chen	Nanjing University of Aeronautics and Astronautics, China
Tao Chen	China University of Geosciences, China
Tianwen Chen	Hong Kong University of Science and Technology, Hong Kong SAR
Tong Chen	University of Queensland, Australia
Weitong Chen	University of Adelaide, Australia
Yi-Hui Chen	Chang Gung University, Taiwan
Yile Chen	Nanyang Technological University, Singapore
Yi-Ling Chen	National Taiwan University of Science and Technology, Taiwan
Yi-Shin Chen	National Tsing Hua University, Taiwan
Yi-Ting Chen	National Yang Ming Chiao Tung University, Taiwan
Zheng Chen	Osaka University, Japan
Zhengzhang Chen	NEC Laboratories America, USA
Zhiyuan Chen	UMBC, USA
Zhong Chen	Southern Illinois University, USA
Peng Cheng	East China Normal University, China
Abdelghani Chibani	Université Paris-Est Créteil, France
Jingyuan Chou	University of Virginia, USA
Lingyang Chu	McMaster University, Canada
Kun-Ta Chuang	National Cheng Kung University, Taiwan

Robert Churchill	Georgetown University, USA
Chaoran Cui	Shandong University of Finance and Economics, China
Alfredo Cuzzocrea	Università della Calabria, Italy
Bi-Ru Dai	National Taiwan University of Science and Technology, Taiwan
Honghua Dai	Zhengzhou University, China
Claudia d'Amato	University of Bari, Italy
Chuangyin Dang	City University of Hong Kong, China
Mrinal Das	IIT Palakkad, India
Debanjan Datta	Virginia Tech, USA
Cyril de Runz	Université de Tours, France
Jeremiah Deng	University of Otago, New Zealand
Ke Deng	RMIT University, Australia
Zhaohong Deng	Jiangnan University, China
Anne Denton	North Dakota State University, USA
Shridhar Devamane	KLE Institute of Technology, India
Djellel Difallah	New York University, USA
Ling Ding	Tianjin University, China
Shifei Ding	China University of Mining and Technology, China
Yao-Xiang Ding	Zhejiang University, China
Yifan Ding	University of Notre Dame, USA
Ying Ding	University of Texas at Austin, USA
Lamine Diop	EPITA, France
Nemanja Djuric	Aurora Innovation, USA
Gillian Dobbie	University of Auckland, New Zealand
Josep Domingo-Ferrer	Universitat Rovira i Virgili, Spain
Bo Dong	Amazon, USA
Yushun Dong	University of Virginia, USA
Bo Du	Wuhan University, China
Silin Du	Tsinghua University, China
Jiuding Duan	Allianz Global Investors, Japan
Lei Duan	Sichuan University, China
Walid Durani	LMU München, Germany
Sourav Dutta	Huawei Research Centre, Ireland
Mohamad El-Hajj	MacEwan University, Canada
Ya Ju Fan	Lawrence Livermore National Laboratory, USA
Zipei Fan	Jilin University, China
Majid Farhadloo	University of Minnesota, USA
Fabio Fassetti	Università della Calabria, Italy
Zhiquan Feng	National Cheng Kung University, Taiwan

Len Feremans	Universiteit Antwerpen, Belgium
Edouard Fouché	Karlsruher Institut für Technologie, Germany
Dongqi Fu	UIUC, USA
Yanjie Fu	University of Central Florida, USA
Ken-ichi Fukui	Osaka University, Japan
Matjaž Gams	Jožef Stefan Institute, Slovenia
Amir Gandomi	University of Technology Sydney, Australia
Aryya Gangopadhyay	UMBC, USA
Dashan Gao	Hong Kong University of Science and Technology, China
Wei Gao	Nanjing University, China
Yifeng Gao	University of Texas Rio Grande Valley, USA
Yunjun Gao	Zhejiang University, China
Paolo Garza	Politecnico di Torino, Italy
Chang Ge	University of Minnesota, USA
Xin Geng	Southeast University, China
Flavio Giobergia	Politecnico di Torino, Italy
Rosalba Giugno	Università degli Studi di Verona, Italy
Aris Gkoulalas-Divanis	Merative, USA
Djordje Gligorijevic	Temple University, USA
Daniela Godoy	UNICEN, Argentina
Heitor Gomes	Victoria University of Wellington, New Zealand
Maciej Grzenda	Warsaw University of Technology, Poland
Lei Gu	Nanjing University of Posts and Telecommunications, China
Yong Guan	Iowa State University, USA
Riccardo Guidotti	Università di Pisa, Italy
Ekta Gujral	University of California Riverside, USA
Guimu Guo	Rowan University, USA
Ting Guo	University of Technology Sydney, Australia
Xingzhi Guo	Stony Brook University, USA
Ch. Md. Rakin Haider	Purdue University, USA
Benjamin Halstead	University of Auckland, New Zealand
Jinkun Han	Georgia State University, USA
Lu Han	Nanjing University, China
Yufei Han	Inria, France
Daisuke Hatano	RIKEN, Japan
Kohei Hatano	Kyushu University/RIKEN AIP, Japan
Shogo Hayashi	BizReach, Japan
Erhu He	University of Pittsburgh, USA
Guoliang He	Wuhan University, China
Pengfei He	Michigan State University, USA

Yi He	Old Dominion University, USA
Shen-Shyang Ho	Rowan University, USA
William Hsu	Kansas State University, USA
Haoji Hu	University of Minnesota, USA
Hongsheng Hu	CSIRO, Australia
Liang Hu	Tongji University, China
Shizhe Hu	Zhengzhou University, China
Wei Hu	Nanjing University, China
Mengdi Huai	Iowa State University, USA
Chao Huang	University of Hong Kong, Hong Kong SAR
Congrui Huang	Microsoft, China
Guangyan Huang	Deakin University, Australia
Jimmy Huang	York University, Canada
Jinbin Huang	Hong Kong Baptist University, Hong Kong SAR
Kai Huang	Hong Kong University of Science and Technology, China
Ling Huang	South China Agricultural University, China
Ting-Ji Huang	Nanjing University, China
Xin Huang	Hong Kong Baptist University, Hong Kong SAR
Zhenya Huang	University of Science and Technology of China, China
Chih-Chieh Hung	National Chung Hsing University, Taiwan
Hui-Ju Hung	Pennsylvania State University, USA
Nam Huynh	JAIST, Japan
Akihiro Inokuchi	Kwansei Gakuin University, Japan
Atsushi Inoue	Eastern Washington University, USA
Nevo Itzhak	Ben-Gurion University, Israel
Tomoya Iwakura	Fujitsu Laboratories Ltd., Japan
Divyesh Jadav	IBM T. J. Watson Research Center, USA
Shubham Jain	Visa Research, USA
Bijay Prasad Jaysawal	National Cheng Kung University, Taiwan
Kishlay Jha	University of Iowa, USA
Taoran Ji	Texas A&M University - Corpus Christi, USA
Songlei Jian	NUDT, China
Gaoxia Jiang	Shanxi University, China
Hansi Jiang	SAS Institute Inc., USA
Jiaxin Jiang	National University of Singapore, Singapore
Min Jiang	Xiamen University, China
Renhe Jiang	University of Tokyo, Japan
Yuli Jiang	Chinese University of Hong Kong, Hong Kong SAR
Bo Jin	Dalian University of Technology, China

Ming Jin	Monash University, Australia
Ruoming Jin	Kent State University, USA
Wei Jin	University of North Texas, USA
Mingxuan Ju	University of Notre Dame, USA
Wei Ju	Peking University, China
Vana Kalogeraki	Athens University of Economics and Business, Greece
Bo Kang	Ghent University, Belgium
Jian Kang	University of Rochester, USA
Ashwin Viswanathan Kannan	Amazon, USA
Tomi Kauppinen	Aalto University School of Science, Finland
Jungeun Kim	Kongju National University, Korea
Kyoung-Sook Kim	National Institute of Advanced Industrial Science and Technology, Japan
Primož Kocbek	University of Maribor, Slovenia
Aritra Konar	Katholieke Universiteit Leuven, Belgium
Youyong Kong	Southeast University, China
Olivera Kotevska	Oak Ridge National Laboratory, USA
P. Radha Krishna	NIT Warangal, India
Adit Krishnan	UIUC, USA
Gokul Krishnan	IIT Madras, India
Peer Kröger	CAU, Germany
Marzena Kryszkiewicz	Warsaw University of Technology, Poland
Chuan-Wei Kuo	National Yang Ming Chiao Tung University, Taiwan
Kuan-Ting Lai	National Taipei University of Technology, Taiwan
Long Lan	NUDT, China
Duc-Trong Le	Vietnam National University, Vietnam
Tuan Le	New Mexico State University, USA
Chul-Ho Lee	Texas State University, USA
Ickjai Lee	James Cook University, Australia
Ki Yong Lee	Sookmyung Women's University, Korea
Ki-Hoon Lee	Kwangwoon University, Korea
Roy Ka-Wei Lee	Singapore University of Technology and Design, Singapore
Yue-Shi Lee	Ming Chuan University, Taiwan
Dino Lenco	INRAE, France
Carson Leung	University of Manitoba, Canada
Boyu Li	University of Technology Sydney, Australia
Chaojie Li	University of New South Wales, Australia
Cheng-Te Li	National Cheng Kung University, Taiwan
Chongshou Li	Southwest Jiaotong University, China

Fengxin Li	Renmin University of China, China
Guozhong Li	King Abdullah University of Science and Technology, Saudi Arabia
Huaxiong Li	Nanjing University, China
Jianxin Li	Beihang University, China
Lei Li	Hong Kong University of Science and Technology (Guangzhou), China
Peipei Li	Hefei University of Technology, China
Qian Li	Curtin University, Australia
Rong-Hua Li	Beijing Institute of Technology, China
Shao-Yuan Li	Nanjing University of Aeronautics and Astronautics, China
Shuai Li	Cambridge University, UK
Shuang Li	Beijing Institute of Technology, China
Tianrui Li	Southwest Jiaotong University, China
Wengen Li	Tongji University, China
Wentao Li	Hong Kong University of Science and Technology (Guangzhou), China
Xin-Ye Li	Bytedance, China
Xiucheng Li	Harbin Institute of Technology, China
Xuelong Li	Northwestern Polytechnical University, China
Yidong Li	Beijing Jiaotong University, China
Yinxiao Li	Meta Platforms, USA
Yuefeng Li	Queensland University of Technology, Australia
Yun Li	Nanjing University of Posts and Telecommunications, China
Panagiotis Liakos	University of Athens, Greece
Xiang Lian	Kent State University, USA
Shen Liang	Université Paris Cité, France
Qing Liao	Harbin Institute of Technology (Shenzhen), China
Sungsu Lim	Chungnam National University, Korea
Dandan Lin	Shenzhen Institute of Computing Sciences, China
Yijun Lin	University of Minnesota, USA
Ying-Jia Lin	National Cheng Kung University, Taiwan
Baodi Liu	China University of Petroleum (East China), China
Chien-Liang Liu	National Yang Ming Chiao Tung University, Taiwan
Guiquan Liu	University of Science and Technology of China, China
Jin Liu	Shanghai Maritime University, China
Jinfei Liu	Emory University, USA
Kunpeng Liu	Portland State University, USA

Ning Liu	Shandong University, China
Qi Liu	University of Science and Technology of China, China
Qing Liu	Zhejiang University, China
Qun Liu	Louisiana State University, USA
Shenghua Liu	Chinese Academy of Sciences, China
Weifeng Liu	China University of Petroleum (East China), China
Yang Liu	Wilfrid Laurier University, Canada
Yao Liu	University of New South Wales, Australia
Yixin Liu	Monash University, Australia
Zheng Liu	Nanjing University of Posts and Telecommunications, China
Cheng Long	Nanyang Technological University, Singapore
Haibing Lu	Santa Clara University, USA
Wenpeng Lu	Qilu University of Technology, China
Simone Ludwig	North Dakota State University, USA
Dongsheng Luo	Florida International University, USA
Ping Luo	Chinese Academy of Sciences, China
Wei Luo	Deakin University, Australia
Xiao Luo	UCLA, USA
Xin Luo	Shandong University, China
Yong Luo	Wuhan University, China
Fenglong Ma	Pennsylvania State University, USA
Huifang Ma	Northwest Normal University, China
Jing Ma	Hong Kong Baptist University, Hong Kong SAR
Qianli Ma	South China University of Technology, China
Yi-Fan Ma	Nanjing University, China
Rich Maclin	University of Minnesota, USA
Son Mai	Queen's University Belfast, UK
Arun Maiya	Institute for Defense Analyses, USA
Bradley Malin	Vanderbilt University Medical Center, USA
Giuseppe Manco	Consiglio Nazionale delle Ricerche, Italy
Naresh Manwani	IIIT Hyderabad, India
Francesco Marcelloni	Università di Pisa, Italy
Leandro Marinho	UFCG, Brazil
Koji Maruhashi	Fujitsu Laboratories Ltd., Japan
Florent Masseglia	Inria, France
Mohammad Masud	United Arab Emirates University, United Arab Emirates
Sarah Masud	IIIT Delhi, India
Costas Mavromatis	University of Minnesota, USA

Maxwell McNeil	University at Albany SUNY, USA
Massimo Melucci	Università degli Studi di Padova, Italy
Alex Memory	Johns Hopkins University, USA
Ernestina Menasalvas	Universidad Politécnica de Madrid, Spain
Xupeng Miao	Carnegie Mellon University, USA
Matej Miheli	University of Zagreb, Croatia
Fan Min	Southwest Petroleum University, China
Jun-Ki Min	Korea University of Technology and Education, Korea
Tsunenori Mine	Kyushu University, Japan
Nguyen Le Minh	JAIST, Japan
Shuichi Miyazawa	Graduate University for Advanced Studies, Japan
Songsong Mo	Nanyang Technological University, Singapore
Jacob Montiel	Amazon, USA
Yang-Sae Moon	Kangwon National University, Korea
Sebastian Moreno	Universidad Adolfo Ibáñez, Chile
Daisuke Moriwaki	CyberAgent, Inc., Japan
Tsuyoshi Murata	Tokyo Institute of Technology, Japan
Charini Nanayakkara	Australian National University, Australia
Mirco Nanni	Consiglio Nazionale delle Ricerche, Italy
Wilfred Ng	Hong Kong University of Science and Technology, Hong Kong SAR
Cam-Tu Nguyen	Nanjing University, China
Canh Hao Nguyen	Kyoto University, Japan
Hoang Long Nguyen	Meharry Medical College, USA
Shiwen Ni	Chinese Academy of Sciences, China
Jian-Yun Nie	Université de Montréal, Canada
Tadashi Nomoto	National Institute of Japanese Literature, Japan
Tim Oates	UMBC, USA
Eduardo Ogasawara	CEFET-RJ, Brazil
Kouzou Ohara	Aoyama Gakuin University, Japan
Kok-Leong Ong	RMIT University, Australia
Riccardo Ortale	Consiglio Nazionale delle Ricerche, Italy
Arindam Pal	CSIRO, Australia
Eliana Pastor	Politecnico di Torino, Italy
Dhaval Patel	IBM T. J. Watson Research Center, USA
Martin Pavlovski	Yahoo Inc., USA
Le Peng	University of Minnesota, USA
Nhan Pham	IBM T. J. Watson Research Center, USA
Thai-Hoang Pham	Ohio State University, USA
Chengzhi Piao	Hong Kong Baptist University, Hong Kong SAR
Marc Plantevit	EPITA, France

Mario Prado-Romero	Gran Sasso Science Institute, Italy
Bardh Prenkaj	Sapienza Università di Roma, Italy
Jianzhong Qi	University of Melbourne, Australia
Buyue Qian	Xi'an Jiaotong University, China
Huajie Qian	Columbia University, USA
Hezhe Qiao	Singapore Management University, Singapore
Biao Qin	Renmin University of China, China
Zengchang Qin	Beihang University, China
Tho Quan	Ho Chi Minh City University of Technology, Vietnam
Miloš Radovanović	University of Novi Sad, Serbia
Thilina Ranbaduge	Australian National University, Australia
Chotirat Ratanamahatana	Chulalongkorn University, Thailand
Chandra Reddy	IBM T. J. Watson Research Center, USA
Ryan Rossi	Adobe Research, USA
Morteza Saberi	University of Technology Sydney, Australia
Akira Sakai	Fujitsu Laboratories Ltd., Japan
David Sánchez	Universitat Rovira i Virgili, Spain
Maria Luisa Sapino	Università degli Studi di Torino, Italy
Hernan Sarmiento	UChile & IMFD, Chile
Badrul Sarwar	CloudAEye, USA
Nader Shakibay Senobari	University of California Riverside, USA
Nasrin Shabani	Macquarie University, Australia
Ankit Sharma	University of Minnesota, USA
Chandra N. Shekar	RGUKT RK Valley, India
Chih-Ya Shen	National Tsing Hua University, Taiwan
Wei Shen	Nankai University, China
Yu Shen	Peking University, China
Zhi-Yu Shen	Nanjing University, China
Chuan Shi	Beijing University of Posts and Telecommunications, China
Yue Shi	Meta Platforms, USA
Zhenwei Shi	Beihang University, China
Motoki Shiga	Tohoku University, Japan
Kijung Shin	KAIST, Korea
Kai Shu	Illinois Institute of Technology, USA
Hong-Han Shuai	National Yang Ming Chiao Tung University, Taiwan
Zeren Shui	University of Minnesota, USA
Satyaki Sikdar	Indiana University, USA
Dan Simovici	University of Massachusetts Boston, USA
Apoorva Singh	IIT Patna, India

Bikash Chandra Singh	Islamic University, Bangladesh
Stavros Sintos	University of Illinois at Chicago, USA
Krishnamoorthy Sivakumar	Washington State University, USA
Andrzej Skowron	University of Warsaw, Poland
Andy Song	RMIT University, Australia
Dongjin Song	University of Connecticut, USA
Arnaud Soulet	Université de Tours, France
Ja-Hwung Su	National University of Kaohsiung, Taiwan
Victor Suciu	University of Wisconsin, USA
Liang Sun	Alibaba Group, USA
Xin Sun	Technische Universität München, Germany
Yuqing Sun	Shandong University, China
Hirofumi Suzuki	Fujitsu Laboratories Ltd., Japan
Anika Tabassum	Oak Ridge National Laboratory, USA
Yasuo Tabei	RIKEN, Japan
Chih-Hua Tai	National Taipei University, Taiwan
Hiroshi Takahashi	NTT, Japan
Atsuhiro Takasu	National Institute of Informatics, Japan
Yanchao Tan	Fuzhou University, China
Chang Tang	China University of Geosciences, China
Lu-An Tang	NEC Laboratories America, USA
Qiang Tang	Luxembourg Institute of Science and Technology, Luxembourg
Yiming Tang	Hefei University of Technology, China
Ying-Peng Tang	Nanjing University of Aeronautics and Astronautics, China
Xiaohui (Daniel) Tao	University of Southern Queensland, Australia
Vahid Taslimitehrani	PhysioSigns Inc., USA
Maguelonne Teisseire	INRAE, France
Ya-Wen Teng	Academia Sinica, Taiwan
Masahiro Terabe	Chugai Pharmaceutical Co. Ltd., Japan
Kia Teymourian	University of Texas at Austin, USA
Qing Tian	Nanjing University of Information Science and Technology, China
Yijun Tian	University of Notre Dame, USA
Maksim Tkachenko	Singapore Management University, Singapore
Yongxin Tong	Beihang University, China
Vicenç Torra	University of Umeå, Sweden
Nhu-Thuat Tran	Singapore Management University, Singapore
Yash Travadi	University of Minnesota, USA
Quoc-Tuan Truong	Amazon, USA

Yi-Ju Tseng	National Yang Ming Chiao Tung University, Taiwan
Turki Turki	King Abdulaziz University, Saudi Arabia
Ruo-Chun Tzeng	KTH Royal Institute of Technology, Sweden
Leong Hou U	University of Macau, Macau SAR
Jeffrey Ullman	Stanford University, USA
Rohini Uppuluri	Glassdoor, USA
Satya Valluri	Databricks, USA
Dinusha Vatsalan	Macquarie University, Australia
Bruno Veloso	FEP - University of Porto and INESC TEC, Portugal
Anushka Vidanage	Australian National University, Australia
Herna Viktor	University of Ottawa, Canada
Michalis Vlachos	University of Lausanne, Switzerland
Sheng Wan	Nanjing University of Science and Technology, China
Beilun Wang	Southeast University, China
Changdong Wang	Sun Yat-sen University, China
Chih-Hang Wang	Academia Sinica, Taiwan
Chuan-Ju Wang	Academia Sinica, Taiwan
Guoyin Wang	Chongqing University of Posts and Telecommunications, China
Hongjun Wang	Southwest Jiaotong University, China
Hongtao Wang	North China Electric Power University, China
Jianwu Wang	UMBC, USA
Jie Wang	Southwest Jiaotong University, China
Jin Wang	Megagon Labs, USA
Jingyuan Wang	Beihang University, China
Jun Wang	Shandong University, China
Lizhen Wang	Yunnan University, China
Peng Wang	Southeast University, China
Pengyang Wang	University of Macau, Macau SAR
Sen Wang	University of Queensland, Australia
Senzhang Wang	Central South University, China
Shoujin Wang	Macquarie University, Australia
Sibo Wang	Chinese University of Hong Kong, Hong Kong SAR
Suhang Wang	Pennsylvania State University, USA
Wei Wang	Fudan University, China
Wei Wang	Hong Kong University of Science and Technology (Guangzhou), China
Weicheng Wang	Hong Kong University of Science and Technology, Hong Kong SAR

Wei-Yao Wang	National Yang Ming Chiao Tung University, Taiwan
Wendy Hui Wang	Stevens Institute of Technology, USA
Xiao Wang	Beihang University, China
Xiaoyang Wang	University of New South Wales, Australia
Xin Wang	University of Calgary, Canada
Xinyuan Wang	George Mason University, USA
Yanhao Wang	East China Normal University, China
Yuanlong Wang	Ohio State University, USA
Yuping Wang	Xidian University, China
Yuxiang Wang	Hangzhou Dianzi University, China
Hua Wei	Arizona State University, USA
Zhewei Wei	Renmin University of China, China
Yimin Wen	Guilin University of Electronic Technology, China
Brendon Woodford	University of Otago, New Zealand
Cheng-Wei Wu	National Ilan University, Taiwan
Fan Wu	Central South University, China
Fangzhao Wu	Microsoft Research Asia, China
Jiansheng Wu	Nanjing University of Posts and Telecommunications, China
Jin-Hui Wu	Nanjing University, China
Jun Wu	UIUC, USA
Ou Wu	Tianjin University, China
Shan-Hung Wu	National Tsing Hua University, Taiwan
Shu Wu	Chinese Academy of Sciences, China
Wensheng Wu	University of Southern California, USA
Yun-Ang Wu	National Taiwan University, Taiwan
Wenjie Xi	George Mason University, USA
Lingyun Xiang	Changsha University of Science and Technology, China
Ruliang Xiao	Fujian Normal University, China
Yanghua Xiao	Fudan University, China
Sihong Xie	Lehigh University, USA
Zheng Xie	Nanjing University, China
Bo Xiong	Universität Stuttgart, Germany
Haoyi Xiong	Baidu, Inc., China
Bo Xu	Donghua University, China
Bo Xu	Dalian University of Technology, China
Guandong Xu	University of Technology Sydney, Australia
Hongzuo Xu	NUDT, China
Ji Xu	Guizhou University, China

Tong Xu	University of Science and Technology of China, China
Yuanbo Xu	Jilin University, China
Hui Xue	Southeast University, China
Qiao Xue	Nanjing University of Aeronautics and Astronautics, China
Akihiro Yamaguchi	Toshiba Corporation, Japan
Bo Yang	Jilin University, China
Liangwei Yang	University of Illinois at Chicago, USA
Liu Yang	Tianjin University, China
Shaofu Yang	Southeast University, China
Shiyu Yang	Guangzhou University, China
Wanqi Yang	Nanjing Normal University, China
Xiaoling Yang	Southwest Jiaotong University, China
Xiaowei Yang	South China University of Technology, China
Yan Yang	Southwest Jiaotong University, China
Yiyang Yang	Guangdong University of Technology, China
Yu Yang	City University of Hong Kong, Hong Kong SAR
Yu-Bin Yang	Nanjing University, China
Junjie Yao	East China Normal University, China
Wei Ye	Tongji University, China
Yanfang Ye	University of Notre Dame, USA
Kalidas Yeturu	IIT Tirupati, India
Ilkay Yildiz Potter	BioSensics LLC, USA
Minghao Yin	Northeast Normal University, China
Ziqi Yin	Nanyang Technological University, Singapore
Jia-Ching Ying	National Chung Hsing University, Taiwan
Tetsuya Yoshida	Nara Women's University, Japan
Hang Yu	Shanghai University, China
Jifan Yu	Tsinghua University, China
Yanwei Yu	Ocean University of China, China
Yongsheng Yu	Macquarie University, Australia
Long Yuan	Nanjing University of Science and Technology, China
Lin Yue	University of Newcastle, Australia
Xiaodong Yue	Shanghai University, China
Nayyar Zaidi	Monash University, Australia
Chengxi Zang	Cornell University, USA
Alexey Zaytsev	Skoltech, Russia
Yifeng Zeng	Northumbria University, UK
Petros Zerfos	IBM T. J. Watson Research Center, USA
De-Chuan Zhan	Nanjing University, China

Aoying Zhou	East China Normal University, China
Bing Zhou	Sam Houston State University, USA
Nianjun Zhou	IBM T. J. Watson Research Center, USA
Qinghai Zhou	UIUC, USA
Xiangmin Zhou	RMIT University, Australia
Xiaoping Zhou	Beijing University of Civil Engineering and Architecture, China
Xun Zhou	University of Iowa, USA
Jonathan Zhu	Wheaton College, USA
Ronghang Zhu	University of Georgia, China
Xingquan Zhu	Florida Atlantic University, USA
Ye Zhu	Deakin University, Australia
Yihang Zhu	University of Leicester, UK
Yuanyuan Zhu	Wuhan University, China
Ziwei Zhu	George Mason University, USA

External Reviewers

Zihan Li	University of Massachusetts Boston, USA
Ting Yu	Zhejiang Lab, China

Sponsoring Organizations

Accton

ACSI

Appier

Chunghwa Telecom Co., Ltd

DOIT, Taipei

ISCOM

Metaage

NSTC

Pegatron

Quanta Computer

TWS

Wavenet Co., Ltd

Contents – Part IV

Financial Data

Information Retrieval and Search

Financial Data

Look Around! A Neighbor Relation Graph Learning Framework for Real Estate Appraisal

Chih-Chia Li(✉), Wei-Yao Wang, Wei-Wei Du, and Wen-Chih Peng

Department of Computer Science, National Yang Ming Chiao Tung University, Hsinchu, Taiwan

{licc.cs09,wwdu.cs10,wcpengcs}@nycu.edu.tw, sf1638.cs05@nctu.edu.tw

Abstract. Real estate appraisal is a crucial issue for urban applications, aiming to value the properties on the market. Recently, several methods have been developed to automatize the valuation process by taking the property trading transaction into account when estimating the property value to mitigate the efforts of hand-crafted design. However, existing methods 1) only consider the real estate itself, ignoring the relation between the properties. Moreover, naively aggregating the information of neighbors fails to model the relationships between the transactions. To tackle these limitations, we propose a novel Neighbor **Re**lation **Gra**ph Learning Fra**m**ework (**ReGram**) by incorporating the relation between target transaction and surrounding neighbors with the attention mechanism. To model the influence between communities, we integrate the environmental information and the past price of each transaction from other communities. Since the target transactions in different regions share some similarities and differences of characteristics, we introduce a dynamic adapter to model the different distributions of the target transactions based on the input-related kernel weights. Extensive experiments on the real-world dataset with various scenarios demonstrate that ReGram robustly outperforms the state-of-the-art methods.

Keywords: Real estate appraisal · Graph neural networks · Dynamic adapters

1 Introduction

Property technology (proptech) has developed proprietary systems by bringing properties and their owners from offline to online and has stimulated several productive studies for digital marketing, *e.g.*, virtual tours, and online appraisal, especially due to the COVID-19 situation [6]. In this paper, we focus on one of the proptech applications, real estate appraisal, which come to play a vital role, not only for society but also for the government, as it reduces the burden of triad relationships (*e.g.*, buyers, sellers, and owners) when bargaining for the

D.-N. Yang et al. (Eds.): PAKDD 2024, LNAI 14648, pp. 3–16, 2024.
https://doi.org/10.1007/978-981-97-2238-9_1

proper value of a property. For example, real estate appraisal provides an objective price to prevent buyers from being deliberately sold overpriced properties by sellers. On the other hand, the price of real estate can be viewed as one of the development metrics in urban areas; thus real estate appraisal benefits the government in terms of planning urban development. Therefore, it is essential to develop automatic algorithms for eliminating unfair trades and boosting economic prosperity.

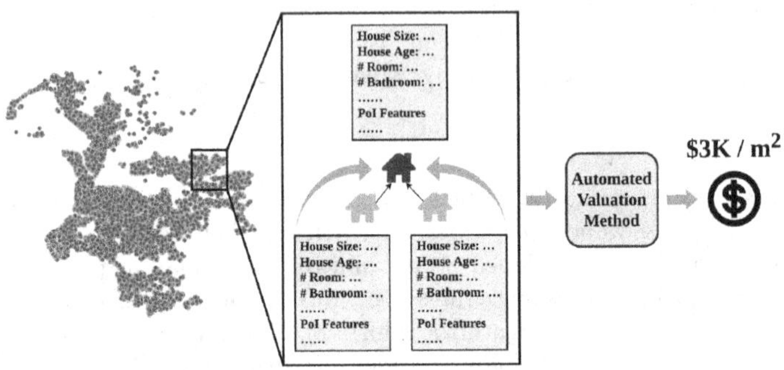

Fig. 1. Illustration of a real scenario in Taipei with AVMs for real estate appraisal.

Generally, there are several domains in the real world that are also tasked with modeling information according to such geographic location for analysis, *e.g.*, rental pricing [18]. A key solution for taking the surrounding information into account is to adopt a graph neural network (GNN) (Fig. 1), which has drawn significant attention due to its capability of modeling non-Euclidean data like social networks [17] and financial trading [7,19]. In real estate appraisal, the only existing approach exploits GNN by considering the neighbor information from different perspectives and dividing the final prediction as multitask learning based on the urban district of the target transaction to model value distribution in different regions [20]. Despite the above progress, there are three shortcomings in the previous work. First, there is some irrelevant information of real estate neighbors (*e.g.*, number of rooms), which would not directly affect the price per unit area of other transactions, even if they are close. Therefore, naively fusing whole features of neighbor transactions will hamper the model performance due to the noise. Second, the values of the properties are significantly influenced by the geographic location, which can represent the regional economic situation of the target transaction. For instance, if a property is located in a wealthy area, the corresponding value is more likely to be expensive. However, existing works ignore that the value of surrounding properties will affect the target transaction, which neglects the impact of the geographic location. Third, when sharing some similarities and differences of characteristics with the transactions in different regions, the existing works fail to consider that there are few transactions in some regions (*e.g.*, remote areas), which causes underfitting and worsens the performance. Therefore, we believe that real estate appraisal with modeling neighbor information is still an unsolved but essential problem.

To tackle the above challenges, we present a novel Neighbor **Re**lation **Gra**ph Learning Framework (**ReGram**) for real estate appraisal. For the first issue, we aggregate the relation between the target transactions and their neighbors by learning the weighted neighbor relationship. For the second issue, in order to consider the impact of geographic location, we introduce a preliminary appraisal of real estate based on the neighbors' prices, which provides surrounding price information for the target property. For the third issue, we introduce a dynamic adaptor to consider the discrepancies between each target transaction for modeling the distribution of the real estate value.

In summary, the main contributions of our paper are as follows:

1. We propose ReGram, a novel neighbor relation graph learning framework to appraise real estate by leveraging the neighborhood relations and their corresponding prices, which can be applied to other geographic urban applications (*e.g.*, rental pricing).
2. A preliminary appraisal of the target transaction based on the neighbors' prices is introduced to provide surrounding price information. Moreover, we propose a dynamic adaptor to flexibly model the targeting price based on various characteristics.
3. Extensive experiments show that ReGram achieves a state-of-the-art performance on the real-world real estate appraisal dataset with various scenarios.

2 Preliminaries

Related Work. To estimate the value of properties, traditional approaches conduct the process of valuation based on domain experts [3,15], but lack the ability to appraise automatically. In recent years, several automated valuation methods (AVMs) have been proposed by adopting machine learning [1,2,14] and deep learning techniques [9,12] but ignoring spatially proximal real estate [8]. Besides, several works have utilized artificial neural networks or big data approach [5,9,12,13] to appraise the real estate. However, most of the previous works ignored the spatio-temporal dependencies among real estate transactions. The information of peer-dependency were utilized with k-nearest neighbors by sampling a fixed number of similar transactions, and generating sequences from target transactions and nearby transactions for estimating the price of real estate [4]. Still, the sampling process treated each feature as being equally important, which causes sampling noise from unrelated features.

Recently, MugRep, which is the only existing method exploiting GNN in real estate appraisal, was proposed with a multitask hierarchical graph based framework by constructing the graphs at the transaction level and community level [20]. To predict the value of real estate, MugRep learned the prediction weights independently by separating the tasks via urban districts. However, MugRep hampers the model prediction due to considering irrelevant information by aggregating all features of neighbors directly, and fails to take the neighbor transactions' price into account, which is critical information to estimate the value of the target real estate. Moreover, they predict the value based on district divisions, which ignores the correlation between each region and is prone to incorrect prediction in regions with insufficient transactions. Our novel approach, in contrast, considers the relation between the target transactions and their neighbors as a weighted relation representation. Besides, we integrate the neighbors' prices to take the surrounding appraisal information of the target property into account, and dynamically model the distributions based on the target transactions.

Definitions of Real Estate Appraisal

Definition 1: Real Estate Transaction. Consider a set of property trading records as real estate transactions $\boldsymbol{S}$ in chronological order; a real estate transaction at the t-th transaction denotes $\boldsymbol{s_t} = \{s_t^e, s_t^o\} \in \boldsymbol{S}$ where the environment and object features are s_t^e and s_t^o.

Definition 2: Real Estate Value. Consider a set of property unit prices as real estate value $\boldsymbol{P}$ in chronological order; a real estate value at the t-th transaction denotes $\boldsymbol{p_t} \in \boldsymbol{P}$, where p_t is the value of the real estate transaction.

Definition 3: Target Transaction $\boldsymbol{s_{t+1}} = \{s_{t+1}^e, s_{t+1}^o\}$ is the real estate that will be appraised.

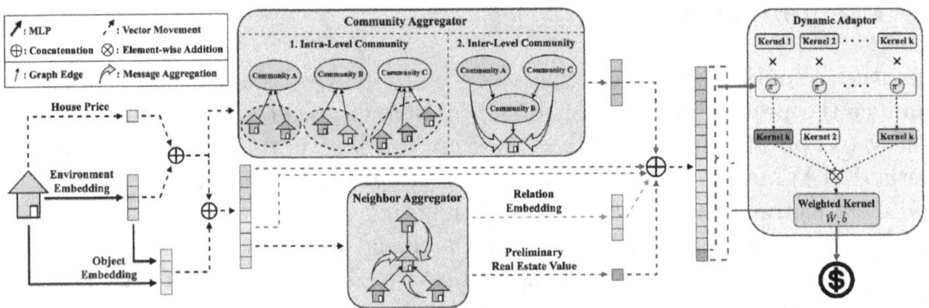

Fig. 2. An overview of the proposed ReGram. The community aggregator encodes the transactions with intra-level and inter-level communities. The neighbor aggregator considers the neighbor information of the target transaction to generate the relation representation and preliminary appraisal. The dynamic adaptor generates input-related weights for appraising the real estate.

Definition 4: Environment Feature represents the surrounding environment information of the property. In this paper, environment features are PoI (point of interest), geographic information (longitude, latitude), land usage, and house age.

Definition 5: Object Feature represents the information about the property such as the size of the house, the number of rooms in the house, and the floor number of the house, etc.

Definition 6: PoI Feature consists of two types of features. The first is the number of PoI with a Euclidean distance. For example, there are two schools around 5km from the property. The other is the minimum distance to each PoI category. For example, the shortest distance to the hospital or gas station.

Definition 7: Community. A community is a hypernode containing multiple transactions that are built in the same region at the same time. In this paper, two transactions belong to the same community if and only if they meet the following conditions: (1) The completion dates (month and year) of construction are the same. (2) The distance between two transactions is less than 500 m [20].

Definition 8: Problem Definition. Given a target transaction $\boldsymbol{s}_{t+1}$ with its previous transactions $\boldsymbol{S}_{[0..t]}$ and corresponding unit price $\boldsymbol{P}_{[0..t]}$, the task is to estimate the price $\boldsymbol{p}_{t+1}$.

3 Approach

Figure 2 illustrates an overview of the proposed model. The graph construction is divided into the transaction-level subgraph, intra-level community, and inter-level community for modeling various relationships between transactions and communities. After the graph construction, real estate transactions were first embedded by the transaction projection to get transaction representations. Then we utilized the neighbor aggregator to produce the relation embedding and a preliminary real estate value from surrounding neighbor transactions, and adopted the community aggregator to get the community-level representation. Afterwards, the dynamic adaptor generated the weighted kernels based on the above contexts and appraised the target transaction.

3.1 Graph Construction

We construct the transaction-level subgraph, intra-level community, and inter-level community similar to [20]. However, previous work heavily relies on plentiful and various data (*e.g.*, check-in, and user-trip) for constructing the community; thus, it is hard to generalize to different situations. Therefore, we introduce the community node using a simple yet effective method by discussing with domain experts, and only used the PoI feature to construct the community-level subgraph, which demonstrates the robustness shown in the experiments.

Transaction-Level Subgraph. To model the relation between transactions, we build the transaction-level subgraph, where each node is a real estate trading transaction. For each edge, the edge between two nodes needs to satisfy all conditions:

1. **Numeric Difference**. The distance between two properties is less than 500 m, the trading date difference is no more than 1 year unless they are both in the same month, and the difference of house age is less than 10 years.
2. **Property Characteristic**. The building type and the property's main purpose (4 and 1622 types, respectively) of the two transactions are required to be the same. For instance, if two properties are both apartment and for commercial-used, then these properties will satisfy this condition.
3. **Characteristic Indicator**. There are also indicators to characterize the property, including the small house indicator, shop indicator, and first-floor indicator. The two properties need to have the same indicator results to match this condition. For example, the two transactions have the same characteristic indicator if both are small and first-floor houses but are not shops.

Intra-Level Community. To capture the influence from surrounding communities, the objective of the intra-level community is to model the relation between transaction and community, where each node represents a transaction or a community. We formulate the community nodes and construct the relation between transaction and community according to Definition 7. Besides, we filter out the transactions for which the trading time is far from the current target transactions (two months was set in this paper) to avoid considering irrelevant information to the current target time.

Inter-Level Community. We designed an inter-level community for modeling the relation between communities, where each node in the inter-level community represents the community. Specifically, we first averaged the PoI features of transactions in that community to indicate the general living environment in that area. Then we iterated all community pairs, connecting the communities if the l2 distance of the pair was the top 0.1% smallest of all the pairs [20].

3.2 Transaction Encoding

To model the contextualized information of transactions, we divided the t-th transaction features into environment feature s_t^e (Definition 4) and object feature s_t^o (Definition 5). The environment embedding e_t was obtained by $e_t = W_{e_2}\sigma(W_{e_1}s_t^e)$, where $W_{e_1} \in \mathbb{R}^{2d_m \times d_e}$ and $W_{e_2} \in \mathbb{R}^{d_m \times 2d_m}$ are learnable matrices, and $\sigma(\cdot)$ is the Mish activation function [16].

For the object embedding, we first followed the process of environment embedding to transform the object feature, and we fused with the environment embedding to obtain the object embedding o_t for considering the interactions between the property characteristics and the surrounding environment: $o_t = W_x\sigma([o_t' \oplus e_t]); o_t' = W_{o_2}\sigma(W_{o_1}s_t^o)$, where $W_{o_1} \in \mathbb{R}^{d_m \times d_o}, W_{o_2} \in \mathbb{R}^{d_m \times d_m}$ and $W_x \in \mathbb{R}^{d_m \times 2d_m}$ are learnable matrices, and $\oplus$ is the concatenation operator.

To consider the relation of information and the price of the transaction, we concatenated the environment embedding e_t, object embedding o_t, and its price p_t as the transaction embedding $x_t : x_t = [e_t \oplus o_t \oplus p_t]$, where $x_t \in \mathbb{R}^{d_x}$ and $d_x = 2d_m + 1$. It is noted that we set the house price of the target transaction as zero since the target price was the value we wanted to estimate.

3.3 Neighbor Aggregator

The neighbor aggregator aims to consider the relations between the target transaction and its neighbors with two steps: relation modeling and relation aggregation.

Relation Modeling. We generated the transaction relation $r_{(t+1)t'}$ by projecting the concatenation of the target transaction embedding x_{t+1} and the neighbor transaction embedding $x_{t'}$, where t' indicates the index of the neighbor transaction (Definition 1): $r_{(t+1)t'} = W_r[x_{t+1} \oplus x_{t'}]$, where $W_r \in \mathbb{R}^{d_m \times 2d_x}$ is a learnable kernel weight.

After generating the transaction relation, the attention mechanism is adopted to characterize the importance of each neighbor transaction for the target transaction. Specifically, the i-th head coefficient is computed as follows:

$$\beta^i_{(t+1)t'} = \text{LeakyReLU}(w^i_a \sigma(r_{(t+1)t'})), \tag{1}$$

where $w_a \in \mathbb{R}^{1 \times d_m}$ is a learnable weight. To extend the potential important neighbors, we computed the attention weight $\alpha^i_{(t+1)t'}$ with softmax temperature τ as follows:

$$\alpha^i_{(t+1)t'} = \frac{\exp(\frac{\beta^i_{(t+1)t'}}{\tau})}{\sum_{k \in N_{t+1}} \exp(\frac{\beta^i_{(t+1)k}}{\tau})}, \tag{2}$$

where N_{t+1} is the set of the target transaction's neighbor indexes.

For the multi-head attention, we took the average of all the attention heads to enhance the capability from different perspectives:

$$\alpha_{(t+1)t'} = \frac{1}{H} \sum_{i=1}^{H} \alpha^i_{(t+1)t'}, \tag{3}$$

where H is the number of attention heads.

On the other hand, we propose the delta value to reflect the potential price difference between the neighbor transaction and target transaction. The computation of the delta value $d_{(t+1)t'}$ is as follows:

$$d_{(t+1)t'} = w_d \sigma(r_{(t+1)t'}), \tag{4}$$

where $w_d \in \mathbb{R}^{1 \times d_m}$ is a learnable weight.

Relation Aggregation. After computing the attention weight and delta value, the relation aggregation produces a preliminary estimated real estate value and the weighted relation embedding from nearby neighbors.

Formally, the neighbor's price is fused with the delta value and aggregate corresponding importance to generate the preliminary value $\tilde{p}_{t+1}$:

$$\tilde{p}_{t+1} = \sum_{t' \in N_{t+1}} \alpha_{(t+1)t'}(p_{t'} + d_{(t+1)t'}). \tag{5}$$

On the other hand, the relation embedding is obtained by aggregating with the attention weights of the neighbors as:

$$r_{t+1} = \sum_{t' \in N_{t+1}} \alpha_{(t+1)t'} r_{(t+1)t'}. \tag{6}$$

3.4 Community Aggregator

To further consider the information from a higher level perspective, we designed a community aggregator to integrate the representation from neighbor communities. Specifically, the environmental representation was computed as $\tilde{x}_{t'} = [e_{t'} \oplus p_{t'}]$, where $\tilde{x}_{t'} \in \mathbb{R}^{d_{\tilde{x}}}$ and $d_{\tilde{x}} = d_m + 1$.

In the community aggregator, we calculated the attention weight $\alpha^i_{t'}$ for each transaction in the community to quantify the corresponding impact:

$$\beta_{t'} = v_u \tanh(W_{u_1} \tilde{x}_{t'}), \alpha^j_{t'} = \frac{\exp(\beta_{t'})}{\Sigma_{k \in C_i} \exp(\beta_k)}, \tag{7}$$

where $W_{u_1} \in \mathbb{R}^{d_m \times d_{\tilde{x}}}$, $v_u \in \mathbb{R}^{1 \times d_m}$, and C_j is the transaction set in community j. The community embedding is computed by aggregating the environmental representation of each transaction with the attention weight:

$$u^j_{t+1} = \mathrm{ReLU}(W_{u_2}(\sum_{t' \in C_j} \alpha^j_{t'} \tilde{x}_{t'})), \tag{8}$$

where $W_{u_2} \in \mathbb{R}^{d_m \times d_{\tilde{x}}}$ is a learnable weight.

Then we propagated the embedding of the target transaction's community neighbors to the target transaction itself. Instead of passing through the community of the target transaction, we focused on the connection between the target itself and the community neighbors to enable the flexibility for community modeling for the other target transactions in the same community. Specifically, we calculated the attention weight for each community to engage with the influential community for the target transaction based on the concatenation of the target transaction's object embedding o_{t+1} and the community embedding u^j_{t+1}:

$$\gamma^j_{t+1} = v_c \tanh(W_{c_1}[o_{t+1} \oplus u^j_{t+1}]), \alpha^j_{t+1} = \frac{\exp(\gamma^j_{t+1})}{\Sigma_{k \in N^c_{t+1}} \exp(\gamma^k_{t+1})}, \tag{9}$$

where $W_{c_1} \in \mathbb{R}^{d_m \times 2d_m}$ and $v_c \in \mathbb{R}^{1 \times d_m}$ are learnable weights, and N^c_{t+1} is the set of the target transaction's community neighbors.

Finally, the neighbor community embedding c_{t+1} was aggregated by:

$$c_{t+1} = \mathrm{ReLU}(W_{c_2}(\sum_{i \in N^c_{t+1}} \alpha^i_{t+1} u^i_{t+1})), \tag{10}$$

where $W_{c_2} \in \mathbb{R}^{d_m \times d_m}$ is the learnable weight.

3.5 Dynamic Adaptor

Generally, prices of properties are influenced differently based on multiple factors. For example, the demand of public transportation affects the real estate price in metropolitan areas, while this demand has fewer effects in remote areas since people mostly drive by themselves. A potential solution is to separate the prediction based on different districts. However, this ignores the correlation between each region and is hard to learn for areas with insufficient transactions. Therefore, we propose the dynamic adaptor to adaptively learn the distribution of different target transactions to predict the real estate price.

Specifically, multiple learnable kernel weights are designed in the dynamic adaptor and are used to appraise the target transaction based on the input-generated weighted kernel. To obtain the input-related importance of the kernels, we used the object embedding o_{t+1}, environment embedding e_{t+1}, neighbor relation embedding r_{t+1} and neighbor community embedding c_{t+1} of the target transaction to generate the target representation $h'_{t+1} \in \mathbb{R}^{4d_m} = [o_{t+1} \oplus e_{t+1} \oplus r_{t+1} \oplus c_{t+1}]$. Then we computed the attention weight of the k-th kernel weight:

$$z_{t+1}^k = w_k \sigma(W_{r_1} h'_{t+1}), \pi_{t+1}^k = \frac{\exp(\frac{z_{t+1}^k}{\tau})}{\sum_{l=1}^{K} \exp(\frac{z_{t+1}^l}{\tau})}, \tag{11}$$

where $W_{r_1} \in \mathbb{R}^{d_m \times 4d_m}$, $w_k \in \mathbb{R}^{1 \times d_m}$, and K is the number of kernels. It is noted that we also applied the softmax temperature to use near-uniform attention in early training epochs, which can better optimize multiple kernels simultaneously and avoid focusing on only one kernel before fully training the kernels.

To generate the input-related weights $\tilde{W}_{t+1} \in \mathbb{R}^{1 \times (4d_m+1)}, \tilde{b}_{t+1} \in \mathbb{R}^{1 \times 1}$ for the target transaction, we performed the aggregation of kernels based on these attention weights:

$$\hat{W}_{t+1} = \sum_{k=1}^{K} \pi_{t+1}^k \tilde{W}_k, \quad \hat{b}_{t+1} = \sum_{k=1}^{K} \pi_{t+1}^k \tilde{b}_k, \tag{12}$$

where $\tilde{W}_k \in \mathbb{R}^{1 \times (4d_m+1)}, \tilde{b}_k \in \mathbb{R}^{1 \times 1}$ are learnable kernel weights of the k-th kernel.

Finally, we adopted linear regression with the weights of the aggregated kernel, and integrated the concatenation of the target representation h'_{t+1}, preliminary estimation $\tilde{p}_{t+1}$ as the final representation h_{t+1} to predict the target appraisal value $\hat{p}_{t+1}$:

$$h_{t+1} = [h'_{t+1} \oplus \tilde{p}_{t+1}], \hat{p}_{t+1} = \hat{W}_{t+1} h_{t+1} + \hat{b}_{t+1}. \tag{13}$$

Training Objective. We minimized the mean square error loss to learn the prediction of real estate:

$$Loss = \frac{1}{|S|} \sum_{s_{t+1} \in S} (\hat{p}_{t+1} - p_{t+1})^2, \tag{14}$$

where S is the set of target transactions and p_{t+1} is the ground truth value.

4 Experiments

In the experiment, we aim to study three research questions: **RQ1:** How does ReGram perform compared with different groups of models for real estate appraisal? **RQ2:** How does neighbor price affect prediction results? **RQ3:** How does each component of ReGram contribute to the effectiveness?

Table 1. The numbers of transactions.

	Train	Validation	Test	Total
New Taipei	89,964	4,349	2,474	96,787
Taipei	34,311	1,815	1,176	37,302
Taoyuan	61,592	2,074	1,170	64,836
Taichung	40,916	1,669	988	43,573
Tainan	17,710	586	423	18,719
Kaohsiung	38,335	1,875	1,142	41,352
Total	282,828	12,368	7,373	302,569

4.1 Experimental Setup

Dataset Description. Since there is no public real estate transaction dataset, the experiments were conducted on the collected dataset in Taiwan special municipalities, including New Taipei, Taipei, Taoyuan, Taichung, Tainan, and Kaohsiung[1]. We collected the transaction data from Taiwan Real Estate Transaction Website[2] and the PoI data from E.Sun Bank. The target transactions in the dataset ranged from 2015/7/1 to 2021/6/30, of which we used 3 months of transactions as validation data (2021/1 - 2021/3), 3 months of transactions as testing data (2021/4 - 2021/6), and the others as training data. Table 1 shows the statistics of the corresponding transactions.

Implementation Details. We performed standard normalization on numeric data, and used one-hot encoding for categorical data city by city. For the object feature, the dimension d_o ranged from 500 to 700, and the dimension of the environment feature d_e was 1750. The hyperparameters were tuned based on the validation set for each model. The dimensions of both object and environment embeddings d_m were set to 256, and we set the numbers of dynamic kernels K to 8 and attention heads H to 8. Besides, we took the softmax temperature $\tau = 30$, and performed batch normalization [10] before the attention calculation in Eqs. 2 and 11. To learn the weight of our model, Adam optimizer [11] was employed with a learning rate of 0.001 and a batch size of 64 for 50 epochs. All the experiments were conducted on a GeForce RTX 2080 Ti 11GB GPU, and are reported the average and standard deviation of these 5 testing scores. The best result in the experiments is in boldface, while the second-best result is underlined.

Baselines. To evaluate the performance of ReGram, we compared with the following models: The **Machine Learning** (ML) based methods include (1) LR, (2) KNN, (3) SVR, and (4) LGBM. The **Deep Neural Network** based model (DNN) consists of multilayer perceptrons. The **Graph-based Neural Network** (GNN) includes the following (1) The Graph Convolutional Network (GCN)

[1] We note that both Beijing and Chengdu datasets used in [20] have not been released.
[2] https://lvr.land.moi.gov.tw/.

aggregates the neighbors' features in equal weight, (2) the Graph Attention Network (GAT) aggregates the neighbors' features in the weight computed by node features, (3) MugRep is the SOTA model in the real estate appraisal with a transaction module and a hierarchical community module in the multitask learning manner. For fair comparisons, the embedding by our transaction projection without the price information was also used in DNN, GCN, GAT, and MugRep. For the ML baselines, we report the testing score due to the non-stochasticity. We trained the model in 5 different random seeds for the DNN and GNN models.

Evaluation Metrics. To evaluate the performance, we used Mean Absolute Percentage Error (MAPE) following [20], which can be viewed as the prediction accuracy of the estimated value[3].

Table 2. Overall performance in MAPE. + NP is denoted as appraising the preliminary value of the target transaction by referencing neighbors' price information.

	New Taipei	Taipei	Taoyuan	Taichung	Tainan	Kaohsiung	Average
LR	98.24	412.06	14.17	15.75	18.98	15.57	95.80
SVR	33.27	23.01	27.69	24.59	27.13	27.46	27.19
KNN	8.67	10.23	10.04	13.14	15.65	11.99	11.62
LGBM	8.37	9.48	10.46	13.16	16.12	**10.70**	11.38
DNN	8.03±0.20	9.16±0.09	9.07±0.24	11.38±0.09	17.13±0.16	11.34±0.15	11.02±0.16
GCN	28.38±1.19	25.91±0.17	26.75±0.37	28.29±1.51	29.77±0.55	27.21±0.89	27.72±0.78
+ NP	26.80±1.29	25.05±0.26	26.06±0.30	28.52±1.08	29.15±0.56	26.46±0.64	27.01±0.69
GAT	10.42±0.09	12.07±0.06	12.68±0.25	13.42±0.17	18.18±0.40	12.28±0.11	13.18±0.18
+ NP	9.71±0.09	11.56±0.18	11.95±0.15	13.20±0.14	16.11±0.32	12.52±0.10	12.51±0.16
MugRep	10.86±0.26	11.70±0.15	11.25±0.28	13.20±0.33	16.83±1.17	12.23±0.43	12.68±0.44
+ NP	7.97±0.14	9.45±0.56	8.86±0.22	11.70±0.15	14.52±0.21	11.79±0.37	10.72±0.28
ReGram	**7.15±0.15**	**8.31±0.06**	**8.33±0.18**	**11.24±0.10**	**14.36±0.39**	10.92±0.12	**10.05±0.17**

4.2 Overall Performance

Table 2 reports the experiment results of our model and baselines, which shows that our model consistently outperformed other baselines with at least 8.8% improvement in terms of MAPE. MugRep performed the best among the graph-based baselines since it utilizes the community-level information and original transaction feature. However, the non-graph based methods achieved better performance compared to the graph-based methods in some municipalities. This indicates that naively aggregating the transaction features also takes some noise into account, while our relation aggregator eliminates the aggregation issue by considering the relation between the target transaction and surrounding important neighbors. Moreover, another reason for the inferior performance of MugRep may be the lack of multi-source data in the dataset, which also shows the demand for a large variety of data for MugRep and the robustness of our proposed method. It is worth noting that KNN achieves competitive performance

[3] Hit-rates@10% and 20% are reported in Appendix due to the space limit.

by only averaging the price of nearest neighbors. This demonstrates the importance and necessity of considering the price information of neighbors, which is also attributed in our model.

In order to further testify the effect of neighbors' prices, we aligned the settings of the graph-based baselines with ReGram by appraising the preliminary value of the target transaction by referencing neighbors' price information (denoted as + NP). It is observed that MugRep, GAT, and GCN reach 15%, 5%, and 3% significant MAPE improvement after taking the price information of neighbors into account, which shows the effectiveness of using neighbors' prices and generalizability to be applied in any graph-based model. Nonetheless, ReGram still outperforms these models even when the neighbors' price information is considered in the baseline models.

4.3 Ablation Study

To understand the contributions of each module, the neighbor aggregator is replaced with GAT using the same settings to generate relation embedding. Moreover, we changed our dynamic adaptor to a single adaptor, which uses one linear kernel weight to appraise the value for all the regions, and applied region-aware multitask learning based on the district to learn independent adaptors. Table 3 shows the results in terms of the average MAPE. We summarize the observations as follows.

Table 3. Ablation study in MAPE.

Neighbor		Community	Adaptor	Average
price	relation			
		✓	Dynamic	10.54±0.21
	✓	✓	Dynamic	10.12±0.15
	GAT	✓	Dynamic	10.68±0.27
✓		✓	Dynamic	10.13±0.16
✓	✓		Dynamic	10.15±0.15
✓	✓	✓	Single	10.15±0.11
✓	✓	✓	Region	10.35±0.18
✓	✓	✓	Dynamic	**10.05±0.17**

As expected, removing any one component from ReGram degrades the prediction performance. In addition, it is also observed that adding the neighbor aggregator boosts the performance by 4.6%, which demonstrates the effectiveness of our relation modeling. Besides, the performance is significantly improved by 4% if we only consider either the weighted relation embedding or the price information (the MAPE is improved from 10.54 to 10.12 and 10.13, respectively). Furthermore, the performance results are inferior when applying GAT to model the relation embedding (the MAPE is degraded from 10.54 to 10.68 compared to ReGram without the neighbor aggregator), which also implies that directly aggregating the neighbors' features introduces some irrelevant noise (e.g., neighbors' house area, floor number) which hampers the performance, while our neighbor aggregator aggregates the relation between target transactions and their neighbors to alleviate this issue.

When we replaced a single fully connected layer (i.e., one kernel) in the final adaptor with a multi-adaptor based on region-aware separation as in [20], the performance relatively deteriorated from 10.15 to 10.35, indicating that a region-aware adaptor is not suitable if the information is insufficient. However, our dynamic adaptor with multiple kernels predicting the value of real estate based on the different transactions again demonstrates robust capability.

5 Conclusion

In this paper, we present a neighbor relation graph learning framework, ReGram, for tackling the challenging real estate appraisal problem. We constructed the community node from the transaction data and built the real estate network at the transaction level and community level to model the relation between the properties in various aspects. To tackle the issue of directly aggregating the features of the neighbors, we propose the neighbor aggregator to aggregate the information by modeling the relation between the target transaction and their neighborhoods, and predict the preliminary value of real estate. Moreover, we introduce the dynamic adaptor by adaptively learning the distribution of different target transactions to consider the similarities and discrepancies between each target transaction. Comprehensive experiments on the real-world dataset demonstrate the effectiveness of our model compared to state-of-the-art baselines in different scenarios. For future research, we aim to explore the model explainability for real estate appraisal, which is an important topic in financial applications.

References

1. Ahn, J.J., Byun, H.W., Oh, K.J., Kim, T.Y.: Using ridge regression with genetic algorithm to enhance real estate appraisal forecasting. Expert Syst. Appl. **39**(9), 8369–8379 (2012)
2. Azimlu, F., Rahnamayan, S., Makrehchi, M.: House price prediction using clustering and genetic programming along with conducting a comparative study. In: GECCO Companion, pp. 1809–1816. ACM (2021)
3. Baum, A., Baum, C.M., Nunnington, N., Mackmin, D.: The income approach to property valuation. Estates Gazette (2013)
4. Bin, J., Gardiner, B., Li, E., Liu, Z.: Peer-dependence valuation model for real estate appraisal. Data-Enabled Discov. Appl. **3**(1), 2 (2019)
5. Bin, J., Gardiner, B., Liu, Z., Li, E.: Attention-based multi-modal fusion for improved real estate appraisal: a case study in Los Angeles. Multimedia Tools Appl. **78**(22), 31163–31184 (2019)
6. Bruvels, A., Piasentin, R.C., Sutton, K., Zimmerman, K.: Proptech: property technology, the new frontier in real property (2022). https://mcmillan.ca/insights/proptech-property-technology-the-new-frontier-in-real-property-part-1-introduction/
7. Cheng, D., Yang, F., Xiang, S., Liu, J.: Financial time series forecasting with multi-modality graph neural network. Pattern Recogn. **121**, 108218 (2022)
8. Fu, Y., Xiong, H., Ge, Y., Yao, Z., Zheng, Y., Zhou, Z.H.: Exploiting geographic dependencies for real estate appraisal: a mutual perspective of ranking and clustering. In: Proceedings of the 20th ACM SIGKDD International Conference on Knowledge Discovery and Data Mining, pp. 1047–1056 (2014)
9. Ge, C., Wang, Y., Xie, X., Liu, H., Zhou, Z.: An integrated model for urban subregion house price forecasting: a multi-source data perspective. In: ICDM, pp. 1054–1059. IEEE (2019)
10. Ioffe, S., Szegedy, C.: Batch normalization: accelerating deep network training by reducing internal covariate shift. In: International Conference on Machine Learning, pp. 448–456. PMLR (2015)

11. Kingma, D.P., Ba, J.: Adam: a method for stochastic optimization. arXiv preprint arXiv:1412.6980 (2014)
12. Law, S., Paige, B., Russell, C.: Take a look around: using street view and satellite images to estimate house prices. ACM Trans. Intell. Syst. Technol. **10**(5), 54:1–54:19 (2019)
13. Lee, S., Kim, J., Huh, J.: Land price forecasting research by macro and micro factors and real estate market utilization plan research by landscape factors: big data analysis approach. Symmetry **13**(4), 616 (2021)
14. Lin, H., Chen, K.: Predicting price of Taiwan real estates by neural networks and support vector regression. In: Proceedings of the 15th WSEAS International Conference on Systems, pp. 220–225 (2011)
15. McCluskey, W., Borst, R.A.: An evaluation of MRA, comparable sales analysis, and artificial neural networks (ANNs) for the mass appraisal of residential properties in Northern Ireland. Assess. J. **4**(1), 47–55 (1997)
16. Misra, D.: Mish: a self regularized non-monotonic activation function. arXiv preprint arXiv:1908.08681 (2019)
17. Sankar, A., Liu, Y., Yu, J., Shah, N.: Graph neural networks for friend ranking in large-scale social platforms. In: WWW, pp. 2535–2546. ACM/IW3C2 (2021)
18. Ye, P., et al.: Customized regression model for airbnb dynamic pricing. In: KDD, pp. 932–940. ACM (2018)
19. Yin, T., Liu, C., Ding, F., Feng, Z., Yuan, B., Zhang, N.: Graph-based stock correlation and prediction for high-frequency trading systems. Pattern Recogn. **122**, 108209 (2022)
20. Zhang, W., et al.: Mugrep: a multi-task hierarchical graph representation learning framework for real estate appraisal. In: KDD, pp. 3937–3947. ACM (2021)

Multi-time Window Ensemble and Maximization of Expected Return for Stock Movement Prediction

Kanghyeon Seo[1], Seungjae Lee[1], Woo Jin Cho[2], Yoojeong Song[2], and Jihoon Yang[1(✉)]

[1] Sogang University, Seoul, Republic of Korea
{seokh,gobs10910,yangjh}@sogang.ac.kr

[2] Semyung University, Chungbuk, Republic of Korea
{ggggg7,yjsong}@semyung.ac.kr

Abstract. This paper proposes a novel end-to-end model that predicts stock movements, **MERTE**: **M**aximization of **E**xpected **R**eturns in multi-**T**ime window **E**nsemble for stock movement prediction. MERTE is based on three main ideas: 1) *an ensemble framework to capture multiple time-based momentums*; 2) *consolidating the expected return of trading to a loss function*; and 3) *learning correlations between the stocks without pre-defined knowledge*. MERTE consists of several base learners with the same neural network structure, but each receives an input of a different time-sequential length. The base learner specializes in learning the time momentum inherent in its given time window, and it also learns trading performance throughout our proposed loss function. The base learner consists of two attention mechanisms to learn correlations and dynamics of the stock movements without any domain knowledge. Experimental results report that MERTE outperforms baseline models, yielding superior trading gains on almost all six real-world datasets.

Keywords: Stock movement prediction · Financial data-mining · Deep learning

1 Introduction

Forecasting stock movement (*rise or fall*) is one of the significant research topics in FinTech since the results could be applied in many ways, such as risk management, investment strategy, and capital allocation. Most contemporary studies using deep learning (DL) techniques employ recurrent neural networks (RNNs) [4,8,13] as sub-modules to capture sequential properties of the stock movements. The RNNs are usually consolidated by attention mechanisms [7,22,23] or graph neural networks [18,21] (GNNs).

Although the studies demonstrated prospective results, they commonly have three limitations: First, the studies with the RNNs use the input of fixed time window length [7,10,21–23]. It is necessary to consider multiple time windows

D.-N. Yang et al. (Eds.): PAKDD 2024, LNAI 14648, pp. 17–29, 2024.
https://doi.org/10.1007/978-981-97-2238-9_2

for accurate stock movement prediction since the stocks move along with the diverse time-based momentums. This is supported by research in financial market domains [1,3]. They commonly state that diverse time-based momentums affect stock movements. Since the time windows of input for RNNs cover the time-based momentums, a model that utilizes the diverse time windows is required to capture the various momentums to improve the predictive results. Second, objective functions (i.e., a loss function) presented in the previous works are for the prediction task only by using cross-entropy loss [22,23] or hinge loss [7]. Because the ultimate goal of stock movement prediction is to maximize trading gains, the model also should be designed to improve trading performance. The better predictive performance could guarantee higher trading gains; however, accurate predictions on large price changes directly determine the result of trading gain. Thus, a new loss function that considers the trading gains is required. Third, the studies using GNNs usually employ pre-defined external datasets to capture edge connections [10,20]. It is labor-instensive to collect the dataset and learn the preprocessing model. Therefore, a more sophisticated structure of GNN is required to capture the correlations between the stocks by using price-based features only. The efficient market hypothesis supports this [6], insisting that the stock prices already contain all available information.

Against the background, we present our research question: **How can we accurately predict stock movements and maximize trading gains while considering multiple time-based momentums without any prior information about the relationship between the stocks?** In response, we propose a novel stock movement prediction model, named **MERTE**: **M**aximization of **E**xpected **R**eturns in a **T**ime-based **E**nsemble for stock movement predictions. MERTE has multiple base learners with the same neural network architectures. Each base learner takes the input of a different time window and is trained independently to be a specialized predictor with maximizing expected trading returns. We introduce a novel loss function for each base learner's training that accounts for predictive and trading performance. We also show the influence of the proposed loss function on the base learner's learning procedure during backpropagation.

2 Related Works

The DL technologies pave the way for representing complex relations within data. Recent works have utilized various attention mechanisms [5,7,22,23] and GNNs [10,11,18,21]. The studies using GNNs aim to learn the stock correlations for better predictive performance. Most of them employ graph convolutional neural networks [10,18,21] with the RNNs for temporal relations of the stocks. The other work [11] constructs knowledge graphs to build more comprehensive and sophisticated relations of the stocks. On the other hand, attention mechanisms are used for various purposes. For example, Ding *et al.* [5] proposed an improved basic transformer model by inserting a multi-scale Gaussian prior, an orthogonal regularization, and a trading gap splitter to enhance the capacity of capturing

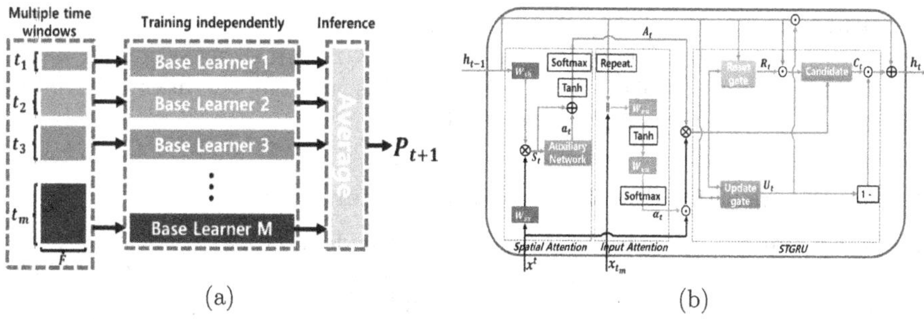

(a) (b)

Fig. 1. Overview of proposed model: (a) a whole description of the proposed time-based ensemble and (b) an architecture of each base learner composed of *spatial attention* to learn self-adaptive adjacency matrix, *input attention* to adaptively weigh on significant features, and *STGRU* to learn spatial and temporal dependencies between stocks.

long-term dependency. Zhang *et al.* [23] propose an integrated transformer model and multiple attention mechanisms to extract meaningful features and to learn long-term dependencies. Yoo *et al.* [22] introduce a novel data-axis transformer model that learns dynamic and asymmetric correlations between the stocks in an end-to-end way. This work is the first to deploy the transformer model to capture the correlational connections of the stock as GNNs did.

Table 1. Notation summary

Notation	Description	Notation	Description
$X_{t_m}(\in \mathcal{R}^{t_m \times N \times F})$	input for mth base learner	N	the number of stocks
F	input feature dimension	$f(\cdot;\theta_m)$	mth base learner with trainable parameter θ_m
$T = \{t_1, ..., t_m\}$	a set of input time windows $t_1 < ... < t_m$	M	the number of base learners
$\hat{Y}_{t+1}(\in \mathcal{R}^N)$	estimated movements of N stocks: $\{0 : fall, 1 : rise\}$	$closing_t^i$	closing price of stock i at time t
$P_{t+1}(P_{t+1}^i)$	estimated probabilities of N stocks (stock i)	$\otimes, \oplus$	matrix product, addition
$\odot$	Hadamard product	$\|$	matrix concatenation in a column-wise way

As we argue in the introduction, we propose MERTE to address three deficiencies in the previous works: singly fixed input of time-sequential length, using the loss function for the prediction task only, and employing the domain-specific data for constructing the correlations between the stocks.

3 Proposed Model

3.1 Problem Formulation

Table 1 presents the notations used in this paper. We aim to forecast N stocks' movements of the subsequent day's closing prices compared to the current day's. We define the problem as:

$$\hat{Y}_{t+1} = round(\frac{1}{M}\sum_{m=1}^{M} f(x_{t_m};\theta_m)).$$

The output of $f(x_{t_m};\theta_m)$ is N forecasted probabilities of the movements. We average the results by M base learners and apply $round(\cdot)$ operation element-wise at the inference stage. We believe that this average operation alleviates performance degradations due to unseen data.

3.2 Multi-time Window Ensemble Classifier

Overall, we aim to train each base learner independently to capture the diverse momentums of stock movements. Fig. 1-(a) shows an overview of MERTE consisting of M base learners with multiple inputs of its time windows $(\{t_1, t_2, ..., t_m\})$. x_{t_m} is fed into the mth base learner. Each base learner is then specialized in learning to predict the movements based on the static time window t_m. All base learners are trained and validated independently from the same initialized θ_m. We save each of θ_m that shows the best validation performance throughout the training stage. The saved θ_ms are used for the inference. Our proposed *fitness* value measures the validation performance on validation data, which is defined as:

$$fitness = \frac{\frac{1}{N}\sum_i^N (P_{t+1}^i \times R_{i,t+1}^{long} + (1 - P_{t+1}^i) \times R_{i,t+1}^{short})}{Prediction\ Accuracy}$$

$$R_{i,t+1}^{long} = \frac{closing_{t+1}^i - closing_t^i}{closing_t^i},\ R_{i,t+1}^{short} = \frac{closing_t^i - closing_{t+1}^i}{closing_t^i},$$

where $R_{i,t+1}^{long/short}$ stands for futures trading return at time step $t+1$ by long or short position on stock i. The learner gains profit if the closing price of stock i at time $t+1$ ($closing_{t+1}^i$) becomes higher (lower) than the price at time t with the long (short) position. The long (short) position for stock i is determined if the estimated probability P_{t+1}^i is under (same as or over) 0.5.

A numerator of the fitness denotes an averaged expected return by the portfolio trading on N stocks. The fitness value then becomes a trading performance indicator per predictive accuracy. We propose this criterion to ensure the model maximizes the trading gains.

3.3 Base Learner

Figure 1-(b) describes the architecture of the base learner. The learner has three sub-modules: **Spatial Attention, Input Attention,** and **Spatiotemporal GRU (STGRU)**.

Spatial Attention. We propose it so the learner can learn a self-adaptive adjacency tensor without prior knowledge in an end-to-end way by stochastic gradient descent. It takes the t step of input x_{t_m}: $x_{t_m}^t \in \mathcal{R}^{N \times F}$ and a previous hidden state $h_{t-1} \in \mathcal{R}^{N \times H}$, where H is a feature dimension of the hidden state. By correlating them, the module produces the self-adaptive adjacency matrix at time t, $A_t \in \mathcal{R}^{N \times N}$. A formulation is defined as:

$$S_t = (x_{t_m}^t W_{sx} + B_{sx}) \otimes (h_{t-1} W_{sh} + B_{sx})',\ a_t = S_t W_{aux} + B_{aux}$$
$$A_t = Softmax(Tanh(S_t \oplus a_t)),$$

where Ws and Bs are trainable weights and biases. At first, $x_{t_m}^t$ and h_{t-1} are linearly transformed into their own embedding space of dimension E by each fully connected layer ($W_{sx} \in \mathcal{R}^{F\times E}, W_{sh} \in \mathcal{R}^{H\times E}$). $(\cdot)'$ denotes a transpose. The two embedded matrices are then multiplied to compute a correlation matrix $S_t \in R^{N\times N}$. We compute S_t with $x_{t_m}^t$ and h_{t-1} because we assume that the stocks' correlations are spatial dependent on each other, including themselves. However, two factors naturally affect the next day's movement: a set of information [2] and noise by noise traders [16]. We thus propose an auxiliary network to reflect these factors (i.e., auxiliary information for time t, notated as a_t) by fully connected layer $W_{aux} \in \mathcal{R}^{N\times N}$. The auxiliary information is then added to S_t to adjust the correlations. Quantity and direction of $S_t \oplus a_t$ are determined by the Tanh activation function, after which the softmax function normalizes them in a row-wise way. The self-adaptive adjacency matrix A_t contains the stocks' spatial correlations by reflecting the properties of the stock movements.

Input Attention. To adaptively extract salient input features for prediction and to capture long temporal dependencies at each time step, we propose to adopt the input attention mechanism [12]. We transform it into the GNN approach. Attention weights α_t between the input sequence x_{t_m} and the prior hidden state h_{t-1} are computed using a multi-layer perceptron with $Tanh(\cdot)$ and $Softmax(\cdot)$ activation functions.

The input attention takes two inputs: repeated prior hidden state $h_{t-1}^{rep} \in \mathcal{R}^{N\times F\times H}$ (h_{t-1} is repeated as much as F times) and transposed sequential data $x_{t_m}' \in \mathcal{R}^{N\times F\times t_m}$. The output is the adaptive attention weights on the input features $\alpha_t \in \mathcal{R}^{N\times F}$ at time t. The procedure at each time t is formulated as:

$$e_t = Tanh([h_{t-1}^{rep} \parallel x_{t_m}']W_u + B_u)W_e + B_e,\ \alpha_t = Softmax(e_t),$$

where $W_u \in \mathcal{R}^{(H+t_m)\times t_m}$, $W_e \in \mathcal{R}^{t_m\times 1}$, and Bs are learnable weights and biases. e_t is normalized by the $Softmax(\cdot)$ activation function in a row-wise way.

STGRU. STGRU can effectively model complex relationships among factors influencing stock movements, including spatial dependencies (e.g., interrelations among individual stocks, sectors, or markets) and temporal dependencies (e.g., time-series data such as historical prices or trading volumes).

We utilize the two outputs of the above two sub-modules (A_t, α_t) for STGRU. A_t produced by the temporal attention sub-module indicates the spatial relationship at time step t. Since STGRU is calculated recursively based on the input data's time length, the same computation is performed at the subsequent time $t+1$. A_{t+1} represents the adjacency matrix representing the new relationship between the stocks at time $t+1$. Throughout multiple gates' repeated calculations (reset gate: R_t, update gate: U_t, candidate: C_t, and hidden state: h_t), temporal relationships are learned for the entire input data's time-sequential length. It is important to note that the input features are multiplied by α at each time step, assigning higher weights for the highly relevant features to predictions. Details of the whole process are:

$$R_t = \sigma(A_t \otimes [(\alpha_t \odot x^t) \parallel h_{t-1}] \otimes W_r + B_r),\ U_t = \sigma(A_t \otimes [(\alpha_t \odot x^t) \parallel h_{t-1}] \otimes W_u + B_u)$$
$$C_t = Tanh(A_t \otimes [(\alpha_t \odot x^t) \parallel (R_t \odot h_{t-1})] \otimes W_c + B_c),\ h_t = (1 - U_t) \odot C_t + U_t \odot h_{t-1},$$

where σ is the sigmoid function. Trainable parameters Ws have all the same dimensions $((F + H) \times H)$. Throughout the above recursive operations, along with the time window, the hidden state h_t obtains sequentially embedded spatiotemporal correlations of the stocks.

Predictor. With the last hidden state h_t from STGRU, a predictor network maps it into the estimated probabilities of N stocks' movements P_{t+1} by a non-linear transformation. The predictor has two fully connected layers with the $Tanh(\cdot)$ activation function, which determines quantities and directions of movement information. P_{t+1} is computed as:

$$P_{t+1} = \sigma(Tanh(h_t W_{pred0} + B_{pred0})W_{pred1} + B_{pred1}),\ \hat{Y}_{t+1} = round(P_{t+1}),$$

where Ws: $W_{pred0} \in R^{H \times 2H}, W_{pred1} \in R^{2H \times 1}$ and Bs are trainable parameters. The estimated class labels $\hat{Y}_{t+1}$ are produced by round operation $(round(\cdot))$. The predictor network refines the information of the hidden state by mapping it into different dimension sizes $(H \to 2H \to 1)$.

3.4 Proposed Loss Function for Base Learner

Our proposed loss function for the mth base learner $(L(\theta_m))$ is formulated as:

$$L(\theta_m) = -\frac{1}{N}\sum_{i}^{N}\{y_i log(P^i_{t+1}) + (1 - y_i) log(1 - P^i_{t+1})\}$$
$$- \beta(P^i_{t+1} \times R^{long}_{i,t+1} + (1 - P^i_{t+1}) \times R^{short}_{i,t+1}),$$

where β is a coefficient regulating the influence of the expected return. P^i_{t+1} is the estimated probability of stock i. We devise this loss function because the expected return adjusts the model's predictive learning. For example, although the model has a large cross-entropy loss value, the overall loss value can be reduced if the expected return term is large enough.

We analyze the influence of the expected return on the base learner's learning procedure throughout a chain rule. Let us define the expected return on stock i at $t + 1$ to π^i_{t+1}. A gradient of each learner on the learnable parameter θ_m is computed as:

$$\frac{\partial L}{\partial \theta_m} = \frac{\partial CE}{\partial \theta_m} - \beta\frac{\partial \pi^i_{t+1}}{\partial \theta_m} = \frac{\partial CE}{\partial P^i_{t+1}}\frac{\partial P^i_{t+1}}{\partial \theta_m} - \beta(R^{long}_{i,t+1}\frac{\partial P^i_{t+1}}{\partial \theta_m} - R^{short}_{i,t+1}\frac{\partial P^i_{t+1}}{\partial \theta_m})$$
$$= (P^i_{t+1} - Y^i_{t+1})\frac{\partial P^i_{t+1}}{\partial \theta_m} - 2\beta R^{long}_{i,t+1}\frac{\partial P^i_{t+1}}{\partial \theta_m} = (P^i_{t+1} - Y^i_{t+1} \underline{- 2\beta R^{long}_{i,t+1}})\frac{\partial P^i_{t+1}}{\partial \theta_m}.$$

Note that $R^{long}_{i,t+1} = -R^{short}_{i,t+1}$. Y^i_{t+1} is a ground truth on stock i. $CE(\cdot)$ denotes the cross-entropy loss function.

We figure out that the underlined term $-2\beta R_{i,t+1}^{long}(= 2\beta R_{i,t+1}^{short})$ is a factor that contributes to the gradient of the base learner's total loss by adjusting the estimated probability P_{t+1}^{i}. This indicates that the trading performance (regarding the expected return) directly affects the base learner's learning procedure with the regulation coefficient β.

4 Experiments

4.1 Experimental Setup

We report the experimental results after averaging five independent runs.

Table 2. Preprocessing of input features.

Input Feature	Normalization
$n_{open}, n_{high}, n_{low}$ n_{close} n_{adj_close}	$n_{\{open,high,low\}} = \frac{\{open,high,low\}_t}{close_t} - 1$ $n_{close} = \frac{close_t}{close_{t-1}} - 1$ $n_{adj\_close} = \frac{adj\_close_t}{adj\_close_{t-1}} - 1$
n_{ma2}, n_{ma3} n_{ma5}, n_{ma8} n_{ma13}	$n_{mak} = \frac{\sum_{i=0}^{k-1} adj\_close_{t-i}/k}{adj\_close_t} - 1$

Dataset and Input Feature. We downloaded the daily price data (*open, high, low, close, and adjusted close (adj_close)*) using yfinance Python package. We collected the six stock datasets of different sectors: **Energy** (01/03/14–10/20/21), **Financials** (01/03/14–11/03/21), **Health Care** (11/01/13–07/30/21), **Industrials** (04/02/13–06/30/22), **Information Technology (IT)** (01/02/14–12/31/21), and **Materials** (01/03/13–03/31/22). Each dataset has 50 different stocks with the highest market capitalization worldwide. We split the period into 8:1:1 for training, validation, and testing.

Following [7,22], we normalized the input features as shown in Table 2. Here, we determined window sizes of the moving average ($n_{ma\#}$) to {2,3,5,8,13}, which are Fibonacci numbers to reflect the recent movement trends of the stocks [17].

Baseline Model. We selected the following baseline models to compare with MERTE. **RNN-based:** Adv-ALSTM [7] and CTV-TabNet [15]. **Transformer-based:** MTC-R [14] and DTML [22]. **GNN-based:** THGNN [19] and HATR [18]. We omitted their brief introductions due to the space limit. Refer to their original papers for the introductions.

Table 3. Predictive performance comparison in terms of ACC (%) and MCC. The bold-types denote the best, whereas the underlines are the next best.

Method	Energy		Financials		Health Care		Industrials		IT		Materials	
	ACC	MCC	ACC	MCC	ACC	MCC	ACC	MCC	ACC	MCC	ACC	MCC
Adv-ALSTM	53.6	0.06	53.2	0.04	52.6	0.05	51.8	0.03	52.3	0.02	52.1	0.03
CTV-TabNet	53.8	0.07	53.5	0.04	52.3	0.03	50.9	0.02	52.6	0.03	52.9	0.03
DTML	55.1	<u>0.10</u>	**55.0**	**0.07**	54.3	0.04	**55.9**	**0.12**	<u>54.7</u>	<u>0.08</u>	52.5	0.05
MTC-R	54.5	0.08	<u>54.3</u>	0.05	54.0	0.03	51.6	0.06	54.2	0.04	51.3	0.02
HATR	54.9	0.08	54.1	0.05	55.1	<u>0.08</u>	53.5	0.08	54.1	0.04	<u>54.2</u>	<u>0.08</u>
THGNN	<u>55.2</u>	<u>0.10</u>	**55.0**	<u>0.06</u>	<u>55.3</u>	<u>0.08</u>	54.0	<u>0.09</u>	54.6	0.05	53.0	0.06
MERTE (Ours)	**57.2**	**0.14**	**55.0**	**0.06**	**56.4**	**0.11**	<u>55.3</u>	**0.12**	**55.9**	**0.09**	**54.9**	**0.10**

Hyperparameter Search. We employed the Adam optimizer [9] for our proposed loss function. We searched the hyperparameters: an initial learning rate $\in \{0.01, 0.001, 0.0001\}$, $\beta \in \{0.01, 1, 2, 5, 10\}$, epochs $\in \{30, 50, 100\}$, a mini-batch size $\in \{64, 128, 256\}$, and the set of time windows $T \in \{[5, 10, 15], [5, 10, 15, 20, 25]\}$. With the same set of hyperparameters, all base learners were trained independently. Each base learner saved its best parameters that show the highest fitness value on the validation data during the training stage. The overall testing results are the average of each result produced by all base learners. For the baselines, we searched their hyperparameters based on their original papers.

4.2 Predictive Performance Comparison

Table 3 reports predictive results in terms of accuracy (ACC) and the Matthews correlation coefficient (MCC), which are widely used for the stock movement task [7,14,15,18,22]. MERTE yielded the highest prediction results on five of six datasets (MERTE showed the second best performance on Industrials), up to 2%p of ACC improvement on Energy than the following best baseline.

Lines 1 and 2 of Table 3 show the results by the RNN-based baselines, which are the lowest on all datasets. Although CTV-TabNet leveraged the same approach as ours, i.e., the various input time-sequential lengths, it did not surpass MERTE. This is because CTV-TabNet did not learn the correlations between the stocks. Lines 3 and 4 report the predictive performance by the transformer-based baselines. Among the two baselines, DTML demonstrated similar yet inferior results on almost all datasets compared to MERTE. This is because DTML is designed to capture the correlations between the stocks and employ them for its accurate predictions, which is different from MTC-R. Since the key to the GNN-based baselines presented in lines 5 and 6 is to learn the correlations between stocks and use them for prediction, they show better prediction results than those shown in lines 1, 2, and 4, which do not learn the correlations. This indicates that using the correlation between stocks is vital in predicting stock movements. The GNN-based baselines also performed similarly to DTML but did not outperform MERTE.

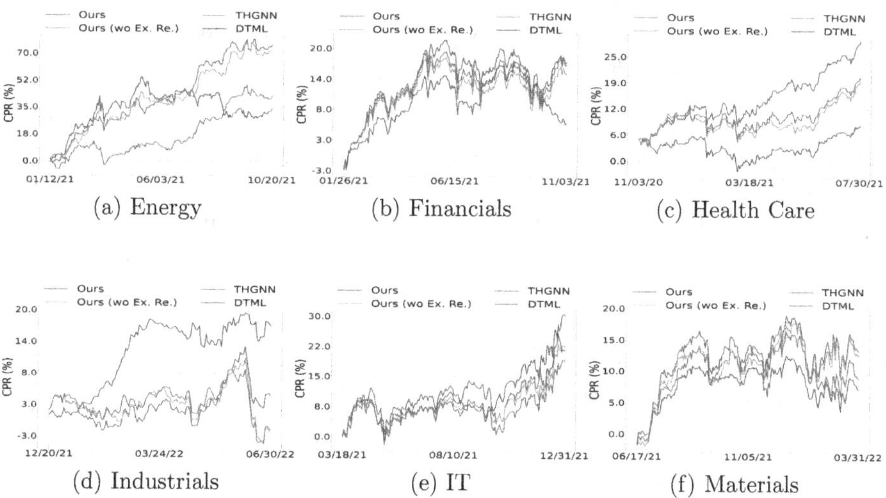

Fig. 2. CPRs (%) of the futures trading of the stock portfolio.

This is because MERTE captures the various stock momentums inherent in the time-sequential length each base learner receives as input, and each base learner uses the auxiliary network to infer the two real-world factors that affect stock movements: information and noise traders.

4.3 Trading Performance Comparison

To evaluate the predictive performance, we conducted *a daily futures trading of the stock portfolio*. At the beginning of the trading day, the model estimates the probabilities of the next day's 50 stock movements. It sets positions on the 50 stocks according to the probabilities (i.e., long position if $prob. \geq 0.5$ or short position if $prob. < 0.5$). After the end of the stock market, the model closes the positions and computes daily returns. The model gains profit if the closing price rises from the long position and vice versa from the short position. The model again builds the portfolio based on the estimated probabilities of the next day's movements. Refer to [14] for details.

Figure 2 exhibits cumulative portfolio returns (CPRs) during the testing periods. To confirm the influence of our proposed function on the trading, we also present a variant of MERTE: Ours (wo Ex. Re.), which refers to MERTE trained without the expected return in the loss function. We observed that MERTE gained superior CPR flows on almost all datasets, except Industrials, compared to the subsequent best predictive baselines. DTML exhibited the highest CPR on Industrials due to its accurate prediction performance, shown in Table 3.

Generally, the trading results can be attributed to accurate stock movement predictions; however, high prediction accuracy does not always guarantee a high CPR. For example, in the financials, DTML and THGNN had the same ACC of 55.0%, as shown in Table 3, but THGNN earned a higher CPR than DTML. This is because the model's accurate ability to predict days with large differences in price changes primarily determines trading profit. In other words, despite having the same ACC as DTML, THGNN achieved a higher CPR than DTML by accurately predicting movements on days with large price changes.

Each base learner of MERTE also learns trading performance in the learning process through our proposed loss function. When comparing the CPRs of Ours (wo Ex. Re.) and Ours, we found that the CPRs of Ours are higher in almost all datasets. These observations confirm that our proposed loss function improves the trading performance.

Table 4. Ablation study. The bold-types represent the highest values in each dataset and the area separated by underlines.

Method	Energy		Financials		Health Care		Industrials		IT		Materials	
	ACC	MCC	ACC	MCC	ACC	MCC	ACC	MCC	ACC	MCC	ACC	MCC
wo Ex. Re	**55.9**	**0.12**	**54.8**	**0.05**	**55.1**	**0.07**	53.9	0.07	**55.1**	**0.07**	53.7	0.06
wo Aux. Net	54.5	0.08	54.0	0.05	54.3	0.04	**54.4**	**0.10**	54.9	0.05	**54.3**	**0.08**
BL (5)	54.3	0.07	54.1	0.04	56.0	**0.10**	54.0	0.07	**55.3**	**0.07**	54.0	0.05
BL (10)	54.5	0.07	**55.2**	**0.06**	55.7	0.09	**54.6**	**0.10**	55.0	0.06	53.8	0.04
BL (15)	55.5	0.09	54.8	0.05	55.6	0.08	53.9	0.06	54.9	0.05	53.9	0.04
BL (20)	**55.7**	**0.10**	54.3	0.05	**56.1**	**0.10**	53.4	0.05	55.1	0.06	**54.1**	**0.06**
BL (25)	55.4	0.08	53.9	0.04	55.8	0.08	53.5	0.06	54.2	0.04	53.0	0.02
MERTE (Ours)	**57.2**	**0.14**	**55.0**	**0.06**	**56.4**	**0.11**	**55.3**	**0.12**	**55.9**	**0.09**	**54.9**	**0.10**

4.4 Ablation Study

To evaluate the effectiveness of our proposed modules, we generated the following variants: MERTE without expected return term (wo Ex. Re.) and MERTE without auxiliary network (wo Aux. Net.). Moreover, we also evaluated each base learner's predictive performance with the time-sequential length of its input (notated as BL (#) in lines 3 to 7 of Table 4).

From lines 1 and 2 of Table 4, we figured out that each of MERTE's modules improved the predictive performance since none surpassed MERTE's results. Furthermore, we also observed that no base learner could outperform MERTE except on the Financials dataset. This is because the base learner was trained on a limited length of the input view, which resulted in restricted predictive results. MERTE mitigated this issue and empirically demonstrated better performance using the multi-time ensemble method. Lastly, we investigated the relation between computational cost and ACC by comparing the performance of

BLs and MERTE. Since BLs are partial models of MERTE, we found that they have less computational costs than MERTE. However, BLs showed lower ACCs than MERTE on almost all datasets, indicating that MERTE improves ACCs at the expense of high computational costs.

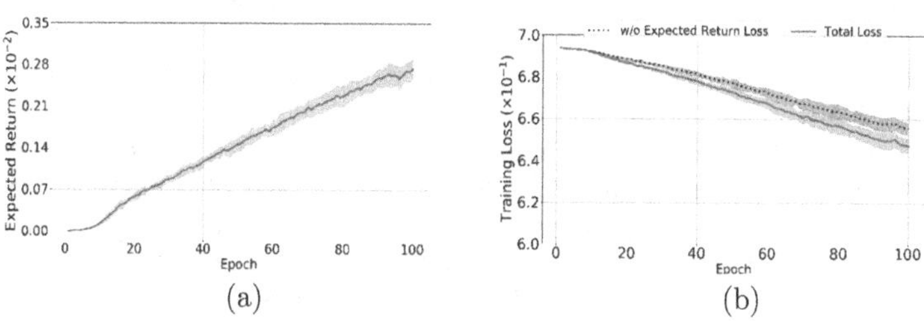

Fig. 3. (a) flow of expected return values during training and (b) loss curves with/without the expected returns on Energy.

4.5 Visualization of Proposed Loss Function

We visualized the expected return value's impact on the training loss. Figure 3-(a) displays the flow of the expected return, whereas Fig. 3-(b) compares the loss values without the expected return term (w/o Expected Return Loss) and our proposed loss (Total Loss). The expected return showed continuous increments during the training stage. These increments directly affected the total loss, reducing the w/o expected return loss curve shown in Fig. 3-(b). From these observations, we found that with our proposed loss function, the base learner of MERTE could acquire the trading capacity in its training process, and it also reduced the loss value for the prediction performance.

5 Conclusion

MERTE is proposed to forecast stock movements using the multi-time window ensemble of the base learners with the newly proposed loss function. It shows improved performance on both the predictive task and the stock portfolio futures tradings, compared to the baselines on almost all six real-world datasets. We also confirm that each proposed module contributes to MERTE's performance. The expected return term consolidated to the loss function and the multi-time window ensemble module are shown to enhance the prediction results. We will develop our approach for a portfolio trading algorithm in future work.

Acknowledgements. This work was supported by Korea Health Industry Development Institute(KHIDI) grant funded by the Korea government(MSIT) (No. H122C1983, Development of National Mental Health Trend Monitoring and Management Platform) and the National Research Foundation of Korea(NRF) grant funded by the Korea government(MSIT) (No. 2022R1F1A1065155).

References

1. Asness, C.S., Moskowitz, T.J., Pedersen, L.H.: Value and momentum everywhere. J. Financ. **68**(3), 929–985 (2013)
2. Barber, B.M., Odean, T.: All that glitters: the effect of attention and news on the buying behavior of individual and institutional investors. Rev. Financ. Stud. **21**(2), 785–818 (2008)
3. Barroso, P., Santa-Clara, P.: Momentum has its moments. J. Financ. Econ. **116**(1), 111–120 (2015)
4. Cho, K., et al.: Learning phrase representations using RNN encoder-decoder for statistical machine translation. In: Proceedings of the 2014 Conference on Empirical Methods in Natural Language Processing, EMNLP, A meeting of SIGDAT, a Special Interest Group of the ACL, Doha, Qatar, pp. 1724–1734. ACL (2014)
5. Ding, Q., Wu, S., Sun, H., Guo, J., Guo, J.: Hierarchical multi-scale gaussian transformer for stock movement prediction. In: IJCAI, pp. 4640–4646 (2020)
6. Fama, E.F.: The behavior of stock-market prices. J. Bus. **38**(1), 34–105 (1965)
7. Feng, F., Chen, H., He, X., Ding, J., Sun, M., Chua, T.S.: Enhancing stock movement prediction with adversarial training. In: Proceedings of the Twenty-Eighth International Joint Conference on Artificial Intelligence, pp. 5843–5849 (2019)
8. Hochreiter, S., Schmidhuber, J.: Long short-term memory. Neural Comput. **9**(8), 1735–1780 (1997)
9. Kingma, D.P., Ba, J.: Adam: a method for stochastic optimization. arXiv preprint arXiv:1412.6980 (2014)
10. Li, W., Bao, R., Harimoto, K., Chen, D., Xu, J., Su, Q.: Modeling the stock relation with graph network for overnight stock movement prediction. In: Proceedings of the Twenty-Ninth International Conference on International Joint Conferences on Artificial Intelligence, pp. 4541–4547 (2021)
11. Long, J., Chen, Z., He, W., Wu, T., Ren, J.: An integrated framework of deep learning and knowledge graph for prediction of stock price trend: an application in Chinese stock exchange market. Appl. Soft Comput. **91**, 106205 (2020)
12. Qin, Y., Song, D., Cheng, H., Cheng, W., Jiang, G., Cottrell, G.W.: A dual-stage attention-based recurrent neural network for time series prediction. In: Proceedings of the 26th International Joint Conference on Artificial Intelligence, Melbourne, Australia, pp. 2627–2633 (2017)
13. Rumelhart, D.E., Hinton, G.E., Williams, R.J.: Learning internal representations by error propagation. In: Rumelhart, D.E., Mcclelland, J.L. (eds.) Parallel Distributed Processing: Explorations in the Microstructure of Cognition, Volume 1: Foundations, pp. 318–362. MIT Press, Cambridge (1986)
14. Seo, K., Yang, J.: Exploring multi-time context vector and randomness for stock movement prediction. In: 2022 IEEE International Conference on Big Data (Big Data), pp. 1114–1123. IEEE (2022)
15. Seo, K., Yang, J.: Exploring candlesticks and multi-time windows for forecasting stock-index movements. In: Proceedings of the 38th ACM/SIGAPP Symposium on Applied Computing, pp. 1100–1109 (2023)
16. Shleifer, A., Summers, L.H.: The noise trader approach to finance. J. Econ. Perspect. **4**(2), 19–33 (1990)
17. Song, Y., Lee, J.W., Lee, J.: Development of intelligent stock trading system using pattern independent predictor and turning point matrix. Comput. Econ. 1–12 (2020)

18. Wang, H., Li, S., Wang, T., Zheng, J.: Hierarchical adaptive temporal-relational modeling for stock trend prediction. In: IJCAI, pp. 3691–3698 (2021)
19. Xiang, S., Cheng, D., Shang, C., Zhang, Y., Liang, Y.: Temporal and heterogeneous graph neural network for financial time series prediction. In: Proceedings of the 31st ACM International Conference on Information & Knowledge Management, pp. 3584–3593 (2022)
20. Xu, Y., Cohen, S.B.: Stock movement prediction from tweets and historical prices. In: Proceedings of the 56th Annual Meeting of the Association for Computational Linguistics (Volume 1: Long Papers), Melbourne, Australia, pp. 1970–1979. Association for Computational Linguistics (2018)
21. Ye, J., Zhao, J., Ye, K., Xu, C.: Multi-graph convolutional network for relationship-driven stock movement prediction. In: 2020 25th International Conference on Pattern Recognition (ICPR), pp. 6702–6709. IEEE (2021)
22. Yoo, J., Soun, Y., Park, Y.C., Kang, U.: Accurate multivariate stock movement prediction via data-axis transformer with multi-level contexts. In: Proceedings of the 27th ACM SIGKDD Conference on Knowledge Discovery & Data Mining, KDD 2021, pp. 2037–2045. Association for Computing Machinery, New York (2021)
23. Zhang, Q., Qin, C., Zhang, Y., Bao, F., Zhang, C., Liu, P.: Transformer-based attention network for stock movement prediction. Expert Syst. Appl. 117239 (2022)

MOT: A Mixture of Actors Reinforcement Learning Method by Optimal Transport for Algorithmic Trading

Xi Cheng(✉), Jinghao Zhang, Yunan Zeng, and Wenfang Xue

Institute of Automation, Chinese Academy of Sciences, Beijing, China
{xi.cheng,jinghao.zhang,yunan.zeng}@cripac.ia.ac.cn,
wenfang.xue@ia.ac.cn

Abstract. Algorithmic trading refers to executing buy and sell orders for specific assets based on automatically identified trading opportunities. Strategies based on reinforcement learning (RL) have demonstrated remarkable capabilities in addressing algorithmic trading problems. However, the trading patterns differ among market conditions due to shifted distribution data. Ignoring multiple patterns in the data will undermine the performance of RL. In this paper, we propose MOT, which designs multiple actors with disentangled representation learning to model the different patterns of the market. Furthermore, we incorporate the Optimal Transport (OT) algorithm to allocate samples to the appropriate actor by introducing a regularization loss term. Additionally, we propose Pretrain Module to facilitate imitation learning by aligning the outputs of actors with expert strategy and better balance the exploration and exploitation of RL. Experimental results on real futures market data demonstrate that MOT exhibits excellent profit capabilities while balancing risks. Ablation studies validate the effectiveness of the components of MOT.

Keywords: Algorithmic trading · Reinforcement learning · Optimal transport

1 Introduction

The goal of algorithmic trading is to maximize long-term profits while keeping risks within an acceptable range [21]. Compared to the traditional approach of relying on the expert judgment of trading timing, algorithmic trading is highly automated and efficient.

Traditional technical analysis methods include mean reversion [10], momentum investing [11], multi-factor models [4], etc. However, financial market data is non-stationary with a low signal-to-noise ratio. Expert-designed technical analysis methods can't generate profits under diverse market conditions. Deep learning methods excel at capturing intricate price patterns and enhance models' performance [15,28,29]. However, the process from supervised models' output to actual investment still requires the construction of strategies, which introduces expert knowledge and subjectivity. RL methods don't require carefully designed strategies by humans. They take market information as states and output trading decisions directly, which makes it easy to incorporate the unique financial constraints (e.g. transaction costs and slippage) into environments. RL has achieved SOTA in many quantitative investment tasks [16,19,30].

D.-N. Yang et al. (Eds.): PAKDD 2024, LNAI 14648, pp. 30–42, 2024.
https://doi.org/10.1007/978-981-97-2238-9_3

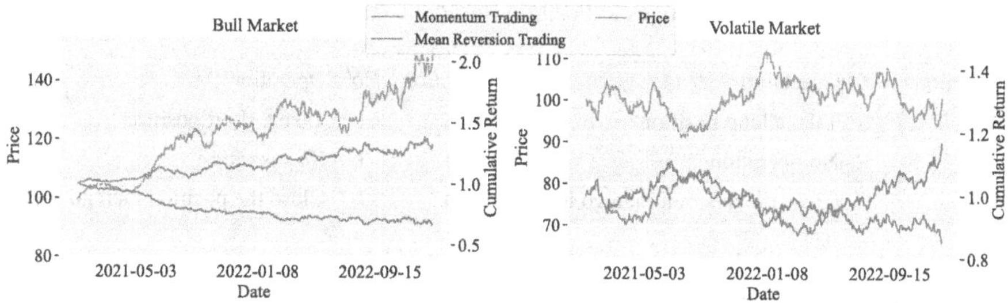

Fig. 1. Profit of strategies in different market conditions. A bull market is suitable for momentum trading, while a volatile market is suitable for mean reversion trading.

However, these methods rely on the assumption that financial data always follow the same distribution. Data patterns often switch in real scenarios. E.g. the most common way to classify market patterns is into two categories: stable (momentum) and volatile (reversal) markets, which require two categories of strategies [14]. These two phenomena are not independent but intertwined with each other. As shown in Fig. 1, when bullish forces > bearish forces or are evenly balanced, the market is in a stable upward (bull) or a volatile state respectively. Momentum trading strategy models the momentum effect of stable market and mean reversion strategy models the reversal effect of volatile market. The same strategy can yield significantly different returns in different market conditions. Inspired by a mixture of experts [5], we propose MOT, which models multiple actors with disentangled representation learning and extracts various pattern information in RL. To allocate samples to agents appropriately, we introduce the Allocation Module with Optimal Transport (OT) regularization loss.

Previous research [16] has introduced imitation learning to RL, allowing agents to learn information from expert knowledge. However, in the early stages of imitation learning, the sampled action used in the training process is not from the agent's generation but is directly given by the expert which is stored in the buffer. As a result, the true output action of the agent differs significantly from the action stored in the buffer. To solve this problem, MOT introduces a pre-training method based on supervised learning to imitation learning. We expect the output generated by the agent to be closer to the expert strategy in imitation learning so the model can be initialized in a better stage.

The training process of MOT can be divided into three stages: first, the Pretrain Module uses supervised learning to train only the actor with expert strategy. Then we use the expert strategy to fill the buffer and train the RL model by imitation learning. After that, MOT uses multiple actors to model different market patterns and uses OT to solve the problem of pattern allocation. The contributions are summarized as follows:

1) MOT is the first that introduces OT algorithm to RL for mining various trading patterns. Allocation Module allocates different samples to appropriate actors.
2) MOT is also the first study that addresses the imitation learning gap between the actor's output and the buffer. MOT introduces a supervised Pretrain Module before imitation learning, which allows the real actor's output to be closer to the expert strategy.
3) Experiments show MOT has great profitability in different market modes while balancing risks. Further studies confirm the effectiveness of three components of MOT.

Table 1. Changes in Position Based on Trading Signals

Po	Action	Po'	Operation	Po	Action	Po'	Operation
0	1	1	Take a long position	0	−1	−1	Take a short position
1	1	1	No operation	−1	−1	−1	No operation
−1	1	1	Close the position then go long	1	−1	−1	Close the position then go short

2 Problem Formulation

The algorithmic trading problem can be represented as Markov Decision Process (MDP) $\mathcal{M} = \langle \mathcal{S}, \mathcal{A}, \mathcal{P}, \mathcal{R}, \gamma \rangle$, where $\mathcal{S}$ represents the state space provided by the environment, $\mathcal{A}$ represents the action space, $\mathcal{P} : \mathcal{S} \times \mathcal{A} \times \mathcal{S} \to [0, 1]$ is the probability function of the conditional state transitions, $\mathcal{R} : \mathcal{S} \times \mathcal{A} \to \mathbb{R}$ is the reward function, and $\gamma \in (0, 1)$ is the discount factor. The specific definition of the five-tuple for MDP is as follows:

The State Space $\mathcal{S}$: The state $\mathbf{S}_t = [\mathbf{S}_t^m; \mathbf{S}_t^a] \in \mathcal{S}$. The account indicators $\mathbf{S}_t^a = [s_t^{a_1}, s_t^{a_2}, ...]$ describe the trader's positions, account cash balance, margin, returns, and other related information of the trader's account. $\mathbf{P}_t = [p_t^o, p_t^h, p_t^l, p_t^c, v_t^o, v_t^a]$ represents the Opening-High-Low-Closing (OHLC) prices, trading volume, and trading value. $\mathbf{Q}_t = [q_t^1, q_t^2, ..., q_t^i]$ are derived from $\mathbf{P}_t$ and technical analysis. The market indicators $\mathbf{S}_t^m = [\mathbf{P}_t; \mathbf{Q}_t]$ include the volume-price data $\mathbf{P}_t$ and the technical indicators $\mathbf{Q}_t$.

The Action Space $\mathcal{A}$: The action $a_t \in \{-1, 1\}$ represents the trading signal output by the agent. −1 corresponds to short selling and 1 corresponds to a long position. We define the agent to trade in units of contracts. The actual execution of trades depends on the trading signal and the trader's existing positions. The specific changes in position and action are summarized in Table 1, where Po means position.

The Transition Function $\mathcal{P}$: We assume that the actions of individual traders do not affect the overall asset price in the market. This implies that the observation transition function of market indicators is independent of trading behavior, i.e. $\mathcal{P}(\mathbf{S}_{t+1}^m | \mathbf{S}_t) = \mathcal{P}(\mathbf{S}_{t+1}^m | \mathbf{S}_t, a_t)$. However, the observation transition function of account prices is influenced by trading behavior, i.e. $\mathcal{P}(\mathbf{S}_{t+1}^a | \mathbf{S}_t) \neq \mathcal{P}(\mathbf{S}_{t+1}^a | \mathbf{S}_t, a_t)$.

The Reward $\mathcal{R}$: We choose the closing price p_t^c to calculate profit r_t. To better simulate real market, we set transaction fee rate μ[1] and slippage σ[2]. The profit r_t is defined as $r_t = (p_t^c - p_{t-1}^c - 2\sigma) \cdot a_{t-1} - \mu \cdot p_t^c \cdot |\Delta po|$, where $\Delta po = Po' - Po$. When setting rewards, it is inappropriate to consider only the profit without taking into account the risk. The Sharpe ratio is the most widely used indicator for balancing risk and returns [24], defined as $SR = \frac{mean(r_t)}{std(r_t)}$. To measure the impact of the profit on SR each step, we adopt the Differential Sharpe Ratio (DSR) [17] as the reward. Considering that the adjacent data is more important than distant previous data in algorithmic trading, DSR employs the smoothing technique of Exponential Moving Average (EMA). DSR_t is defined as:

[1] Transaction costs are charged as a percentage of the contract.

[2] Slippage refers to the difference between the expected and the actual execution price.

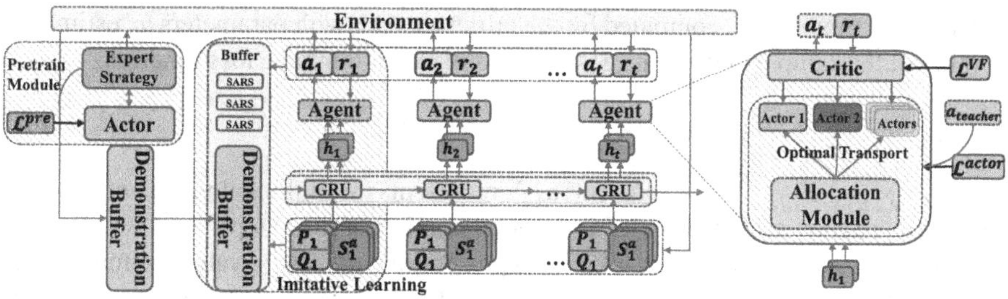

Fig. 2. The architecture of MOT. First, we pretrain the actor using the expert strategy and then proceed with imitation learning. We model different market patterns using multiple actors and allocate samples to the actors using the Allocation Module.

$$DSR_t = \frac{B_{t-1}\Delta A_t - \frac{1}{2}A_{t-1}\Delta B_t}{(B_{t-1} - A_{t-1}^2)^{\frac{3}{2}}}, \tag{1}$$

representing the impact of each new profit r_t on SR after applying EMA. A_t is the first moment and B_t is the second moment of profits r_t estimated by EMA. We utilize the DSR_t as the reward $\mathcal{R}$. If the account money is insufficient, the trading will be terminated in advance, and we simulate this by setting a margin threshold.

3 Methodology

The overview of MOT is present in Fig. 2. First, to ensure alignment between the actions in Demonstration Buffer and the actual outputs of the actor, we introduce Pretrain Module. Second, we leverage imitation learning to initialize the RL algorithm. Third, we use multiple actors with disentangled representation learning and model various market conditions. Last, Allocation Module allocates samples to different actors by OT algorithm.

3.1 Imitation Learning

In RL-based algorithmic trading, the initial exploration phase is often inefficient and yields low profits. Imitation learning leverages expert knowledge and provides the actor with a favorable starting point. We employ PPO [23] as the backbone to address the MDP problem. To capture the temporal patterns of states $\mathbf{S}_t$, we utilize Gated Recurrent Units (GRU) [1] to obtain the hidden representation $h_t = GRU(h_{t-1}, \mathbf{S}_t)$ of states $\mathbf{S}_t$. h_t is then fed into the actor and critic networks as inputs.

The actor network aims to find the optimal policy π by maximizing the advantage function. The input is the environment state $\mathbf{S}_t$ and the output is the action a_t. To ensure sufficient exploration by the agent, we add noise ε to the output of the actor network. The actual executed action $a_t = \pi^{\theta}(h_t) + \varepsilon$, where ε represents the noise, π is the policy given by actor network with parameters θ. The trading experience trajectories (SARS: state, action, reward, new state) are stored in the buffer $\mathcal{B}$. After sampling, we update the gradients of the actor network and the critic network using the data from $\mathcal{B}$.

The value function V, computed by the critic network with parameters ω, estimates the value of the sample under state $\mathbf{S}_t$. It is optimized through the loss function:

$$\mathcal{L}^{VF}(\omega) = \mathbb{E}\left[(V_\omega(\mathbf{S}_t) - V_t)^2\right], \tag{2}$$

where $V_t = \sum_{t'=t}^{T-1} \mathbb{E}[\gamma^{T-t'-1} DSR_{t'}(\mathbf{S}_{t'}, a_{t'})]$ represents the empirical value of the accumulated future rewards DSR and T is the total number of time steps.

Let $\delta_t^V = DSR_t + \gamma V(\mathbf{S}_{t+1}) - V(\mathbf{S}_t)$ represent the advantage value estimation. In our research, the advantage function is computed by generalized advantage estimator (GAE) [23]: $\hat{A}_t^{GAE(\gamma,\lambda)} = \sum_{k=t}^{T-1} (\gamma\lambda)^{k-t} \delta_k^V$, where γ is the discount factor, λ represents the trade-off between variance and bias.

PPO introduces a surrogate objective function to measure the similarity between the updated policy and the previous policy. The policy ratio formula is $\frac{\pi_\theta(a_t|\mathbf{S}_t)}{\pi_{\theta_{\text{old}}}(a_t|\mathbf{S}_t)}$. $\pi_{\theta_{\text{old}}}$ and π_θ represents the original and updated policy respectively. The objective function $\mathcal{L}^{CLIP}(\theta)$ for policy update is as Eq. 3, ϵ is the clipping threshold.

We employ the commonly used Dual Thrust [13] as the expert strategy to provide demonstration actions. We store the demonstration trajectory SARS in Demonstration Buffer (DB) and train the agent using samples from DB. The training of the actor-critic network in imitation learning follows the same approach as the PPO algorithm, with the only difference being that the training data is from DB. Subsequently, the actor-critic network continues to train by PPO method, as shown in Eq. 2 and Eq. 3:

$$\mathcal{L}^{CLIP}(\theta) = \mathbb{E}\left[\min(\frac{\pi_\theta(a_t|\mathbf{S}_t)}{\pi_{\theta_{old}}(a_t|\mathbf{S}_t)}\hat{A}_t, clip(r_t(\theta), 1-\epsilon, 1+\epsilon)\hat{A}_t)\right]. \tag{3}$$

3.2 Pretrain Module

The Pretrain Module is used to align the actions in the buffer $\mathcal{B}$ with the outputs of the actor. As mentioned before, it can be observed that a_{expert} in DB is directly provided by Dual Thrust strategy rather than generated by the actor network π_θ. Therefore, when using the demonstration data for gradient descent of the network, there is a significant discrepancy between the distribution of the actor network's output action π_θ and the action a_{expert} [16]. This has a negative impact on the stability of the RL network.

To address this issue, we aim to align the output action $a_t = \pi_\theta$ of the actor network with the expert-provided action a_{expert} by training the actor network using supervised learning. The loss function is defined as Eq. 4:

$$\mathcal{L}^{pre} = CrossEntropy(a_{expert}, \pi_\theta(h_t)) \tag{4}$$

Pretrain Module accelerates the actor's understanding of the task by mimicking expert strategies and enhances the actor's ability to effectively engage in the imitation learning process. Pretrain Module is positioned before imitation learning as Fig. 2.

3.3 Multiple Actors

We employ multiple actors to model strategies in different patterns. Futures data is derived from the trading activities of numerous participants and reflects different trading

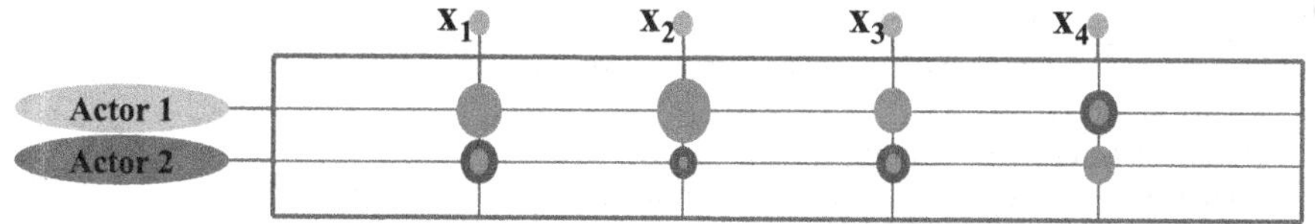

Fig. 3. OT refers to assigning x to the actor with the minimum L_{err}^{ij} while achieving a balanced allocation proportion, $\frac{x \ to \ Actor \ 1}{x \ to \ Actor \ 2} \approx \frac{w_1}{w_2}$. The pink circles represent L_{err}^{ij}.

patterns [22]. Ignoring multiple patterns will reduce the performance of models [8]. All k actors of MOT are constructed in the same manner, as depicted in Fig. 2 and Eq. 3. For convenience, we illustrate how the model is trained with $k = 2$.

To integrate the outputs of the two actors, we use an Allocation Module to assign weights to them. Regarding the construction of the Allocation Module, we first consider what inputs should be provided to it. The historical sequence of futures states $\mathbf{S}_t$ plays a significant role in determining the current market patterns. Additionally, the historical decision errors of different actors represent their decision-making performance and also influence the current sample allocation. We use GRU to extract latent feature representations from $\mathbf{S}_t^i$, denoted as $\hat{h}_t^i = GRU(\hat{h}_{t-1}^i, \mathbf{S}_t^i)$, where i means i-th sample. As the calculation of sample decision errors, we provide posterior teacher actions on the training set. The teacher action $a_{teacher} = 1$ when the futures close price p_t^c increases in the next time step and -1 otherwise. Let a^{i1} and a^{i2} represent the action output by actor 1 and actor 2. The sample decision error $\mathbf{e}_t^i$ is then computed as $[a_{teacher\ t}^i - a_t^{i1}, a_{teacher\ t}^i - a_t^{i2}]$. To avoid introducing future information, we utilize the previous error $\mathbf{e}_{t-1}^i$. Subsequently, we concatenate $\hat{h}_t^i$ and embedding of error sequence $\mathbf{d}_{t-1}^i = GRU(\mathbf{d}_{t-2}^i, \mathbf{e}_{t-1}^i)$ and feed them into a fully connected layer to predict the allocation results, denoted as $\mathbf{b}_t^i = FC(\hat{h}_t^i, \mathbf{d}_{t-1}^i)$. In different patterns Allocation Module should have different attention for the two actors in Eq. 5, where q_t^i represents the allocation weights, and a_t^i represents the final action. To ensure the discrete differentiability of the Allocation Module, we utilize the gumbel-softmax method [9] to compute Eq. 5. It is worth noting that the allocation of samples is not binary, but rather a soft allocation ranging $0 < \mathbf{q}_t^i < 1$.

$$\mathbf{q}_t^i = softmax(\mathbf{b}_t^i), \ a_t^i = \mathbf{q}_t^{i^T} [a_t^{i1}, a_t^{i2}], \tag{5}$$

However, if the actors want to learn different patterns, the representations should be as dissimilar as possible. Inspired by disentangled representation learning, we take the inputs x of the actors' last layers as the representations and design a disentangled loss to enable the agent to learn different patterns, $\mathcal{L}^{dis} = \sum_{i=1}^{N} x_{i1} \cdot x_{i2}$.

3.4 Optimal Transport Regularization

However, the model lacks a mechanism to ensure the effective allocation of samples to actors. Sometimes, the majority of samples are assigned to one actor. We incorporate OT techniques to ensure that the Allocation Module assigns more appropriate samples to each actor, thereby capturing diverse patterns more accurately.

Algorithm 1. Training process of MOT

```
Initialize actor network parameters θ_0, critic network parameters ω_0 and epochs K
Obtain the expert strategy
Pretrain the actor by L^{pre} in Equation 4
Add the expert strategy to DB and train by imitation learning, get the dual policies π_{θ_j}(a|S),
  j = 1, 2
for k = 0, 1, 2, ... do
    Collect the trajectory τ_t = (S_t, a_t, DSR_t, S_{t+1})_{t=0}^{T-1} by allocating the policy in Equation 5
    Compute advantages Â_t by current value V_{ω_t}(S_t)
    Compute the policy ratio π_{θ_t}(a_j|S_t) / π_{θ_{t-1}}(a_j|S_t)
    Compute the loss L^{OT} and L^{dis} in Equation 7
    Update the policy network by maximizing the clipped objective using L^{actor}(θ) in Equation 7 (both for actor 1 and actor 2)
    Update the critic network by minimizing loss L^{VF}(ω) in Equation 2
end for
```

We need to consider two main requirements. Firstly, the Allocation Module should allocate the samples to the actor with the smallest decision error. In other words, if $|a^i_{teacher\ t} - a^{i1}_t| > |a^i_{teacher\ t} - a^{i2}_t|$, we tend to assign the sample to actor 2. Secondly, the allocation of samples to the actors should be proportional to their respective patterns.

Below, we formally define the allocation problem. Assume we utilize N samples in each epoch of PPO's gradient descent process. Based on the error vector, we can construct an error matrix denoted as $L_{err} \in [N \times 2]$. Each element L^{ij}_{err} in it represents the decision error of the i-th sample on the j-th actor, given by $L^{ij}_{err} = a^i_{teacher} - a^{ij}$. Corresponding to L_{err} is the allocation matrix $M \in [N \times 2]$, where each element $M^{ij} \in \{0, 1\}$. The value of 1 in the allocation matrix M indicates that Allocation Module assigns the i-th sample to the j-th actor, while the value of 0 indicates no allocation.

The OT method is particularly suitable for solving allocation problem. OT involves determining an optimal allocation of resources from one location to another while minimizing overall cost or distance. It is also commonly employed to measure the difference between two probability distributions. Our research aims to find the optimal allocation scheme that minimizes L_{err}. The specific formulation of the problem is as follows,

$$
\begin{aligned}
&\underset{M}{min}\ (L \cdot M) \\
&s.t. \frac{\sum_{i=1}^{N} M^{i1}}{N} = w_1, \quad \frac{\sum_{i=1}^{N} M^{i2}}{N} = w_2, \quad M^{i1} + M^{i2} = 1, \forall i = 1, 2, \dots, N,
\end{aligned}
\tag{6}
$$

where w_1 and w_2 represent the proportions corresponding to different modes (assumed to be $\frac{1}{2}$). We employ the Sinkhorn method to solve the OT problem [2]. Figure 3 provides a visual explanation of the problem we aim to address.

To align the distribution of the output $\mathbf{q}^i$ from the allocation module with M^i of the OT problem, we incorporate a cross-entropy loss term. Considering Allocation Module as part of actors, Eq. 3 can be expanded to Eq. 7, λ_O is the hyperparameter. The third

Table 2. Experimental Results (↑ indicates the higher the better, ↓ indicates the opposite)

Methods	ARR (↑)	VO (↓)	ASR (↑)	MDD (↓)	CR (↑)	SoR (↑)
Long Hold	−2.598	0.261	−0.638	113.121	−0.001	−0.080
Short Hold	3.163	0.259	0.782	0.894	0.041	0.093
Dual Thrust	10.130	0.253	2.628	0.033	3.962	0.365
GRU	11.342(1.12)	0.242(0.00)	3.004(0.31)	0.016(0.02)	4.280(0.23)	0.399(0.05)
iRDPG	14.453(0.98)	0.254(0.01)	3.955(0.18)	0.023(0.03)	5.881(3.21)	0.537(0.03)
PPO	12.245(0.23)	0.243(0.00)	3.223(0.05)	0.022(0.02)	4.281(0.23)	0.436(0.01)
MOT-ND	15.322(1.25)	0.246(0.01)	4.252(0.24)	**0.005(0.01)**	**7.277(3.51)**	0.587(0.07)
MOT-NO	17.236(1.05)	0.248(0.01)	4.447(0.18)	0.026(0.01)	5.558(0.75)	0.529(0.08)
MOT	**20.379(0.85)**	**0.228(0.00)**	**5.395(0.26)**	0.011(0.02)	6.582(0.66)	**0.605(0.05)**

term is L^{OT}. The pseudocode for the MOT is shown in Algorithm 1.

$$\mathcal{L}^{actor}(\theta) = \mathcal{L}^{CLIP}(\theta) + \mathcal{L}^{dis} + \lambda_O \sum_{k=1}^{2} M_t^{ik} log(\mathbf{q}_t^{ik}). \tag{7}$$

4 Experiments

4.1 Dataset

We utilize the IF stock index futures dataset whose underlying asset is the CSI 300 Index. The dataset provides minute-level trading data of contracts. Each minute bar includes OHLC, trading volume, etc. The total trading duration in a day is 240 min. We collected it from ricequant.com[3] and divided the data into a training set from 2015-12-31 to 2018-05-08 and a test set from 2018-05-09 to 2019-05-09.

4.2 Baselines, Evaluation Metrics and Hyperparameters

Baselines: Long Position Hold (buy futures and hold), Short Position Hold (borrow contracts and hold), Dual Thrust [13] (a technical analysis trading strategy commonly used for intraday trading), GRU [1] (a variant of RNNs[4]), PPO [23] (a RL method that improves stability by preventing large policy changes[5]), iRDPG [16] (SOTA: an off-policy algorithm that incorporates expert strategy and behavior cloning).

Evaluation Metrics: We will measure the model's performance by Accumulated Rate of Return (ARR, the overall profitability), Volatility (VO, measures by standard deviation of profit r), Annualized Sharpe Ratio (ASR, annualized version of SR), Maximum

[3] A well-known Chinese quantitative trading platform, https://www.ricequant.com/.

[4] We chose it as a baseline because we employed the GRU method in the Pretrain Module before imitation learning. The results of GRU demonstrate the performance of the Pretrain Module.

[5] We enhance PPO using imitation learning mentioned in Methodology Section.

Drawdown (MDD, the maximum decline of an asset's value from its peak to the lowest over a period), Calmar Ratio (CR=$\frac{ARR}{MDD}$, risk-adjusted ARR based on MDD) and Sortino Ratio (SoR=$\frac{mean(r)}{std(min(r,0))}$, excess return per unit of downside risk).

Hyperparameters: We set transaction fee rate μ = 2.3 × 10^5 and slippage σ = 0.2. Insufficient account assets may trigger a forced liquidation. We set the margin threshold as 70% and initial capital $C = 50000\ CNY$. We repeated 6 experiments for each model.

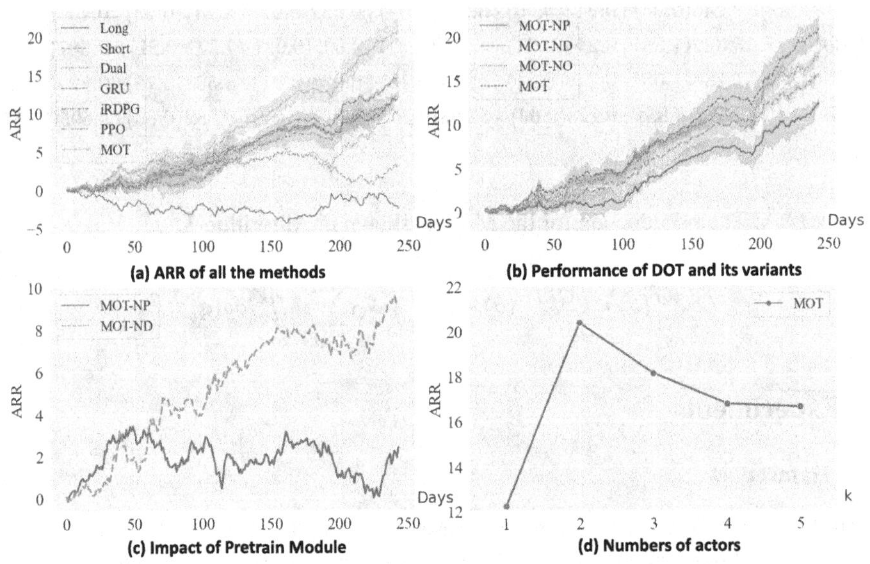

Fig. 4. Performance of different models in terms of ARR

4.3 Experimental Results

Table 2 provides a summary of the results. Figure 4(a) depicts ARR of all the methods. From Table 2, MOT outperforms other methods in terms of profit and risk-reward balance. ARR is the most crucial indicator, and our model achieves the highest ARR. The ARR of PPO is about 1.0 higher than that of GRU, indicating that PPO exhibits greater robustness. The ASR, CR, and SoR are composite metrics that consider both risk and return. Deep learning methods (last 6 rows in Table 2) outperform the technical indicator models (first 3 rows in Table 2) in these three metrics, which suggests the former better represents complex states under high-noise conditions. MOT performs second in terms of MDD, indicating that MOT only requires a short time period to recover from losses. RL models outperform time-series models, as the latter primarily focuses on predicting price trends without considering the high costs caused by incorrect predictions. Since greater risk leads to greater returns, profits are higher when there are significant price fluctuations. So the correlation among most methods is very high.

4.4 Ablation Study

We conducted ablation experiments to show the effectiveness of its three components. The experimental results and the trend of ARR are depicted in Table 2 and Fig. 4(b).

Overall Performance. MOT-NP applies imitation learning based on PPO without Pretrain Module. MOT-ND is obtained by removing multiple actors from the final model, while MOT-NO eliminates the process of OT. From Fig. 4(b), we observe that ARR curve of MOT remains higher than other variants in most periods. Table 2 shows that MOT performs best in terms of ARR, VO, ASR, and SOR. Among the three modules, OT method contributes the most to the improvement of model performance, followed by the Pretrain Module. MOT-ND excels in MDD metric, indicating that the model without multiple actors' design tends to generate more conservative strategies. While a conservative trading strategy often misses the optimal investment opportunities. Since the calculation of CR relies on MDD, MOT-ND also exhibits higher CR.

Effectiveness of Pretrain Module. The influence of the expert strategy in DB diminishes over time and the benefit of imitation learning is mainly observed in the early stages. For the ablation experiment, we selected the agent trained for 100 epochs after imitation learning. Figure 4(c) illustrates the impact of Pretrain Module on imitation learning and the yellow curve is the model with Pretrain Module. It can be observed that MOT-ND demonstrates a steady increase accompanied by minor fluctuations in profit. In contrast, MOT-NP experiences some declines and doesn't learn well. This indicates that Pretrain Module contributes to the improvement of imitation learning.

Effectiveness of Multiple Actors and OT Modeling. Figure 5 demonstrates the variation in weights assigned to two actors before and after OT modeling. In a relatively volatile period, the model assigns weights more randomly without OT while assigns higher weights to actor 2 with OT. Notably, the introduction of OT leads to higher returns and enhances the ability to capture complex patterns. Figure 4(d) illustrates the impact of actors' number to MOT. MOT achieves the best profitability when $k = 2$ while achieves the worst when $k = 1$. This indicates that only one actor is insufficient to capture all patterns, while an excessive number of actors may lead to redundancy. In our model, the optimal number of actors is 2.

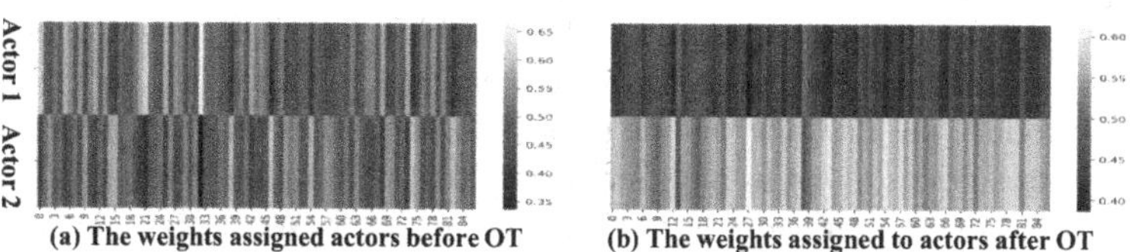

(a) The weights assigned actors before OT (b) The weights assigned to actors after OT

Fig. 5. Effectiveness of OT modeling

5 Related Work

Investment Strategies Based on Expert Knowledge. The early method used expert knowledge to construct heuristic rules [10,20], which can be divided into two categories: fundamental analysis and technical analysis. Fundamental analysis captures diverse factors such as industry trends, company financial statements, and public opinion. This method is more commonly used by long-term investors to find undervalued assets. Popular technical indicators include Relative Strength Index [27], Average Direction Index [6], On-Balance Volume [26], etc. Commonly used investment strategies include momentum trading [7] and mean reversion strategy [20]. However, interrelated technical indicators are correlated with each other, and building them directly from the market introduces too much market noise. Typically, rules constructed based on expert knowledge can only capture trading opportunities under specific market conditions [3].

Investment Strategies Based on RL. In contrast to supervised learning, which still requires expert knowledge to construct strategies, RL can optimize strategies in an end-to-end form. Moody et al. [18] made the first attempt to apply recurrent RL (RRL) algorithm to algorithmic trading. However, traditional RL methods are not well-suited for environments with large state spaces, making it challenging to select market features. Deep RL methods have partially addressed this problem. Si et al. [25] argue that strategies need to consider multiple factors and combine multi-objective optimization with deep RL to address this issue. Oliveira et al. [19] adopts SARSA, which maps states and actions to specific cells in a table to learn the value function. Since insufficient financial data causes overfitting, Jeong et al. [12] divided stocks into groups based on their correlations and introduced transfer learning into the Deep Q-Network (DQN). To shorten the inefficient random exploration phase, iRDPG [16] incorporates technical analysis through imitation learning. Yuan et al. [30] argue that daily frequency data cannot meet the high demands of RL and instead use minute frequency data. And PPO algorithm achieves more stable returns compared to DQN and SAC algorithms.

6 Conclusion

In this paper, we propose MOT, an RL-based model for algorithmic trading problems. Specifically, we model the algorithmic trading problem as MDP and leverage imitation learning to enable the agent to learn from expert knowledge. To better initialize MOT, we introduce the Pretrain Module prior to the imitation learning phase. Considering that futures prices result from different patterns, we employ multiple actors with disentangled representation learning to model the patterns. We design the Allocation Module to integrate the outputs of multiple actors and incorporate OT techniques to guide the learning of the Allocation Module. Experimental results demonstrate that our model achieves superior profitability while controlling the risk, showcasing its robustness in financial markets with complex data patterns. Further ablation studies confirm the effectiveness of the three components of MOT.

Acknowledgements. This work was supported by the National Natural Science Foundation of China (No. 72374201).

References

1. Chung, J., Gulcehre, C., Cho, K., Bengio, Y.: Empirical evaluation of gated recurrent neural networks on sequence modeling. arXiv preprint arXiv:1412.3555 (2014)
2. Cuturi, M.: Sinkhorn distances: lightspeed computation of optimal transport. In: NIPS, vol. 26 (2013)
3. Deng, Y., Bao, F., Kong, Y., Ren, Z., Dai, Q.: Deep direct reinforcement learning for financial signal representation and trading. IEEE TNNLS **28**(3), 653–664 (2016)
4. Fama, E.F., French, K.R.: Multifactor explanations of asset pricing anomalies. J. Financ. **51**(1), 55–84 (1996)
5. Fedus, W., Zoph, B., Shazeer, N.: Switch transformers: scaling to trillion parameter models with simple and efficient sparsity. JMLR **23**(1), 5232–5270 (2022)
6. Gurrib, I., et al.: Performance of the average directional index as a market timing tool for the most actively traded USD based currency pairs. Banks Bank Syst. **13**(3), 58–70 (2018)
7. Hong, H., Stein, J.C.: A unified theory of underreaction, momentum trading, and overreaction in asset markets. J. Financ. **54**(6), 2143–2184 (1999)
8. Houlsby, N., et al.: Parameter-efficient transfer learning for NLP. In: ICML, pp. 2790–2799. PMLR (2019)
9. Jang, E., Gu, S., Poole, B.: Categorical reparameterization with gumbel-softmax. arXiv preprint arXiv:1611.01144 (2016)
10. Jegadeesh, N., Titman, S.: Returns to buying winners and selling losers: implications for stock market efficiency. J. Financ. **48**(1), 65–91 (1993)
11. Jegadeesh, N., Titman, S.: Cross-sectional and time-series determinants of momentum returns. Rev. Financ. Stud. **15**(1), 143–157 (2002)
12. Jeong, G., Kim, H.Y.: Improving financial trading decisions using deep q-learning: predicting the number of shares, action strategies, and transfer learning. Expert Syst. Appl. **117**, 125–138 (2019)
13. Kim, H.J., Shin, K.S.: A hybrid approach based on neural networks and genetic algorithms for detecting temporal patterns in stock markets. Appl. Soft Comput. **7**(2), 569–576 (2007)
14. Li, Z., Tam, V.: A machine learning view on momentum and reversal trading. Algorithms **11**(11), 170 (2018)
15. Lin, H., Zhou, D., Liu, W., Bian, J.: Learning multiple stock trading patterns with temporal routing adaptor and optimal transport. In: 27th ACM SIGKDD, pp. 1017–1026 (2021)
16. Liu, Y., Liu, Q., Zhao, H., Pan, Z., Liu, C.: Adaptive quantitative trading: an imitative deep reinforcement learning approach. In: Proceedings of the AAAI Conference on Artificial Intelligence, vol. 34, pp. 2128–2135 (2020)
17. Moody, J., Saffell, M.: Reinforcement learning for trading. In: NIPS, vol. 11 (1998)
18. Moody, J., Wu, L.: Optimization of trading systems and portfolios. In: Proceedings of the IEEE/IAFE 1997 CIFEr, pp. 300–307. IEEE (1997)
19. de Oliveira, R.A., Ramos, H.S., Dalip, D.H., Pereira, A.C.M.: A tabular sarsa-based stock market agent. In: Proceedings of the First ACM International Conference on AI in Finance, pp. 1–8 (2020)
20. Poterba, J.M., Summers, L.H.: Mean reversion in stock prices: evidence and implications. J. Financ. Econ. **22**(1), 27–59 (1988)
21. Pricope, T.V.: Deep reinforcement learning in quantitative algorithmic trading: a review. arXiv preprint arXiv:2106.00123 (2021)

22. Ritter, J.R.: Behavioral finance. Pac.-Basin Finance J. **11**(4), 429–437 (2003)
23. Schulman, J., Wolski, F., Dhariwal, P., Radford, A., Klimov, O.: Proximal policy optimization algorithms. arXiv preprint arXiv:1707.06347 (2017)
24. Sharpe, W.F.: Mutual fund performance. J. Bus. **39**(1), 119–138 (1966)
25. Si, W., Li, J., Ding, P., Rao, R.: A multi-objective deep reinforcement learning approach for stock index future's intraday trading. In: 2017 10th ISCID, vol. 2, pp. 431–436. IEEE (2017)
26. Tsang, W.W.H., Chong, T.T.L., et al.: Profitability of the on-balance volume indicator. Econ. Bull. **29**(3), 2424–2431 (2009)
27. Wilder, J.W.: New concepts in technical trading systems. Trend Research (1978)
28. Xu, W., et al.: HIST: a graph-based framework for stock trend forecasting via mining concept-oriented shared information. arXiv preprint arXiv:2110.13716 (2021)
29. Xu, W., Liu, W., Xu, C., Bian, J., Yin, J., Liu, T.Y.: Rest: relational event-driven stock trend forecasting. In: Proceedings of the Web Conference 2021, pp. 1–10 (2021)
30. Yuan, Y., Wen, W., Yang, J.: Using data augmentation based reinforcement learning for daily stock trading. Electronics **9**(9), 1384 (2020)

Agent-Based Simulation of Decision-Making Under Uncertainty to Study Financial Precarity

Pegah Nokhiz[1], Aravinda Kanchana Ruwanpathirana[2(✉)], Neal Patwari[3], and Suresh Venkatasubramanian[1]

[1] Brown University, Providence, RI, USA
{pegah_nokhiz,suresh}@brown.edu
[2] University of Utah, Salt Lake City, UT, USA
kanchana.ruwanpathirana@utah.edu
[3] Washington University in St. Louis, St. Louis, MO, USA
npatwari@wustl.edu

Abstract. Financial insecurity in the U.S. is on the rise, accelerated by the growth of the gig economy and the associated income instability, increasing inequality, and the effects of algorithmic decision-making. Such insecurity has been studied within the framework of *precarity* – a concept that captures people's latent uncertainty and precariousness. To alleviate precarity, we must study it. Precarity manifests over time as a sequence of events for an individual. Therefore, we must study *individual* trajectories, rather than the trajectory of aggregate properties of populations or snapshot analysis of an automated decision process. Doing so requires an agent behavior model that can simulate a number of related phenomena simultaneously: how individual consumption reacts to uncertainty in one's financial status, how predictive tools impact income, and how utility-maximizing individuals behave in the long term. In this paper, we develop an agent-based simulation framework with realistic elements to examine the dynamics of precarity. Our model combines different threads of inquiry in economics and incorporates models of consumption, ruin, and investment. Our results illustrate how precarity, if ignored by policy-makers, can exacerbate the ill-effects of automated decision-making. Our framework also allows us to experiment with different strategies to mitigate precarity and evaluate their effectiveness.

1 Introduction

Financial insecurity is a characteristic of modern life in America [30]. A combination of socioeconomic factors, increasing inequality, unstable jobs, and data-driven algorithmic decision-making has left families increasingly vulnerable to financial "shocks" that can have an outsized and often irreversible long-term effect on their finances.

The first two authors contributed equally.

D.-N. Yang et al. (Eds.):.PAKDD 2024, LNAI 14648, pp. 43–56, 2024.
https://doi.org/10.1007/978-981-97-2238-9_4

Latent Socioeconomic Factors. Some individuals and households are more vulnerable financially due to latent socioeconomic factors. According to financial reports [15,30], these latent factors can be demographic – communities of color tend to be more likely to experience negative income shocks – as well as economic, with factors ranging from supporting family, an unstable job, and so on.

Job Instability. The rise of the gig economy has led to an increase in paycheck instability with associated long-term effects. Consider a gig worker versus an office worker with a stable job and salary. While both might start with the same set of observable economic features, i.e., similar assets and income levels, the gig worker's finances (income and employment status) are prone to be more volatile in the long run due to the nature of short-term and unpredictably valued contracts in the gig economy [10].

The Looming Risk of Financial Ruin. A recent CNBC survey revealed that 58% of American households are living paycheck to paycheck [8]. A household that cannot build up savings is one that is vulnerable to (even small) financial shocks. At the margins, this precariousness can tip households over the edge into bankruptcy and even homelessness: a recent study from UCSF indicated that the most commonly reported cause of homelessness was a loss of or reduction in income (with 12% of respondents indicating this) [20].

There is a name for this *precarious* state of being. Precarity [5,6] is a transdisciplinary term that characterizes the latent instability, *precariousness*, and therefore vulnerability of people's lives. Researchers have proposed quantitative measures of precarity [25], and it has been linked to automated decision-making – [24] has shown how precarity can be exacerbated by the compounding effects of repeated algorithmic decisions that take financial variables into account when making predictions that in turn cause future financial shocks.

To study precarity – how it appears, what conditions make one precarious, and how we might mitigate it via interventions – we need to be able to model long-term financial behavior. This is challenging because a) precarity is a property of an individual financial trajectory rather than an aggregate property of a population and b) it requires modeling individual behavior in response to repeated financial shocks rather than examining the effect of a single shock. Agent-based modeling [19,24] is, therefore, the appropriate approach to take; if we can realistically model individual behavior, we can study how individuals respond to financial shocks and what kinds of mitigation strategies might be most effective. Returning to the UCSF study on homelessness in California, one finding was that a majority of homeless Californians had a median income of $960 (significantly lower than the expenses needed to maintain a house), and 70% of them believed that monthly assistance of $300-$500 would have prevented homelessness [20].

Our Work. We develop an agent-based simulation framework to study precarity. Our framework consists of three components: a) A mechanism for agents to maximize utility while subject to constraints under which precarity emerges – the risk of bankruptcy, consumption constraints, and uncertainty about future earnings and asset growth, b) A mechanism to add income shocks into the system, and c) Mechanisms to intervene to mitigate the effects of external shocks.

This framework draws on a number of inter-related threads of work in the economics literature that seek to model human consumption, effects of uncertainty in decision-making, and challenges of avoiding ruin. None of these existing lines of work captures all of the desiderata above; one key contribution of our framework is combining them to show how an agent might maximize utility under all of these constraints.

2 Background: Modeling Consumption

The most important part of our agent-based simulation framework is the process by which agents consume and earn utility. In this section, we describe the standard toolkit from the literature on consumption, and then build our (constrained) framework in Sect. 3. A reader knowledgeable in economic models of consumption may skip this section.

There are numerous models that seek to capture human consumption behavior (which we review in Sect. 6). At their core, all of these models assume that an agent consumes an amount c in order to maximize utility $u(c)$, where $u(\cdot)$ is some concave function. Agents are assumed to receive *income* y_t at time t, as well as maintaining assets x_t. Further, most models assume some form of *discounting*: that an agent prefers to receive utility now rather than later. This is formalized by saying that the actual utility gained by consuming c_t at some future time t is $\beta^t u(c_t)$, where $0 < \beta < 1$ is a *discounting factor.*

In all models, the goal is then to determine how an agent might choose consumption c_t at each time t to maximize their long-term utility, given by $\sum_{t=0}^{\infty} \beta^t u(c_t)$. We note that while utility maximization is not necessarily the only way to model agents of bounded rationality [3], it is the framework with the most extensive machinery and tools for modeling.

2.1 Capturing Uncertainty

The *income fluctuation problem* (IFP) [23,26] investigates how an agent might maximize utility when their income y_t fluctuates stochastically, while also assuming that assets earn a fixed rate of return r. The objective is to maximize:

$$\mathbb{E}\left\{\sum_{t=0}^{\infty} \beta^t u(c_t)\right\} \tag{1}$$

with respect to $\{c_t\}_t$ such that $x_{t+1} = r(x_t - c_t) + y_{t+1}$ and $0 \leq c_t \leq x_t$ where y_t is non-capital labor income, $\beta \in (0,1)$ is the discount factor, and r is the interest rate on assets ($r \geq 1$). The *constant relative risk aversion* (CRRA) utility of the individual is $u(c) = \frac{c^{1-\gamma_c}}{1-\gamma_c}$, which is a standard utility function in economics [22,29] used to capture people's preference for low-risk and lower gain compared to risky results with higher payoffs. Parameter γ_c has a value $\gamma_c > 0$.

3 The Framework: Introducing Real Constraints

In order to model realistic agent behavior so as to capture precarity, our agents must a) try to avoid ruin (i.e., situations when the assets go below zero); b) have a fixed time horizon (i.e., a "time of death"); and c) have minimum required consumption (minimum subsistence) at each time step (for e.g., for basic necessities like food and shelter). Formally, we define ruin as $\exists t < \infty$ such that $x_t \leq 0$. Having a time of death means that the utilities are summed only till some time τ_d. Finally, minimum consumption adds the constraint $c_t \geq c$ for all time t.

These constraints are related. For example, in the IFP model, utility-maximizing agents can avoid ruin (i.e., continue to consume into the future) as long as there are no minimum subsistence constraints (we illustrate this in a lemma in Appendix A). The appendix and the code are available as supplementary material, which can be found at https://github.com/kanchanarp/Simulation-of-Decision-making-under-Uncertainty-to-Study-Precarity. Some models avoid the issue of ruin by allowing agents to go into debt indefinitely as long as eventually all dues are paid. To account for ruin, we do not allow agents to take on debt in our model. Thus, in order to capture the above constraints, we need to make modifications to the base IFP model. To do this, we will draw on a different model of utility-maximizing consumption in the context of investment.

3.1 Background: Modeling Ruin

Bayraktar and Young [4] introduce an investment model with a time of death τ_d, time of ruin τ_0 ($\tau_0 = \inf\{t < \infty \,|x_t \leq 0\}$), as well as a soft constraint that allows the individuals to avoid ruin before the time of death. Their model can be described as follows:

$$\max E\left(\int_0^{\min(\tau_d,\tau_0)} \hat{\beta}^t u(c_t)dt\right) \text{ such that}$$

$$dx_t = [rx_t + (\mu - r)\pi_t - c_t]dt + \sigma\pi_t dZ_t$$

$$\mathcal{P}[\tau_0 \leq \tau_d] \leq \phi(x_0)$$

where x_t is the amount of current assets, π_t is the amount invested in *risky assets* like volatile stock investments, Z_t is a *Brownian motion*, $\phi(x_0)$ is a probability dependent on x_0 which is the initial amount of assets the agent starts with, and $\hat{\beta}$ is a discount factor. r is the rate of return for riskless assets (e.g., savings accounts). μ and σ are the rates for the static return and stochastic return on risky assets, respectively. Given this constrained model, [4] construct an equivalent unconstrained optimization problem and then refer to prior work by [18] to show how this unconstrained problem can be solved. Given their dynamic equations, the unconstrained optimization problem is stated as,

$$V(x_0) = \max E\left(\int_0^{\tau_0} e^{-\gamma t}\hat{\beta}^t u(c_t)dt + Pe^{-\gamma\tau_0}\hat{\beta}^{\tau_0}\right),$$

where $P \leq 0$ is a Lagrange parameter and $E\left(e^{-\gamma\tau_0}\hat{\beta}^{\tau_0}\right)$ encodes the relaxation of $\mathcal{P}[\tau_0 \leq \tau_d] \leq \phi(x_0)$. The function V is called a value function, and γ comes from the time of death distribution which is exponential with parameter γ.

3.2 Our New Model

Recall the objective function (1) from IFP. We consider a continuous relaxation of the model, under a single-asset economy with a constant real interest rate. That is, let x_t, y_t, and c_t be the assets, income, and consumption, respectively, at time t and r be the real interest rate on assets. This yields a dynamic equation $dx_t = [(r-1)\,x_t - rc_t + y_t]\,dt$ and the objective, max $E\left(\int_0^\infty \hat{\beta}^t u(c_t)dt\right)$ where $\hat{\beta}$ is a discount factor.

Combining this with the objective function and ruin constraints introduced by [4] in Sect. 3.1, we obtain the following optimization problem:

$$\begin{aligned} &\max E_{\{c_t|t=0\dots\}}\left(\int_0^{\min(\tau_d,\tau_0)} \hat{\beta}^t u(c_t)dt\right) \\ &\text{s.t. } dx_t = [(r-1)\,x_t - rc_t + y_t]\,dt \\ &\qquad \mathcal{P}[\tau_0 \leq \tau_d] \leq \phi(x_0) \end{aligned} \tag{2}$$

Note that our model is stochastic since the income decision y_t (and therefore c_t and x_t) is stochastic. Therefore, we have expectations over the sequence c_t in our optimization function, instead of just a deterministic formulation. Model (2) is similar in form to the [4] model. *This key insight allows us to borrow their idea of a value function with unconstrained optimization-type formulation.* Let $c(x_t)$ be a function that returns the optimal consumption value given x_t and let $\beta = \gamma + \log(1/\hat{\beta})$. We can see that the value function still remains

$$V(x) = \max E_{\{x_t|t=0\dots\}}\left(\int_0^{\tau_0} e^{-\beta t}u(c(x_t))dt + Pe^{-\beta\tau_0}\right) \tag{3}$$

$$dx_t = [(r-1)\,x_t - rc_t + y_t]\,dt \tag{4}$$

Eliminating the Time-of-Ruin Parameter τ_0. The optimization problem (3) involves a time of ruin τ_0 parameter that we do not have any information about. We need a method to remove τ_0 from the equation to derive a function that does not depend on τ_0. For this purpose, we can use a tool that was used by [18]: the Feynman-Kac formula [16]. Feynman-Kac states that, given a model $dx = \mu(x,t)dt + \sigma(x,t)dW^Q$ (where Q is a Wiener process), for any time T, the function defined $V(x,t)$ by $\frac{\partial V(x,t)}{\partial t} + \mu(x,t)\frac{\partial V(x,t)}{\partial x} + \frac{1}{2}\sigma^2(x,t)\frac{\partial^2 V(x,t)}{\partial x^2} - g(x,t)V(x,t) + f(x,t) = 0$ can be shown to be defined by,

$$\begin{aligned} V(x,t) = E&\left(\int_t^T e^{-\int_t^r g(x,\tau)d\tau} f(x_r,r)dr \mid x_t = x\right) \\ &+ E\left(e^{-\int_t^T g(x,\tau)d\tau} V(x_T,T) \mid x_t = x\right) \end{aligned}$$

Letting $\sigma(x,t) = 0$, $\forall x$ and $\mu(x,t) = ((r-1)x_t - rc_t + y_t)$ and setting $g(x,t) = \beta$, $f(x,t) = u(c(x_t))$ and $V(X_T) = V(0) = P$ (since $T = \tau_0$ in our case) we can get the Eqs. (3) and (4), and Feynman-Kac then gives us,

$$\beta V(x) = \frac{\partial V(x)}{\partial t} + ((r-1)x - rc(x) + y)\, V'(x) + u(c(x))$$

This gives us an optimization problem that is independent of τ_0.

Solving the Optimization Problem. We will now try to solve this equation. We can first remove the $\frac{\partial V(x)}{\partial t}$ to derive an equation that depends on $V'(x)$ so we have a homogeneous equation. Using the model Eq. (4) we can show that, $\frac{\partial V(x)}{\partial t} = \frac{\partial V(x)}{\partial x}\frac{\partial x}{\partial t} = ((r-1)x - rc(x) + y)\, V'(x)$. This gives us,

$$\begin{aligned}\beta V(x) &= u(c(x)) + 2\,((r-1)x - rc(x) + y)\, V'(x)\\ &= u(c(x)) + \frac{(r-1)x - rc(x) + y}{r} u'(c(x))\end{aligned}$$

where the second equation comes from solving for $V'(x)$ by taking the derivative with respect to $c(x)$ (which gives us $V'(x) = u'(c(x))/(2r)$).

Taking the derivative of $\beta V(x_t)$ with respect to x_t, we get,

$$\begin{aligned}\beta V'(x_t) - \frac{1}{2r}\beta u'(c(x_t)) &= u'(c(x_t))c'(x_t)\\ &+ \left(\frac{r-1}{r} - c'(x_t)\right) u'(c(x_t))\\ &+ \left(\frac{r-1}{r}x_t - c(x_t)\right) u''(c(x_t))c'(x_t)\\ &+ \frac{y_t}{r}u''(c(x_t))c'(x_t) - \frac{1}{2r}\beta u'(c(x_t))\\ &= 0\end{aligned}$$

and by setting $u(c(x_t))$ to be the CRRA utility, and using the fact that the $u'(c(x_t)) = c(x_t)^{-\gamma_c}$ and $u''(c(x_t)) = -\gamma_c c(x_t)^{-1-\gamma_c}$, we get

$$c'(x_t) = \frac{\left(r - 1 - \frac{\beta}{2}\right) c(x_t)}{\gamma_c\left((r-1)x_t - rc(x_t) + y_t\right)} \tag{5}$$

Our ultimate goal was to find $c(x_t)$. We can see that given (5), $c(x_t)$ can be obtained by solving a differential equation. Using a symbolic solver to solve the differential equation, we get

$$x_t = k_1 c(x_t)^{\frac{\gamma_c r}{r-1-\frac{\beta}{2}}} + \frac{\gamma_c r}{\frac{\beta}{2} + (\gamma_c - 1)(r-1)} c(x_t) - \frac{y_t}{r} \tag{6}$$

where k_1 is a constant that we need to determine. To determine k_1, we can first use the fact that $V(0) = P$ and that $\beta V(0) = u(c_0) + \frac{(y_0 - rc_0)}{r} u'(c_0)$ to derive the c_0 value which can then be plugged into the Eq. (6) to find the value of k_1. Given k_1, we have a well-defined polynomial that involves $c(x_t)$ which we can solve for any specific x_t using a polynomial solver. Details on how we incorporate minimum subsistence constraints into this formulation are in Appendix B.

Putting it All Together. Agents in the framework are individuals who earn income, possess assets, and decide whether to consume or save. We form income and asset distributions of 10,000 agents using real-world data. They receive instant (temporary) or long-term (permanent) positive or negative shocks to their income (permanent shocks happen every 25 timesteps and temporary shocks happen every 20 timesteps) based on the (positive or negative) decision they obtain from a gradient-boosted binary classifier trained on their income and assets (features) to predict a binary value on if they are doing okay or not financially. The details of the data used, how the shocks are disbursed and the labels can be found in Appendix C. Our implementation using IFP is inspired by https://python.quantecon.org/ifp_advanced.html. In the following sections, we deploy our simulation framework to study the precarity of agents as they consume and earn in the face of financial shocks.

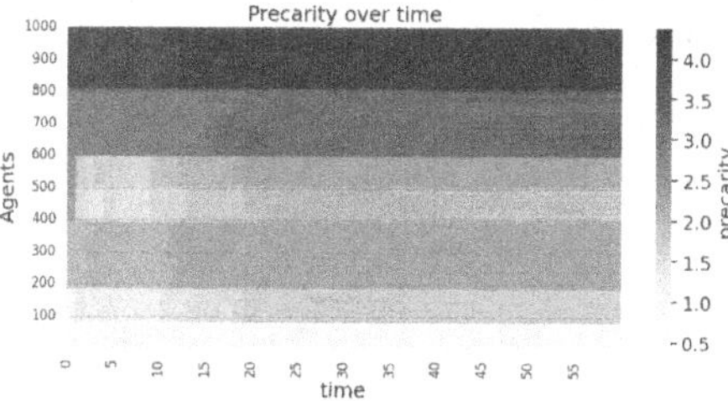

Fig. 1. Long-term precarity analysis of a sample population of 1,000 individuals ordered by their initial instability.

4 Simulation Study: Precarity

Precarity Index. Ritschard et al. [25] were the first to develop a measure of precarity over financial state trajectories. Precarity is about instability in a *sequence* of events rather than as a population aggregate. They quantified it as a combination of three components: the net decline associated with transitions from stronger to weaker financial states, the variability in the sequence measured via the entropy of the distribution of states, and the initial instability of the trajectory. Formally, the *precarity index* $p(\sigma)$, of a sequence is defined as: $p(\sigma) = \lambda r(s_1) + (1 - \lambda)c(\sigma)^{\alpha}(1 + q(\sigma))^{\gamma}$ where the initial instability is $r(s_1)$, the net decline of the sequence is $q(\sigma)$ and the amount of variability is $c(\sigma)$. The weighting hyper-parameters are $\lambda = 0.2$, $\alpha = 1$, and $\gamma = 1.2$ according to [25]. Note that a higher precarity index denotes higher instability and is unfavorable. In this paper, we encode agent trajectories as a sequence of asset distribution deciles. We set the initial instability of an agent to be a function of the number of transitions required to reach the highest asset decile from their current decile and their perceived financial well-being based on real-world data [9].

4.1 Long Term Precarity

In this experiment, our goal is to see how automated decisions and shocks affect long-term population precarity. For a sample population of size 1,000, we run this experiment for 60 timesteps (months).

Analysis. There are two takeaways from Fig. 1. Firstly, we can see the overall increase in precarity is directly linked to people's initial instability – the higher the initial instability is, the more precarity increases, culminating at an unfavorable high precarity value. That is, agents with low precarity (e.g., agents 0 to 100) have a corresponding lower precarity over time. More precarious subpopulations (e.g., agents 200 to 800), however, observe an overall increase in their long-term precarity with respect to their initial instability. Secondly, there is variability within groups with similar initial instability. This indicates that initial conditions alone are inadequate to determine where someone will land in the future.

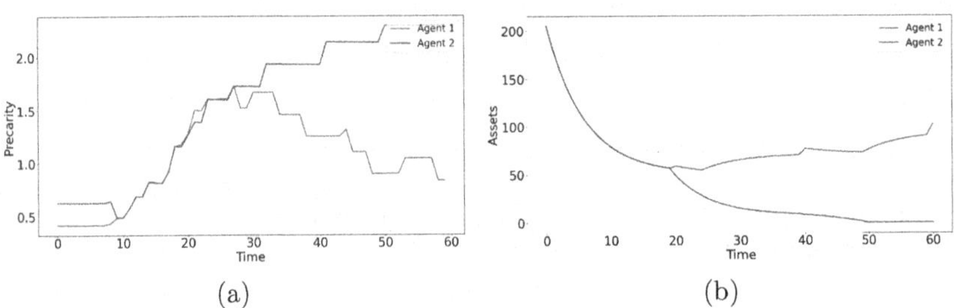

Fig. 2. Gig vs. office worker with different latent initial instability, similar initial assets and starting income distributions. Assets are in thousands of dollars.

4.2 Factors Contributing to Precarity

The first simulation showed the long-term effects of financial fragility, as well as how (up to a point) starting conditions can determine one's financial outcome. In this set of experiments, we go a little deeper into the hidden or latent factors of an individual's financial state (that can lead out varying outcomes) that contribute to overall financial precarity. This is important for policymakers as well as for algorithmic decision-making: if the goal of interventions is to improve financial

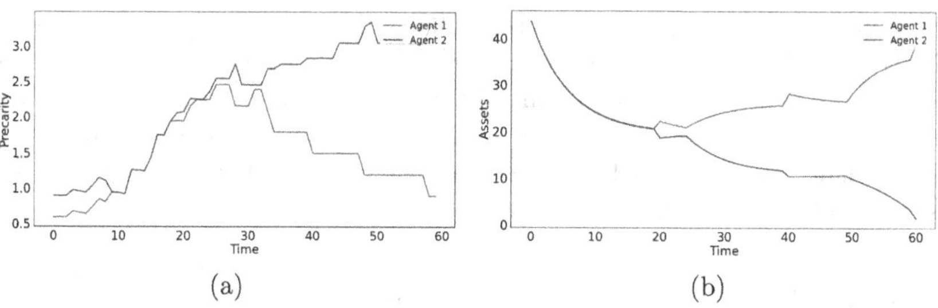

Fig. 3. Two agents with different initial instability and similar initial assets ($43,800) with marginally different initial incomes. Assets are in thousands of dollars.

conditions for individuals, then it is important to "see" the latent factors that can affect individual responses.

The approach we take is to compare pairs of individuals that differ in one key aspect and examine the evolution of their financial state over time. In the first scenario, the latent factor we control is income instability – the **"gig worker vs. office worker"** case. In the second, we look at individuals with marginally different incomes – **"minor income difference"**. In our plots we show results for two such individuals – we repeated these simulations for multiple pairs that fit each scenario and obtained similar results. Our goal is to examine: 1) if hidden differences (in latent instability) can lead to largely different financial outcomes and 2) if marginal differences in observable features (portrayed in the marginal difference in initial income) can lead to vastly different aftermaths.

For the first scenario, we consider agents with two different profiles. Agent one has lower initial instability and agent two has a higher initial instability (even though their observable features i.e., assets and income, are exactly the same for an initial period). We can think of the first individual as an office worker with a stable income and the second individual as a gig worker with more latent instability and unstable income. The simulation environment is the same for both agents. After 20 months, and motivated by the volatile/insecure nature of gig work, Agent 2 experiences a drop in income. Agent 2's income distribution shifts to a lower income level (chosen uniformly at random from the range of low-income values, in our experiments). In the second scenario, the two agents have exactly similar assets. They only have an income difference of $20. All other experimental parameters are exactly similar for both agents.

Analysis. There are four takeaways from this experiment. Firstly, (see Fig. 2), an automated decision-maker that only looks at observable features, e.g., income and asset values (which are the same for both individuals) to assign snapshot decisions cannot account for diverging consequences as a result of the (hidden) instability. Secondly, in Fig. 2, we observe the effects of the magnitude of precarity: large hidden differences in instability could lead to considerably different financial outcomes. Thirdly, in Fig. 3, we observe the extreme difference in the precarity and asset trajectories of agents who are only marginally different in their income values. This illustrates that although their initial finances are very similar, small differences can lead to substantially differential financial outcomes. Lastly, Fig. 3 also shows that one initial negative outcome – the initial decision – can have consequences that get amplified by the subsequent set of automated decisions made for an agent. This illustrates the way in which compounded decisions can have a significant effect on an individual trajectory.

5 Simulation Study: Interventions

Simulation can serve as a sandbox to test interventions that would otherwise be difficult if not impossible to explore in the real world. Fiscal interventions are actions taken by the government in the form of a collection of different subsidies, tax rebates, and unemployment benefits to address different economic circumstances. The Coronavirus Aid, Relief, and Economic Security (CARES) Act is a prime example of such a stimulus package [28].

We measure the effectiveness of interventions in terms of their *durability*. We define durability in two ways. Firstly, and temporally, we measure durability as the number of timesteps the agents would have more money compared to the scenario in which they received no interventions, i.e., their asset value be more than the baseline of the assets they would have gathered without any interventions. Secondly, and financially, we measure durability as the net difference in assets (at the end of the simulation) compared to not getting any help at all.

Tax Incentives. In this setting, we provide the agents with tax breaks according to their corresponding income bracket for tax season 2019–2020 [11] which effectively increases their income to the gross value for that bracket. We wish to examine three different intervention dimensions: when the intervention is made, how long the effect of the intervention lasts (i.e., the durability), and what effect changing the span of the intervention has.

As in the previous section, we compare pairs of agents who differ in one of two ways (income instability or a minor income difference). We explore two lengths of 12 and 6 months of tax breaks with different start points for 50 pairs of agents in each of the two scenarios above, as follows. In an **early** intervention when agents are latently different based on precarity but have similar observable financial features, we assign the suffering second agent one-year (6 months if exploring a 6-month span) tax breaks during the first month. In a **middle** intervention, we provide the agent with a tax break at timepoint 24 for a year (or 6 months). In the **late** intervention we explore another 1-year (or 6-month) tax break starting at timepoint 48 and after several shocks and automated decisions have occurred.

Analysis. Not surprisingly, interventions work. Durability increases for pairs of agents, as depicted (on average) in Table 1. The corresponding after-intervention precarity and asset plots are in Appendix D. There are three specific takeaways from this experiment as well. First, in general, earlier interventions when agents are latently different but have similar observable features result in improved agent durability. Secondly, the earlier interventions help both temporal and financial durability. Lastly, the longer the intervention time span, the more durable the agents will be both in terms of assets and months.

Table 1. Mean intervention durability for 50 pairs of agents in each scenario with early, middle, and late tax breaks (studied separately for 12 months and 6 months). Assets are in thousands of dollars and rounded to the closest $1,000.

Scenario	Months			Assets		
	Early	Mid	Late	Early	Mid	Late
Gig Worker (12 m.)	**41.3**	19.2	4.7	**110**	20	7
Gig Worker (6 m.)	**33.6**	16.8	4.6	**54**	9	5
Income Diff (12 m.)	**33.4**	33.1	11.1	**64**	39	18
Income Diff (6 m.)	24.8	**25.2**	11	**32**	20	13

Direct Subsidies. Direct subsidies are direct payments to help people in financial need. We explore different subsidy values in Appendix E. As with tax relief, subsidies work, and knowing the diverging latent precarious nature of agents beforehand (as early as possible) can help policy-makers, in the long run, both in terms of time and money. Finally, we look at some real-world statistical corroborations in Appendix F to compare the overall trends of our results.

6 Related Work

Within the domain of decision-making, sequential decision-making has been extensively studied [17,21]. In this field, automated algorithms have emerged as indispensable tools. However, their opacity [12] often obscures their inner workings, leading to unpredictability and unexpected outcomes for the agents. Understanding the behavior of agents within a financial decision-making system requires models that are capable of capturing the consumption behavior of agents. Traditional models such as the permanent income hypothesis [13], the life-cycle model [14], and the neoclassical model [7] assume agents can foresee the future with certainty, albeit with varying time horizons. The income fluctuation problem (IFP) is a notable consumption model that integrates uncertainty and dynamic optimization over an infinite time horizon [23,26]. Simulations are commonly employed to study behavior, as demonstrated in works like [24], which utilizes data-driven simulations based on Markov decision processes and IFP. While IFP provides a robust foundation, it lacks certain realistic elements such as considerations for time of death, ruin, and minimum consumption. To address these limitations, our approach draws upon frameworks on the modeling of liability to bankruptcy (ruin) [2], income shock analysis, optimal stimulus allocation to minimize bankruptcy risk [1], and investment models [18]. Some of these frameworks incorporate constraints, including upper bounds on the probability of ruin before death [4]. Additionally, our model integrates insights from various sources, including [27], which examines the concept of lower bounds on consumption (minimum subsistence) within the context of utility-maximizing consumption. By combining diverse dynamics from multiple areas of study, our model aims to provide a realistic representation of decision-making processes.

7 Conclusions

The main contribution of our paper is an agent-based simulation framework for exploring financial insecurity precipitated by algorithmic decision-making through the lens of precarity. There are a number of ways in which we could develop this framework further, e.g., exploring unfairness and other forms of inequity in automated decision-making involving sub-populations with different sensitive attributes, as well as targeted interventions that might mitigate the effects of (biased) decision-making for individuals over long time horizons.

Acknowledgments. This research was supported in part by grants from the MacArthur Foundation and the Ford Foundation.

References

1. Abebe, R., Kleinberg, J., Weinberg, S.M.: Subsidy allocations in the presence of income shocks. In: Proceedings of the AAAI Conference on Artificial Intelligence, vol. 34, no. 05, pp. 7032–7039 (2020). https://doi.org/10.1609/aaai.v34i05.6188. https://ojs.aaai.org/index.php/AAAI/article/view/6188
2. Asmussen, S., Albrecher, H.: Ruin Probabilities, vol. 14. World Scientific, New Jersey (2010)
3. Barberis, N.C.: Thirty years of prospect theory in economics: a review and assessment. J. Econ. Perspect. **27**(1), 173–196 (2013)
4. Bayraktar, E., Young, V.R.: Maximizing utility of consumption subject to a constraint on the probability of lifetime ruin (2012)
5. Benhabib, J., Bisin, A., Zhu, S.: The wealth distribution in Bewley economies with capital income risk. J. Econ. Theory **159**, 489–515 (2015)
6. Butler, J.: Precarious Life: The Powers of Mourning and Violence. Verso Books, London (2006)
7. Bütler, M.: Neoclassical life-cycle consumption: a textbook example. Econ. Theor. **17**(1), 209–221 (2001)
8. Dickler, J.: With inflation stubbornly high, 58% of Americans are living paycheck to paycheck: CNBC survey (2023). https://www.cnbc.com/2023/04/11/58percent-of-americans-are-living-paycheck-to-paycheck-cnbc-survey-reveals.html
9. Division of Consumer and Community Affairs: Economic well-being of US households in 2022 (2023). https://www.federalreserve.gov/publications/files/2022-report-economic-well-being-us-households-202305.pdf
10. Donovan, S.A., Bradley, D.H., Shimabukuru, J.O.: What does the gig economy mean for workers? (2016)
11. El-Sibaie, A.: Tax brackets (2019). https://taxfoundation.org/data/all/federal/2019-tax-brackets

12. Eslami, M., Vaccaro, K., Lee, M.K., Elazari Bar On, A., Gilbert, E., Karahalios, K.: User attitudes towards algorithmic opacity and transparency in online reviewing platforms. In: Proceedings of the 2019 CHI Conference on Human Factors in Computing Systems, CHI 2019, pp. 1–14. Association for Computing Machinery, New York (2019). https://doi.org/10.1145/3290605.3300724
13. Friedman, M.: The permanent income hypothesis. In: A Theory of the Consumption Function, pp. 20–37. Princeton University Press, Princeton (1957)
14. Friedman, M.: Theory of the Consumption Function. Princeton University Press, Princeton (2018)
15. Hardy, B.L., Logan, T.D.: Racial economic inequality amid the COVID-19 crisis. The Hamilton Project **1**, 1–10 (2020)
16. Kac, M.: On distributions of certain Wiener functionals. Trans. Am. Math. Soc. **65**(1), 1–13 (1949)
17. Kannan, S., Roth, A., Ziani, J.: Downstream effects of affirmative action. In: Proceedings of the Conference on Fairness, Accountability, and Transparency, FAT* 2019, pp. 240–248. Association for Computing Machinery, New York (2019). https://doi.org/10.1145/3287560.3287578
18. Karatzas, I., Lehoczky, J.P., Sethi, S.P., Shreve, S.E.: Explicit solution of a general consumption/investment problem. Math. Oper. Res. **11**(2), 261–294 (1986). http://www.jstor.org/stable/3689808
19. Klügl, F., Kyvik Nordås, H.: Modelling agent decision making in agent-based simulation - analysis using an economic technology uptake model. In: Proceedings of the 2023 International Conference on Autonomous Agents and Multiagent Systems, AAMAS 2023, pp. 1903–1911. International Foundation for Autonomous Agents and Multiagent Systems, Richland, SC (2023)
20. Kushel, M., Moore, T.: Towards a new understanding: the California statewide study of people experiencing homelessness (2023)
21. Liu, L.T., Dean, S., Rolf, E., Simchowitz, M., Hardt, M.: Delayed impact of fair machine learning. In: Dy, J., Krause, A. (eds.) Proceedings of the 35th International Conference on Machine Learning. Proceedings of Machine Learning Research, vol. 80, pp. 3150–3158. PMLR, Stockholmsmassan, Stockholm Sweden (2018). https://proceedings.mlr.press/v80/liu18c.html
22. Ljungqvist, L., Sargent, T.J.: Recursive Macroeconomic Theory. MIT Press, Cambridge (2018)
23. Ma, Q., Stachurski, J., Toda, A.A.: The income fluctuation problem and the evolution of wealth. J. Econ. Theory **187**, 105003 (2020)
24. Nokhiz, P., Ruwanpathirana, A.K., Patwari, N., Venkatasubramanian, S.: Precarity: modeling the long term effects of compounded decisions on individual instability. In: Proceedings of the 2021 AAAI/ACM Conference on AI, Ethics, and Society, AIES 2021, pp. 199–208. Association for Computing Machinery, New York (2021). https://doi.org/10.1145/3461702.3462529
25. Ritschard, G., Bussi, M., O'Reilly, J.: An index of precarity for measuring early employment insecurity. In: Ritschard, G., Studer, M. (eds.) Sequence Analysis and Related Approaches. Life Course Research and Social Policies, pp. 279–295. Springer, Cham (2018). https://doi.org/10.1007/978-3-319-95420-2_16
26. Sargent, T., Stachurski, J.: Quantitative economics. Technical report, Citeseer (2014)
27. Shin, Y.H., Lim, B.H.: Comparison of optimal portfolios with and without subsistence consumption constraints. Nonlinear Anal. **74**(1), 50–58 (2011). https://doi.org/10.1016/j.na.2010.08.014

28. Stone, C.: Fiscal stimulus needed to fight recessions (2020). https://www.cbpp.org/research/economy/fiscal-stimulus-needed-to-fight-recessions
29. Wakker, P.P.: Explaining the characteristics of the power (CRRA) utility family. Health Econ. **17**(12), 1329–1344 (2008)
30. Weller, C.: Working-class families are getting hit from all sides. Technical report, Center for American Progress (2018). https://www.americanprogress.org/article/working-class-families-getting-hit-sides/

Information Retrieval and Search

Semantic Completion: Enhancing Image-Text Retrieval with Information Extraction and Compression

Xue Chen[1] and Yi Guo[1,2,3(✉)]

[1] East China University of Science and Technology, Shanghai, China
y30221074@mail.ecust.edu.cn, guoyi@ecust.edu.cn
[2] Business Intelligence and Visualization Research Center, National Engineering Laboratory for Big Data Distribution and Exchange Technologies, Shanghai, China
[3] Shanghai Engineering Research Center of Big Data and Internet Audience, Shanghai, China

Abstract. Image-text retrieval is an essential branch in the field of information retrieval, facing the serious challenge of the cross-modal semantic gap. Although significant progress has been made in recent years, most research has ignored an essential problem: text as image description is incomplete, so the problem of semantic loss between image and text still exists. In this paper, we propose a novel information extraction and compression based image-text retrieval method to alleviate the above problem. The method aims to bridge the semantic gap between the two modalities by generating rich and high-quality semantic descriptions from a set of related sentences via an information extraction and compression module. To validate the effectiveness of the method, we conducted extensive experiments on the Flickr30K and MSCOCO datasets. The experimental results show that our method achieves significant performance improvement in image text retrieval with an appropriate fusion ratio. When the amount of pre-trained images is 4M, the evaluation metrics of our method improve by at least 4.22% and 3.69% compared to the baseline method. This further confirms the advantages and potential of our method in solving the semantic loss problem in image-text retrieval.

Keywords: Image-Text Retrieval · Information Extraction and Compression · Semantic Completion

1 Introduction

In the digital age, information overload has become a reality. Image and text data are growing explosively, so it becomes challenging to retrieve relevant information quickly and accurately from huge amounts of data. Nevertheless, image-text retrieval faces the challenge of cross-modal semantic gap due to the different characteristics and diversity of expressions of images and texts.

To address the above problems, researchers have proposed a series of innovative approaches and made significant research progress. For instance, VL-BERT [19] uses Transformer structure to extract the deep representation of

D.-N. Yang et al. (Eds.): PAKDD 2024, LNAI 14648, pp. 59–71, 2024.
https://doi.org/10.1007/978-981-97-2238-9_5

vision and language, and achieves the fusion of modal features and the optimization of semantic expressiveness through the cross-fertilization of multiple heads of attention. GAN-INT [18], on the other hand, adopts a generative adversarial framework to achieve end-to-end cross-modal low-dimensional embedding learning, and improves the generation and reasoning capabilities of vision and language through the co-training of generators and discriminators. FB-Net [22] employs a two-branch foreground-context fusion network and a multi-scale semantic scanning strategy to effectively utilize information from different parts of the image for a comprehensive understanding of the image content. AGREE [15] proposes a lightweight image-text retrieval method, which achieves cross-modal entity alignment through alignment and reordering and introduces a reordering strategy for zero-sample scenarios, which significantly improves image and text retrieval performance. However, these methods ignore the incompleteness problem of text as an image description, i.e., the expressive power of textual descriptions is limited and cannot completely reproduce or express the full information content in an image. Therefore, further research and solution to this problem is necessary.

In this paper, we propose a semantic completion method based on information extraction and compression (SCIEC) for solving the problem of losing semantic information due to incomplete textual descriptions of image correspondences. The method analyses and models multiple description statements corresponding to an image, capturing the overlap and complementarity of semantic concepts, and generating richer and higher-quality integrated descriptions to bridge the semantic gap. By deeply modeling the correlations between different description statements, the method effectively bridges the shortcomings of a single description. In addition, we explore the optimal fusion ratio between high-quality sentences and original sentences to enhance the discriminative power of text features and further improve the quality of image-text retrieval. The experiments demonstrate the potential and advantages of the approach in generating high-quality descriptions and improving retrieval performance. The main contributions of this paper are summarised as follows: (1) A semantic completion method based on information extraction and compression (SCIEC) is proposed, which significantly improves the quality of image-text retrieval by modeling the correlation between different semantic units to achieve in-depth integration and completion of semantic information. (2) Explore the optimal fusion ratio between high-quality sentences and original sentences to enhance the differentiation of textual features, and improve the accuracy and relevance of retrieval by fusing textual information from both. (3) Extensive experiments were conducted on two benchmark tests, Flickr30K and MSCOCO, and the experimental results proved the potential and superiority of the method.

2 Related Work

Image-text retrieval is a task involving visual and language modalities, and its goal is to achieve mutual retrieval between images and text. Currently, the mainstream image and text retrieval model structures are mainly divided into two types: dual-stream and single-stream structures.

2.1 Dual-Stream Structure

In the dual-stream structure models [3,8,9,20,24], images and text are processed through separate encoders to capture their respective semantic information, and the features of the two modalities are fused through additional fusion layers. The purpose of this structure is to effectively establish semantic associations between images and text to improve the performance of the model in various cross-modal tasks. VSE++ [4] and SCAN [6] are two common dual-stream structure models. VSE++ introduces the concept of hard negative sample mining, which improves the model's ability to understand the semantic association between images and text by selecting pairs of negative samples that are difficult to distinguish. While SCAN pays more attention to image and text alignment tasks. It adopts an unsupervised learning image classification method, and uses the self-attention mechanism to learn the associations between different areas in the image. CLIP [17] uses contrastive learning methods to learn the shared semantic space of images and texts, optimizing the model by maximizing the similarity of positive sample pairs and minimizing the similarity of negative sample pairs. In addition to the above models, there are other models related to semantic association between images and text, such as COTS [13] and LexLIP [14]. COTS combines object detection and natural language processing to improve the semantic association between images and text by considering the contextual information of the target object. LexLIP uses multi-task learning to jointly train image classification and text classification tasks, and cross-learning visual and language information to improve results. These dual-stream structural models effectively establish semantic associations between images and text by introducing innovative methods and technologies, promoting the rapid development of cross-modal information processing.

2.2 Single-Stream Structure

The core idea of the single-stream structure model [1,5,7,10,21] is to process images and text as a whole simultaneously through a shared encoder, and this approach allows the model to better capture the semantic relevance of the two modalities and use the complementary information between them to improve the overall performance. For instance, UNITER [2] proposes a multimodal feature representation learning framework that represents vision and language features in the same space and uses contrast learning to enhance the distinction between different modal features. Unicoder-VL [11], on the other hand, proposes a common vision-language encoder that learns the joint representation of images and text through cross-modal pre-training. Compared to UNITER, it provides a more efficient and scalable way to perform vision and language cross-learning. Oscar [12] and ImageBERT [16] are both pre-training models for vision and language tasks, aiming to improve the learning of joint representations between images and text. The difference is that Oscar emphasizes object and semantic alignment pre-training, while ImageBERT performs cross-modal pre-training with large-scale weakly supervised image-text data. Furthermore, in recent work, VinVL [25] revisits the importance of visual representation in vision-language models. It proposes a new representation learning framework that can effectively combine vision and language features to improve model performance. TCL [23] proposes a vision-language pre-training framework. The model introduces Triple Contrastive Learning, which further improves the model's representation learning capability by simultaneously learning the similarity and dissimilarity of positive and negative samples, thus improving the model's performance. These single-stream structural models effectively improve the semantic relevance between image and text in image-text retrieval through the shared encoder, contrast learning, and vision-language alignment pre-training, and show significant results and advantages.

In contrast, our approach combines the advantages of single and dual-stream structures and addresses the incompleteness of textual descriptions from different perspectives. By introducing IEC modules to fill in the missing semantic information, our approach can effectively compensate for the shortcomings of single descriptions and provide richer and higher-quality integrated descriptions, thus further improving the performance of image-text retrieval tasks.

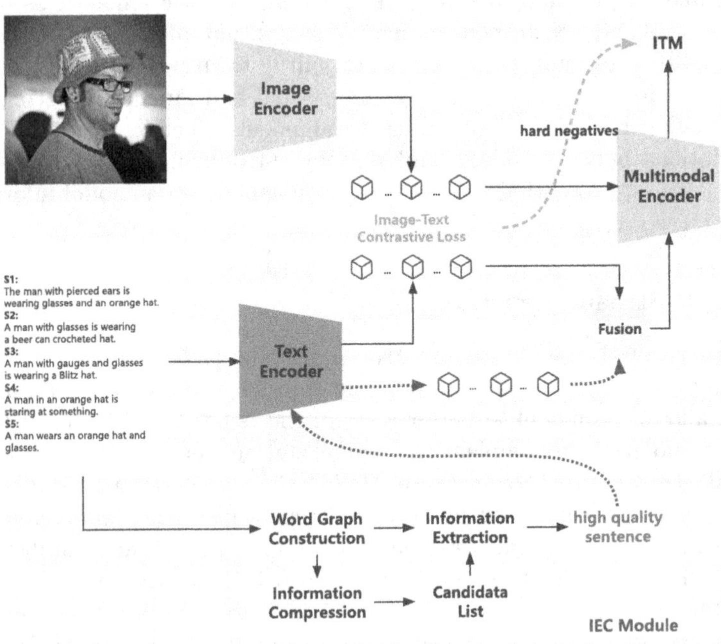

Fig. 1. An Overview of SCIEC.

3 Methodology

3.1 Overview

As shown in Fig. 1, SCIEC consists of an image encoder, a text encoder, and a multi-modal encoder. Before fusion, we use image-text contrast loss to mine hard negative samples and apply them to the image-text matching training task. The IEC module is applied to generate a complete high-quality semantic description and fused with the original sentence to achieve text semantic completion. Finally, the integrated text features and image features are sent to the multi-modal encoder for learning.

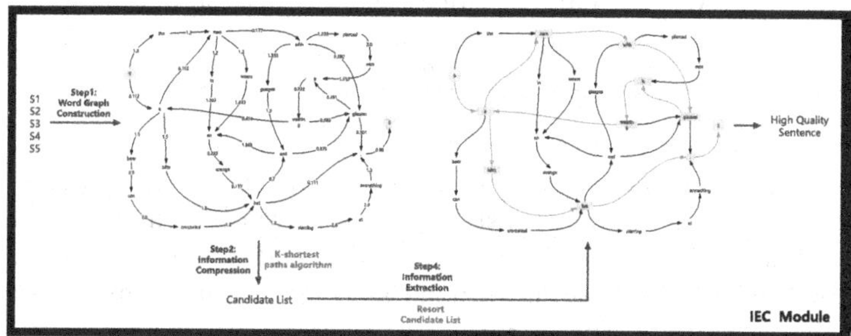

Fig. 2. An Overview of the information extraction and compression module.

3.2 Information Extraction and Compression (IEC)

Given a set of related sentences $S = \{s_1, s_2, ..., s_n\}$ (usually n = 5) to describe the image, in order to obtain a complete and rich image description, we construct a word graph for this set of sentences. By constructing and analyzing the word graph, we can better understand sentence relationships, extract key information, and form coherent and expressive expressions in the description. This method contributes to avoiding redundancy and repetition, while ensuring the completeness and accuracy of the description. Next, we introduce in detail the information compression and information extraction methods based on word graphs. The details of the IEC module are shown in Fig. 2.

Information Compression. This method constructs a directed graph based on the input sentence list S. During the construction process, each sentence is iteratively added to the directed graph according to the following rules:

(1) Node mapping or creation: We process the elements in the sentence in sequence: non-stop words, stop words and punctuation marks. Specifically, for non-stop words, we process them in sequence according to the following

three situations: a. Non-stop words that do not exist in the directed graph; b. Non-stop words that exist in the graph and have clear mapping; c. There are multiple candidate non-stop words in the graph; for non-stop words with unclear mapping, we check the immediate context (previous and following words in the sentence and adjacent nodes in the graph) and select the one with the larger Candidate words that overlap or map more frequently. For stop words, stop words are mapped only if there is overlap among non-stop word neighbors, otherwise a new node is created. For punctuation, punctuation is mapped only if the preceding and following words and adjacent nodes in the sentence are the same.

(2) Edge calculation and addition: After completing node creation and mapping, calculate the edges between each pair of nodes and add them between the mapped words. We calculate the edge weight using the following formula:

$$w(v_i, v_j) = \frac{freq(v_i) + freq(v_j)}{freq(v_i)freq(v_j)\sum_{s \in S} d(s, v_i, v_j)^{-1}} \tag{1}$$

where $freq(v_i)$ refers to the number of words mapped to node v_i, and $d(s, v_i, v_j)$ refers to the offset distance between words v_i and v_j in sentence s. By following these rules for node mapping and creation, and computing edges, we can establish accurate node associations and semantic relationships, thus forming a representation of a directed graph. At this time, the K-shortest path algorithm can yield any number of shortest paths. By normalizing the total path weight across the path length, we perform a round of sorting on the resulting paths.

Information Extraction. It aims to reorder the path list obtained by the above method using the key information extracted from the input sentence list S. Specifically, first, we construct a weighted word graph based on the co-occurrence relationships between words and use the TextRank algorithm to calculate the saliency score of each node. This weighted word graph helps capture the relationships between words and identify keywords. Next, we generate and score candidate keyphrases. These candidate keyphrases are sequences of adjacent words that satisfy specific syntactic patterns. We score each candidate keyphrase, taking into account the saliency scores of the contained words and the length of the candidate keyphrase. The scoring formula is as follows:

$$keyScore(K) = \frac{\sum_{k \in K} TextRank(k)}{length(K) + 1} \tag{2}$$

To generate a more streamlined set of keyphrases, we cluster the generated keyphrases and select the highest-scoring keywords in each cluster as representatives. Finally, we perform a second round of sorting on the information-compressed path list. The ranking process here is based on the normalized length of the path and the sum of the scores of keywords contained on the path. We rerank the paths by multiplying the total path weight over the normalized path length by the sum of the keyword scores contained on the path. The score of a sentence compression cs is given by:

$$totalScore(cs) = \frac{\sum_{v_i,v_j \in path(cs)} w(v_i, v_j)}{length(cs) \sum_{k \in cs} keyScore(k)} \tag{3}$$

Through the above optimization steps, we can extract key information, reasonably sort and organize it, and generate more expressive and accurate sentences.

3.3 Training Tasks

Image-Text Contrastive Learning. It aims to embed image and text representations into a shared semantic space by making the embedding representations of positive samples (relevant image-text pairs) closer in the embedding space while embedding negative samples (irrelevant image-text pairs) to achieve accurate cross-modal image-text matching. SCIEC encodes the image $\boldsymbol{I}$ and text $\boldsymbol{T}$ into embedding vectors E_i and E_t, respectively. For each positive or negative sample, we use the dot product to calculate the similarity between the embedding vectors of the image and text: $S(\boldsymbol{I}, \boldsymbol{T}) =< E_i, E_t >= E_i^T E_t$ We define a loss function with a mini-batch containing M correlation pairs, which includes two parts: image-to-text retrieval and text-to-image retrieval. The goal of this loss function is to maximize the similarity of positive samples and minimize the similarity of negative samples. The specific formula is as follows:

$$L_m^{i2t}(\boldsymbol{I}) = -log\frac{exp(S(I, T_m)/\tau)}{\sum_{m=1}^{M} exp(S(I, T_m)/\tau)} \tag{4}$$

$$L_m^{t2i}(\boldsymbol{T}) = -log\frac{exp(S(T, I_m)/\tau)}{\sum_{m=1}^{M} exp(S(T, I_m)/\tau)} \tag{5}$$

where τ is a learnable temperature parameter. Finally, we define the overall loss of the model as follows: $L_{ita} = (L_{i2t} + L_{t2i})/2$, where L_{i2t} is the comparative loss from image-to-text retrieval, and L_{t2i} is the comparative loss of text-to-image retrieval. By dividing the sum of the two comparative loss functions by two, the average value of the overall loss can be obtained to balance the importance of image-to-text and text-to-image retrieval, and their contributions are relatively evenly included in the overall loss.

Image-Text Matching. To distinguish between image-text pairs that have similar representations but are not related, we use a multimodal encoder. This encoder accepts the image and text as input and generates their joint representation. The predicted output of the image-text matching task is obtained through a fully connected layer and a Softmax function. We calculate the loss for the task using the cross-entropy loss function, based on the labels of the positive and negative samples. Positive samples are labelled as 1 and negative samples are labelled as 0. The key method is to introduce contrast similarity for hard negative example sampling. Specifically, based on the similarity between the image and text, negative text samples with high similarity to the image and negative image samples with high similarity to text are selected in the same batch of data. Such a negative example selection strategy increases the training difficulty and improves the robustness of the model. Suppose we have an image-text pair whose predicted output is y_{pred} and the corresponding true label is y_{true}, then the formula of image-text matching loss can be expressed as:

$$L_{itm} = -y_{true}log(y_{pred}) - (1 - y_{true})log(1 - y_{pred}) \quad (6)$$

The full pre-training objective of SCIEC is: $L = L_{ita} + L_{itm}$.

4 Experiments

4.1 Experimental Settings

Datasets. We used two widely used datasets, Flickr30K and MSCOCO, to train and evaluate the SCIEC model. The Flickr30K dataset contains 31,783 images (29,783 training, 1,000 validation, and 1,000 test images) from the Flickr website, each of which has 5 manually annotated English descriptions. The MSCOCO dataset covers 80 common object categories and contains 123,287 images from different scenarios (113,287 training, 5,000 validation, and 5,000 test images), each with 5 different sentence descriptions. Through training and evaluation on these two datasets, we can conduct a comprehensive analysis and comparison of the performance of the SCIEC model in image-text retrieval tasks.

Evaluation Metrics. In the image and text retrieval task, R@K is an important indicator, which can provide an intuitive measure of the system recall rate, that is, the degree of coverage of the system for relevant images of the query. In our experiment, we used R@1, R@5, and R@10 to evaluate the performance of our proposed method; Use RSum, which is the sum of them, to evaluate the overall retrieval performance. The formula for RSum is as follows:

$$RSum = R@K_{i2t} + R@K_{t2i}(K = 1, 5, 10) \quad (7)$$

Implementation Details. We use ViT-B/16 as image encoder and BERTBase as text encoder and multi-modal encoder. In the process of fusing image and text features, we adopt the cross-attention mechanism of the multi-modal encoder, in which each layer performs cross-attention and fusion of image and text features. We use the PyTorch framework to implement the SCIEC method and choose the AdamW optimizer for model training. By setting the weight decay to 0.02, you can control the regularization of the model and prevent overfitting. We train SCIEC on both datasets with a learning rate of 1e−5, a batch size of 32, and a training epoch of 10. All experiments are performed on a single NVIDIA A6000 GPU.

4.2 Comparison to Baseline

We compare our method with various methods on Flickr30K and MSCOCO datasets, which can be divided into three types: single-stream architecture, dual-stream architecture, and hybrid architecture. We selected six representative SOTA models in recent years for comparison, namely UNITER, VILLA, OSCAR, CLIP, ALIGN, and BLIP.

Table 1 shows the experimental results of our proposed method SCIEC on Flickr30K and MSCOCO datasets respectively. Specifically, (1) the performance of SCIEC is significantly improved when the number of pre-trained images is increased from 4M to 14M. (2) SCIEC outperforms other methods in all recall metrics when the number of pre-trained images for the model is the same. (3) When the number of images used to train SCIEC is 14M: on the Flickr30K dataset, SCIEC's performance is comparable to that of ALIGN trained on a large-scale quantitative set; and on the MSCOCO dataset, SCIEC outperforms ALIGN.

Table 1. Performance of models on Flickr30K and MSCOCO datasets.

Method	#Pre-train Images	Flickr30K							MSCOCO						
		Image-to-Text			Text-to-Image			RSum	Image-to-Text			Text-to-Image			RSum
		R@1	R@5	R@10	R@1	R@5	R@10		R@1	R@5	R@10	R@1	R@5	R@10	
UNITER	4M	87.3	98.0	99.2	75.6	94.1	96.8	550.9	65.7	88.6	93.8	52.9	79.9	88.0	468.9
VILLA	4M	87.9	97.5	98.8	76.3	94.2	96.8	551.5	-	-	-	-	-	-	-
OSCAR	4M	-	-	-	-	-	-	-	70.0	91.1	95.5	54.0	80.8	88.5	479.9
CLIP	400M	88	98.7	99.4	68.7	90.6	95.2	540.6	57.84	81.22	87.78	37.02	61.66	71.5	397.02
ALIGN	1.2B	95.3	99.8	100.0	84.9	**97.4**	98.6	576.0	77.0	93.5	96.9	59.9	83.3	89.8	500.4
BLIP	14M	94.3	99.5	99.9	83.54	96.66	98.32	572.22	75.76	93.8	96.62	57.32	81.84	88.92	494.26
SCIEC	4M	94.4	99.3	99.9	82.8	96.36	98.16	570.92	72.72	92.18	95.94	56.42	81.64	89.04	487.94
SCIEC	14M	**95.3**	**99.8**	**100.0**	**85.48**	97.38	**98.82**	**576.78**	**77.28**	**94.2**	**97.28**	**60.30**	**83.95**	**90.54**	**503.55**

4.3 Ablation Study

To evaluate the contribution of our proposed IEC module to model performance, we performed an ablation study to train the model. We conducted extensive experiments on Flickr30K and MSCOCO datasets and obtained the four combinations in Table 2 by controlling the participation of the IEC module. The results are shown in Table 2. When the IEC module participates in model training, all recall indicators are improved to a certain extent. To present the results more visually, we used graphs. As shown in Fig. 3, we observed the performance changes of the model on the image and text retrieval task by adjusting the number of pre-training images and the participation of the IEC module. It can be clearly seen from Fig. 3 that as the number of pre-trained images increases and the IEC module participates in training, the performance of our model gradually improves. Therefore, the IEC module holds significant importance in enhancing model performance.

Table 2. Performance of the SCIEC model on the Flickr30K and MSCOCO datasets. ΔR shows the difference between SCIEC-base+iec (including the IEC module) and SCIEC-base (without the IEC module) when the number of pre-trained images is equal.

Dataset	Method	#Pre-train	Image-to-Text			Text-to-Image			RSum	ΔR
		Images	R@1	R@5	R@10	R@1	R@5	R@10		
Flickr30K	SCIEC-base	4M	92.6	99.0	99.9	81.46	95.92	97.82	566.7	0
	SCIEC-base+iec	4M	94.4	99.3	99.9	82.8	96.36	98.16	570.92	↑ **4.22**
	SCIEC-base	14M	95.1	99.6	99.9	84.9	97.1	98.66	575.26	0
	SCIEC-base+iec	14M	95.3	99.8	100.0	85.48	97.38	98.82	576.78	↑ **1.52**
MSCOCO	SCIEC-base	4M	72.08	91.48	95.66	55.67	80.79	88.43	484.25	0
	SCIEC-base+iec	4M	72.72	92.18	95.94	56.42	81.64	89.04	487.94	↑ **3.69**
	SCIEC-base	14M	75.76	93.22	96.5	59.56	83.56	90.17	498.77	0
	SCIEC-base+iec	14M	77.28	94.2	97.28	60.30	83.95	90.54	503.55	↑ **4.78**

In addition, we also explored the optimal fusion ratio between the original sentence and high-quality sentences (sentences generated through IEC module). Figure 4 shows the impact of the change in fusion ratio between the original sentence and high-quality sentence on the RMean index of the SCIEC model on the Flickr30K test set when the number of pre-trained images is 4M. The RMean metric is the average of the recall metrics for R@1, R@5, and R@10. As can be seen from Fig. 4, as the original sentence gradually takes a dominant position in the fusion, the performance of the SCIEC model gradually improves and eventually reaches a saturated state. When the fusion ratio reaches 6:4, the model performance begins to exceed the threshold we set (the RMean index of SCIEC-base under the same configuration). As the fusion ratio further increases, the performance of the model gradually stabilizes. Taking these results into consideration, we selected a fusion ratio of 9:1 as the best combination of original sentences and high-quality sentences in the experiment to achieve the best performance of the model.

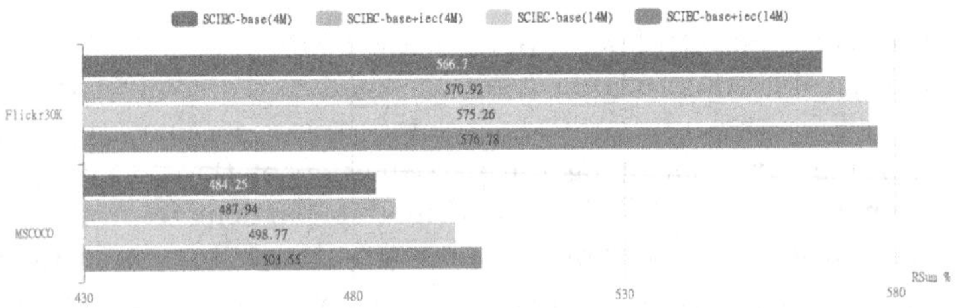

Fig. 3. Performance of SCIEC(4M) models (with and without IEC module) on Flickr30K and MSCOCO datasets.

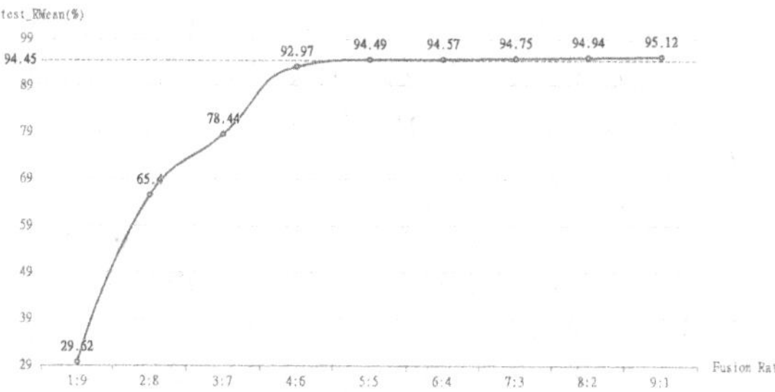

Fig. 4. Performance of SCIEC(4M) with different fusion ratios on Flickr30K dataset.

5 Conclusion

In this paper, we propose a novel image-text retrieval method, SCIEC, which generates rich and complete semantic representations through an information extraction and compression module to achieve the goal of semantic text completion. Subsequently, we align the images and semantically complete text representations and fuse them with a multimodal encoder. Finally, we conduct extensive experiments on the Flickr30K and MSCOCO datasets and compare them with existing methods. The experimental results demonstrate the effectiveness and potential of the SCIEC approach in improving retrieval accuracy and enriching semantic representations.

Acknowledgements. This work was supported by the Science and Technology Program project of Shanghai Municipal Committee of Science and Technology (Grants: 22511104800 and 22DZ1204903).

References

1. Chen, F., Chen, X., Xu, S., Xu, B.: Improving cross-modal understanding in visual dialog via contrastive learning. In: ICASSP 2022 - 2022 IEEE International Conference on Acoustics, Speech and Signal Processing (ICASSP), pp. 7937–7941 (2022). https://api.semanticscholar.org/CorpusID:248218567
2. Chen, Y.-C., et al.: UNITER: UNiversal image-TExt representation learning. In: Vedaldi, A., Bischof, H., Brox, T., Frahm, J.-M. (eds.) ECCV 2020. LNCS, vol. 12375, pp. 104–120. Springer, Cham (2020). https://doi.org/10.1007/978-3-030-58577-8_7. https://api.semanticscholar.org/CorpusID:216080982
3. Dou, Z.Y., et al.: An empirical study of training end-to-end vision-and-language transformers. In: 2022 IEEE/CVF Conference on Computer Vision and Pattern Recognition (CVPR), pp. 18145–18155 (2021). https://api.semanticscholar.org/CorpusID:241033425
4. Faghri, F., Fleet, D.J., Kiros, J.R., Fidler, S.: VSE++: improving visual-semantic embeddings with hard negatives. In: British Machine Vision Conference (2017). https://api.semanticscholar.org/CorpusID:6095318
5. Gan, Z., Chen, Y.C., Li, L., Zhu, C., Cheng, Y., Liu, J.: Large-scale adversarial training for vision-and-language representation learning. ArXiv abs/2006.06195 (2020). https://api.semanticscholar.org/CorpusID:219573512

6. Gansbeke, W.V., Vandenhende, S., Georgoulis, S., Proesmans, M., Gool, L.V.: Learning to classify images without labels. ArXiv abs/2005.12320 (2020). https://api.semanticscholar.org/CorpusID:220347634
7. Huang, Z., Zeng, Z., Huang, Y., Liu, B., Fu, D., Fu, J.: Seeing out of the box: end-to-end pre-training for vision-language representation learning. In: 2021 IEEE/CVF Conference on Computer Vision and Pattern Recognition (CVPR), pp. 12971–12980 (2021). https://api.semanticscholar.org/CorpusID:233169113
8. Jain, A., et al.: MURAL: multimodal, multitask retrieval across languages. ArXiv abs/2109.05125 (2021). https://api.semanticscholar.org/CorpusID:237490989
9. Jia, C., et al.: Scaling up visual and vision-language representation learning with noisy text supervision. In: International Conference on Machine Learning (2021). https://api.semanticscholar.org/CorpusID:231879586
10. Kim, W., Son, B., Kim, I.: ViLT: vision-and-language transformer without convolution or region supervision. In: International Conference on Machine Learning (2021). https://api.semanticscholar.org/CorpusID:231839613
11. Li, G., Duan, N., Fang, Y., Jiang, D., Zhou, M.: Unicoder-VL: a universal encoder for vision and language by cross-modal pre-training. In: AAAI Conference on Artificial Intelligence (2019). https://api.semanticscholar.org/CorpusID:201058752
12. Li, X., et al.: OSCAR: object-semantics aligned pre-training for vision-language tasks. In: Vedaldi, A., Bischof, H., Brox, T., Frahm, J.-M. (eds.) ECCV 2020. LNCS, vol. 12375, pp. 121–137. Springer, Cham (2020). https://doi.org/10.1007/978-3-030-58577-8_8. https://api.semanticscholar.org/CorpusID:215754208
13. Lu, H., Fei, N., Huo, Y., Gao, Y., Lu, Z., Wen, J.: COTS: collaborative two-stream vision-language pre-training model for cross-modal retrieval. In: 2022 IEEE/CVF Conference on Computer Vision and Pattern Recognition (CVPR), pp. 15671–15680 (2022). https://api.semanticscholar.org/CorpusID:248218570
14. Luo, Z., et al.: LexLIP: lexicon-bottlenecked language-image pre-training for large-scale image-text retrieval. ArXiv abs/2302.02908 (2023). https://api.semanticscholar.org/CorpusID:256615217
15. Pena-Pereira, F., Wojnowski, W., Tobiszewski, M.: Agree-analytical greenness metric approach and software. Anal. Chem. **92**, 10076–10082 (2020). https://api.semanticscholar.org/CorpusID:219706533
16. Qi, D., Su, L., Song, J., Cui, E., Bharti, T., Sacheti, A.: ImageBERT: cross-modal pre-training with large-scale weak-supervised image-text data. ArXiv abs/2001.07966 (2020). https://api.semanticscholar.org/CorpusID:210859480
17. Radford, A., et al.: Learning transferable visual models from natural language supervision. In: International Conference on Machine Learning (2021). https://api.semanticscholar.org/CorpusID:231591445
18. Reed, S.E., Akata, Z., Yan, X., Logeswaran, L., Schiele, B., Lee, H.: Generative adversarial text to image synthesis. In: International Conference on Machine Learning (2016). https://api.semanticscholar.org/CorpusID:1563370
19. Su, W., et al.: VL-BERT: pre-training of generic visual-linguistic representations. ArXiv abs/1908.08530 (2019). https://api.semanticscholar.org/CorpusID:201317624
20. Tan, H.H., Bansal, M.: LXMERT: learning cross-modality encoder representations from transformers. In: Conference on Empirical Methods in Natural Language Processing (2019). https://api.semanticscholar.org/CorpusID:201103729
21. Wang, W., Bao, H., Dong, L., Wei, F.: VLMo: unified vision-language pre-training with mixture-of-modality-experts. ArXiv abs/2111.02358 (2021). https://api.semanticscholar.org/CorpusID:241035439

22. Xu, J., Liu, Z., Pei, X., Wang, S., Gao, S.: FB-Net: dual-branch foreground-background fusion network with multi-scale semantic scanning for image-text retrieval. IEEE Access **11**, 36516–36537 (2023). https://api.semanticscholar.org/CorpusID:257897254
23. Yang, J., et al.: Vision-language pre-training with triple contrastive learning. In: 2022 IEEE/CVF Conference on Computer Vision and Pattern Recognition (CVPR), pp. 15650–15659 (2022). https://api.semanticscholar.org/CorpusID:247011309
24. Yao, L., et al.: FILIP: fine-grained interactive language-image pre-training. ArXiv abs/2111.07783 (2021). https://api.semanticscholar.org/CorpusID:244117525
25. Zhang, P., et al.: VinVL: revisiting visual representations in vision-language models. In: 2021 IEEE/CVF Conference on Computer Vision and Pattern Recognition (CVPR), pp. 5575–5584 (2021). https://api.semanticscholar.org/CorpusID:235692795

Fast Edit Distance Prediction for All Pairs of Sequences in Very Large NGS Datasets

A. K. M. Tauhidul Islam[1(✉)] and Sakti Pramanik[2]

[1] Oracle Corp, Austin, USA
tauhid.islam@oracle.com

[2] Michigan State University, East Lansing, MI, USA

Abstract. All the known edit distance calculation algorithms run in near quadratic time with respect to sequence length. For very large number of sequences such as next generation sequencing (NGS) datasets, all pair edit distance calculation based on near quadratic run time may take days or weeks. To solve this performance bottleneck problem, several sub-quadratic run time algorithms have been proposed. Recently, Pramanik et al. [1] has proposed fast reference sequence based edit distance prediction method which addresses this performance bottleneck problem. They are very effective for correctly predicting smaller edit distances (useful, for example, clustering NGS datasets with low threshold) but less effective for larger edit distances. In this paper, we propose faster edit distance prediction method based on a very small number of special reference sequences. These sequences are very effective for predicting close to 100% accuracy. They require several novel techniques based on non-matching sub sequences. We have provided Propositions and Theorems to justify the basis for developing these novel techniques. Using these strategies, we are able to develop a linear time edit distance prediction method with respect to sequence length.

Keywords: Edit distance prediction · NGS sequence similarity

1 Introduction

The edit distance between two arbitrary strings represents their dissimilarity. Given strings A and B, the edit distance is the minimum number of substitutions, insertions and deletions required to convert string A into string B [2]. Despite wide applicability of edit distance measures for dissimilarity between sequences, near quadratic computation complexity of these methods with respect to string lengths has been a performance bottleneck. In recent years, emergence of many large NGS sequence datasets and a need to find novel clusters of similar sequences has reinforced the necessity of faster edit distance calculation methods for huge number of pairs of sequences.

Reducing edit distance computation time complexity has been an active research problem. Several edit distance computation methods of sub-quadratic

D.-N. Yang et al. (Eds.): PAKDD 2024, LNAI 14648, pp. 72–91, 2024.
https://doi.org/10.1007/978-981-97-2238-9_6

time complexity with respect to string length have been proposed [2,6,12]. These methods are still not fast enough for many emerging applications that require all pair distance calculation for large datasests such as large set of NGS sequences. On the other hand, many NGS sequence analysis techniques only require pairs of sequences with high similarity thresholds. For example, clustering NGS sequences at species level requires pairwise edit distances equivalent up to 97% similarity. For these scenarios, faster alternatives of costly edit distance calculation methods are more suitable. Existing edit distance approximation methods [3–5,7,11] further reduce computation time and approximate edit distances within a predefined error bound. For example, the state-of-the-art method [4] runs in near linear time and approximates edit distances within a factor of $2^{O(\sqrt{l}ogl)}$ where l is the sequence length.

In order to further reduce the NGS sequence similarity computation time, several heuristics based methods of linear time complexity [14,15,17] have recently been proposed. These are k-mer (all possible sub-strings of length k) based techniques that compare k-mers between a pair of NGS sequences to measure sequence similarity. These similarity measures are complementary to distance/dissimilarity between sequences and will be used interchangeably throughout rest of the paper. Such sequence similarity does not effectively represent evolutionary based edit distances between the NGS sequences and therefore cannot create edit distance based clusters for NGS data sets.

Recently reference sequence based techniques [1,10] have been proposed for efficient edit distance prediction. For instance, in Pramanik et al. [1], a set of n randomly generated reference sequences is used to convert dataset sequences into feature vectors where each dimension represents the edit distance between the dataset sequence and a reference sequence. Given dataset sequences A and B, a set of reference sequence R, and an edit distance function $ed(A,B)$; $|ed(A,r) - ed(B,r)| \leq ed(A,B)$ where $r \in R$ because of the triangle inequality property of edit distances. Thus, finding the maximum value of $|ed(A,r) - ed(B,r)|$ is the basis for correctly predicting edit distances. This is achieved by finding the Maximum Vector component Difference (MVD), i.e., $max(|ed(A,r) - ed(B,r)|) \leq ed(A,B)$. The MVD based method predicts edit distances nearly accurately in several scenarios. As we increase the number of reference sequences, therefore the number of dimensions in the vector, the MVD gets closer to the actual edit distances. But this is at the cost of higher computation time due to increased processing cost of more reference sequences. Further, MVD based edit distance prediction is less accurate for larger edit distances. However, MVD based approach is still very effective for applications requiring high similarity thresholds.

In this paper, we propose an edit distance prediction method based on a very small number of special reference sequences. Edit distances between these novel reference sequences are very high and, therefore, they are uncorrelated. The proposed approach provides much faster time and works for larger edit distances as well. This is due to a set of innovative techniques developed to solve several challenging problems. These novel methods avoid the limitations of the previous approaches and improve the performance of edit distance prediction significantly. Primary contributions of the paper are:

- We develop a novel linear time edit distance prediction method with respect to sequence length, which is better than the existing methods both in terms of execution time and accuracy of prediction.
- Existing k-mer based similarity measures are popular for their fast execution time but is not based on edit distance. Accuracy of similarity measure with respect to edit distance is poor for these methods. The proposed approach predicts actual edit distance, including the type of edit operations creating the edit distance.
- We conduct extensive experiments and theoretical analysis to justify the superior performance of the proposed method over the existing methods with respect to time and accuracy.
- We have applied the proposed edit distance calculation method for clustering of large NGS sequence datasets. The proposed method based cluster accuracy was very high compared to that based on edit distances. Further, cluster creation time was significantly low for the proposed method compared to that based on edit distances.

Rest of the paper is organized as follows. Section 2 presents the existing sequence dissimilarity comparison approaches. Key ideas of the proposed methods are presented in Sect. 3. Edit distance prediction techniques are presented in Sect. 4. The proposed edit distance prediction method is presented in Sect. 5. Section 6 reports the experimental results. Conclusion is provided in Sect. 7.

2 Related Works

Various popular distance measures such as the edit, the Hamming, and others, are used to measure similarities between NGS sequences. The edit distance is commonly used as the basis for measuring evolutionary relationships between sequences. For NGS sequences, this distance measure is more effective for many applications because it provides the appropriate global similarity. Dynamic programming based edit distance calculation method [2] is of complexity $O(l^2)$ where l is the sequence length. Although, there have been many efforts to reduce such high computation complexity, the well known algorithms for edit distance calculation run in nearly quadratic time complexities [2,6,12]. Many NGS sequence analysis applications require pairs of sequences with high similarity thresholds only. For those applications, faster alternatives of exact edit distance calculation methods are more appropriate.

Several edit distance approximation methods [3–5,7,11] have been proposed to further reduce the time complexity at the cost of accuracy. Andoni et al. [4] proposed a near linear time edit distance approximation method up to a factor of $2^{O(logl)}$. However, the approximation error in terms of percentage is high because the edit distance are integer values. For smaller edit distances, the change in approximation will be significant.

In order to further reduce sequence similarity computation time between large number of sequences, k-mer based similarities were also adopted [8,14,15]. The percentage of matching k-mers is used as a measure of similarities between the sequences. Although such approaches run in linear time with respect to sequence length, the similarity measures do not represent the edit distance. For instance, VSEARCH [15] compares shared distinct k-mers for sequence similarity which also does not represent the edit distance. Local Sensitive Hashing (LSH) [14] based methods compare k-mers in the corresponding positions which is not the case for edit distance with insertion and deletion operations.

Order Min Hash (OMH) [17] creates local sensitive hash of sequences by storing the k-mers as well as their relative orders in a sequence to represent edit distance based similarity. This approach shows superior accuracy for filtering out pairs of sequences with low similarity compared to that of only k-mer similarity based methods. But the OMH based similarity is not an alternative of edit distance calculation methods, rather it can effectively reduce the number of costly pairwise edit distance calculation.

Recently reference sequence based methods [1,10] are proposed to predict edit distances between pairs of sequences in NGS datasets. In this approach, each dataset sequence is converted into a feature vector by computing edit distances of the sequence with respect to a set of reference sequences. Metric distance measures are applied on these feature vectors to predict the edit distances. For example, [1] used the MVD strategy between the corresponding dimensions of the feature vectors for edit distance prediction. The MVD is bounded by the triangle inequality property of the metric distance measures. Prediction accuracy improves with increasing number of reference sequences at the cost of higher computation time. For NGS sequence analysis techniques such as hierarchical clustering for high similarity thresholds, the MVD based method shows superior accuracy over that of the existing k-mer based methods. However, the k-mer based clustering is faster than that of MVD based method. Further, pairwise edit distance prediction accuracy drops sharply with this method for larger edit distances.

In this paper, we use a k-mer based technique to find the disjoint non-matching sub-sequences. We predict edit distances between the sub-sequences using differences of edit distances with respect to single letter reference sequences which is also of linear time complexity with respect to l.

3 Basic Concepts

3.1 Preliminary

S refers to a dataset of N sequences $\{s_1, s_2, \ldots, s_i \ldots, s_N\}$ where s_i is a sequence of letters in the alphabet Σ. Given a sequence A in the dataset, $A[i]$ represents the ith letter of the sequence. $A[i:j]$ represents the sub-string starting at i^{th} position and ending at j^{th} position. $A(i,j)$ represents a sub-string of length j starting at the ith position. Λ refers to a NULL letter to represent insertion and deletion operations.

Definition 1. *Given a string A, an edit operation is a triple (i, x, y) replacing a letter x at the ith position of A with a letter y where $x, y \in (\sum \cup \Lambda)$. The cost of an edit operation is determined by the function $\gamma(i, x, y)$. An edit operation (i, x, y) is (i) a substitution if $x \neq y$ and $x, y \in \sum$, (ii) an insertion if $x = \Lambda$ and $y \in \sum$ (iii) a deletion if $x \in \sum$ and $y = \Lambda$.*

Edit distance between sequences A and B is represented by $ed(A, B)$. R refers to a set of reference sequences. R_l refers to a set of reference sequences where length of each sequence is l and the edit distances among them are also l. Although different sets of sequences satisfy the requirement of R_l, we choose each $r \in R_l$ containing exactly the same letter in Σ because they possess a unique property which is exploited for efficient edit distance prediction. R_l contains sequences for each letter in Σ, therefore $|R_l| = |\Sigma|$. For instance, $R_5 = \{aaaaa, ccccc, ggggg, ttttt\}$ where $\Sigma = \{a, c, g, t\}$ and $l = 5$.

3.2 Key Ideas

Key ideas of our proposed edit distance prediction method are as follows. First, we convert each sequence in S into a feature vector of dimension $|\Sigma|$ where each attribute of the feature vector represents $ed(A, r)$ for an $r \in R_l$. For instance, the feature vector of $A = tctgta$ would be $< 5, 5, 5, 3 >$ for $R_l = \{aaaaaa, cccccc, gggggg, tttttt\}$. The difference between the edit distance of A with a reference sequence and the edit distance of B with the same reference sequence, $A, B \in S$, attempts to capture the edit distance between A and B due to the character of that reference sequence. We sum these differences of distances in the feature vectors to predict $ed(A, B)$. Since an edit operation requires changing a character to another, ideally it should be counted twice in the differences of distances. Let us consider, sequences $A = cgtagcatcagc$, $B = catagcatcatc$ and $C = cggtagcatcac$. Length of the reference sequences is set to $\max(|A|, |B|, |C|) = 12$. The feature vectors of A, B and C are $< 9, 8, 9, 10 >$, $< 8, 8, 11, 9 >$, $< 9, 8, 9, 10 >$ respectively. Sum of the differences of distances between feature vectors of A and B is 4. Hence, $ed(A, B)$ is correctly predicted as 2. On the other hand, the difference of distances between A and C with respect to $r = g^{13}$ (4th dimension) is 0. Our approach fails to capture the edit operations with respect to g because these edit operations are captured by both A and C with respect to the reference sequence. Sum of differences of distances between feature vectors of A and C is 0. Therefore, $ed(A, C)$ is incorrectly predicted as 0. With increasing edit distance between datasets sequences this approach is less likely to predict the edit distances correctly. Given an alphabet Σ, the probability of correct edit distance prediction for two substitution operations is $\frac{(|\Sigma|-1)+(|\Sigma|-2)^2}{|\Sigma|^2}$. This ratio gets higher with larger Σ. In order to improve prediction accuracy, we use heuristics based on fragmenting the sequences into corresponding sub-sequences such that these subsequences have fewer edit operations between them. Since the DNA alphabet consists of only four letters (Nucleotides), we compare differences of distances with respect to sequences of di-nucleotides to improve edit distance prediction accuracy. Details of the heuristics that significantly improve the accuracy of prediction are presented in Sect. 5.

3.3 Selecting the Reference Sequences

An effective set of reference sequences is critical for our proposed edit distance prediction methods. The set of reference sequences are selected so that the following two key properties are satisfied between the reference sequences: 1. length of the sequences must be at least as long as the longest dataset sequence and 2. the edit distances between a pair of the reference sequences must be significantly large.

Edit Distances Between Reference Sequences. Given a dataset sequence $A \in S$, if the reference sequences are correlated then $\forall r \in R, ed(A, r)$ are similar to each other. Our goal is to minimize correlations among the reference sequences in R. Correlation among the reference sequences decreases monotonically with increasing distances among the reference sequences. The proposed single letter reference sequences in R_l provide the maximum edit distances between each of them. We can compute $ed(A, r)$ for $r \in R_l$ in linear time with respect to l using a formula. Correctness of this formula is proved in Theorem 1 based on the concept of *trace*. Given dataset sequences A and B, the set of traces represents all possible combinations of temporally ordered sequences of edit operations to convert A into B. For the sake of completeness, definition and cost function of a *trace* [2] are presented here.

Definition 2. *A trace from sequence A to sequence B is a triple (T, A, B), where T is a set of ordered pairs of integers (i, j) satisfying:*

1. $1 \leq i \leq |A|$ *and* $1 \leq j \leq |B|$
2. *for any distinct pairs* (i_1, j_1) *and* (i_2, j_2) *in* (T, A, B), *(a)* $i_1 \neq i_2$ *and* $j_1 \neq j_2$, *(b)* $i_1 < i_2$ *iff* $j_1 < j_2$.

When sequences A and B are understood, we'll use T to denote (T, A, B). Any $(i, j) \in T$ represents a line between the letters of $A[i]$ and $B[j]$. Each letter can be connected by at most one line and the lines do not cross. The ordered pairs indicate the change operations between the sequences. A *change* operation is either an exact match or a substitution between the corresponding letters. Figure 1 gives a pictorial representation of $(T, A, B) = \{(1,1), (2,2), (3,4), (4,5), (5,6), (6,7), (7,8), (8,9), (9,10), (10,11), (12,12)\}$. The unconnected letters in A are deleted and the unconnected letters in B are inserted into A.

Cost of trace T, denoted as $cost(T)$ is the summation of cost of each change, insertion and deletion operations. Change operation does not contribute to the cost of T if both letters are same. For example, the number of change, insertion and deletion operations in trace T are 11, 1 and 1 respectively. All the change

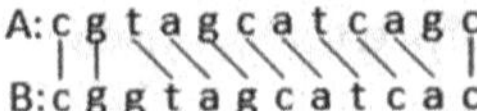

Fig. 1. A trace between sequences A and B

operations have the same letters, so they do not contribute to the cost. Assuming cost of each edit operation is 1, $cost(T) = 0 + 1 + 1 = 2$. The size of a trace is the number of ordered pairs in it. A minimum cost trace between A and B represents $ed(A, B)$ [2].

Theorem 1. *Given a reference sequence, $r \in R_l$ where $r = x^l$, and a dataset sequence A of length l', $ed(r, A) = (l-k)$ where $l \geq l'$ and $k = \Sigma_{i=1}^{l'} 1$, if $A[i] = x$.*

Proof. Let us consider a maximum size trace between r and A, (T_0, r, A), where every letter of A is connected by a line to a letter in x^l. Therefore, the size of T_0 is l'. Thus, the number of change operations in T_0 is l'. Let us assume, $x \in \Sigma$ appears k times in A. Therefore, the cost of change operations for those k pairs of $(i, j) \in T_0$ is zero. In addition, $(l - l')$ deletion operations occur in r. Since $l \geq l'$, there won't be any insertion operation.

$$cost(T_0) = \sum (r[i] \to A[j]) + \sum (\Lambda \to r[i]) + \sum (r[i] \to \Lambda)$$
$$cost(T_0) = (l' - k) + 0 + (l - l') = l - k$$

We will show that T_0 is the minimum cost trace and, therefore, the edit distance between A and r. Assume an arbitrary trace (T_i, r, A) of size l'' where $l'' \leq l'$. Therefore, $(l - l'')$ positions in r and $(l' - l'')$ positions in A do not participate in the change operations and therefore not part of T_i. Letters of such positions in r will be deleted and those in A will be inserted into r.

$$cost(T_i) = (l'' - k) + (l' - l'') + [(l' - l'') + (l - l')]$$
$$cost(T_i) = (l - k) + (l' - l'') = cost(T_0) + (l' - l'')$$
$$cost(T_0) \leq cost(T_i)$$

Hence, T_0 represents the minimum cost trace and therefore $cost(T_0) = l - k$ is the edit distance.

For example, in Fig. 2(a), $|(T_0, A, t^l)|$ is the largest because all letters are connected and the cost of T_0 is minimum. Therefore, assuming cost of each edit operation is 1, $cost(T_0) = ed(A, t^l)$. Here, the number of change, insertion and deletion operations are 11, 1 and 0 respectively. However, 2 of the change operations have the same character. Therefore, $cost(T_0) = (11-2)+1+0 = 10$. On the other hand, in Fig. 2(b) $cost(T_1) = 12 > cost(T_0)$.

Now that we know a simple method for computing edit distance between a data set sequence and a reference sequence, we will need to compute the differences of the edit distances of A and B with respect to the reference sequences to predict edit distance between A and B. Equation 1 shows the relationship between the edit distance and the proposed differences of distances based method. This is based on the triangular property of edit distances.

$$ed(A, B) \geq |ed(A, x^l) - ed(B, x^l)| \text{ where } x \in \Sigma \tag{1}$$

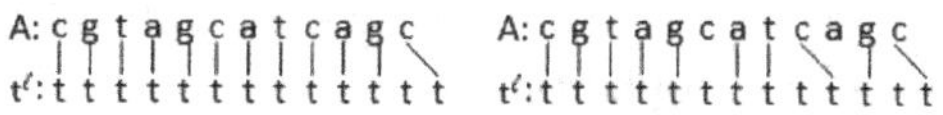

(a) Trace T_0

(b) Trace T_1

Fig. 2. (a) a maximum size trace and (b) a trace with fewer change operations

4 Predicting the Edit Distance

We convert each sequence $A \in S$ into a feature vector of dimension $|\Sigma|$ in metric space where each attribute represents $ed(A, x^l)$ for $x \in \Sigma$. The position of the attributes at A feature vector is an ordered set for the letters in Σ. For instance, the feature vector of $A = tctgta$ is $< 5, 5, 5, 3 >$. We use the differences in values of the corresponding dimensions in the feature vectors to predict edit distance between the NGS sequences.

The following Proposition shows that if the edit distance between a pair of sequences is 1, the difference of distances with respect to a $r \in R_l$ is also 1. Hence, the edit distance will be predicted correctly.

Proposition 1. *If* $ed(A, B) = 1$, $\exists x \in \Sigma$ *s.t.* $|(ed(A, x^l) - ed(B, x^l)| = 1$ *where* $l = \max(|A|, |B|)$.

Proof. Proof of the Proposition is trivial and therefore omitted.

For example, let us consider, $A = tctgta$. The letter t is inserted at $A[4]$ to convert it into $B = tcttgta$. In this case, $\left|ed(A, t^l) - ed(B, t^l)\right| = 1$.

$\left|ed(A, x^l) - ed(B, x^l)\right|$ where $x \in \Sigma$ will be able to predict $ed(A, B) > 1$ with additional heuristics. Each edit operation affects two letters. For example, for a substitution operation $(a \rightarrow g)$, differences of distances with respect to a^l and g^l will be non-zero. Similarly, for $(\Lambda \rightarrow c)$, difference of distance with respect to c^l and Λ i.e., the length difference will be non-zero. Using these properties, we present the following Proposition.

Proposition 2. *For* $ed(A, B) > 1$, *the predicted edit distance based on differences of distances with respect to* $r = x^l \in R_l$ *is bounded as follows,*

$$ed(A, B) \geq \frac{\sum_{i=1}^{|\Sigma|} \left|ed(A, x^l) - ed(B, x^l)\right| + \big||A| - |B|\big|}{2} \tag{2}$$

Proof. Let us consider, a substitution operation $(x \rightarrow y)$ where $x, y \in \Sigma$. The differences of distances of with respect to $x^l, y^l \in R_l$ will be affected for $(x \rightarrow y)$. Similarly, for $(\Lambda \rightarrow y)$, the difference of distances with respect $y^l \in R_l$ and the length difference of the dataset sequences will be affected. Hence, for any edit operation, the edit distance can be found by dividing the sum of differences of distance by two.

On the other hand, for any combination of multiple edit operations, the differences of distances with respect to $x^l \in R_l$ are minimized when x appears in edit operations in both sequences. Similarly, length difference is minimized for multiple insertion and deletion operations.

Therefore, the Inequality holds for $ed(A, B) > 1$.

For example, in Fig. 3(a), differences of distances with respect to g^l, a^l and t^l are 2, 1, 1 respectively. The edit distance is correctly predicted by dividing sum of the differences of distances by two. On the other hand, in Fig. 3(b), the edit distance is predicted incorrectly because the difference of edit distances with respect to g^l is incorrect. Although g is inserted at $B[4]$ followed by deleted in $A[11]$. The difference of distances between A and B with respect to g^l is 0.

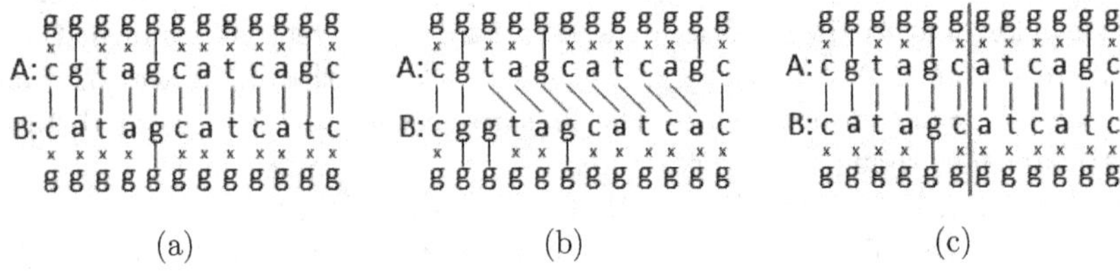

Fig. 3. Edit distance prediction between pairs of sequences using the proposed cost function. In (a), edit distance was correctly predicted. However, in (b), edit distance prediction was not correct. (c) Edit distance prediction for $ed(A, B) > 1$ by dividing the sequences into sub-sequences so that each pair of sub-sequences has only edit operation.

It should also be noted that Proposition 2 is based on the Proposition 1. For instance, in Fig. 3(a), we know from the differences of distances that there are more than one edit operations. If we break the sequences into two pairs of sub-sequences, there is only one edit operation in each pair of sub-sequences (Fig. 3(c)). Thus we can apply Proposition 1 to predict edit distances between the pairs of sub-sequences to determine edit distance between the entire sequences.

For larger edit distances, the differences of distances based predictions will be less accurate. In order to minimize this problem, we develop a strategy to find non-matching (not exact match) sub-sequences between the sequences. Then, we apply Proposition 2 on the non-matching sub-sequences to calculate edit distances between the pairs of sequences. More detail about this technique is presented in Sect. 5.

5 Edit Distance Prediction from Non-matching Sub-sequences

In this section, we present the proposed edit distance prediction method. First, we find non-matching sub-sequences between a pair of sequences. We create a HashTable from the k-mers of a sequence where the k-mers are keys and the positions of the k-mers are associated values. Then we discard positions of shared k-mers in both sequences. We separate the remaining k-mers of each sequence based on consecutive positions and create a sub-sequence for each subset of k-mers. We select corresponding pairs of sub-sequences based on their positions and apply Proposition 2 to predict edit distances between them. Summation of calculated edit distances between corresponding sub-sequences is presented as the predicted edit distance between the sequences. In addition, we predict the number of each type of edit operations. Pseudocode of this technique is presented in Function 1 and Function 2.

FUNCTION 1 *Edit_Distance_Prediction (A,B,k, d)*
Input: Sequence A, Sequence B, k-mer length k, edit distance threshold d
Output:Predicted edit distance, $ed_p(A,B)$, Types of edit operations, $inserts$, $deletes$, $substitutions$
1. $ed_p(A,B) = \infty, inserts = 0$, $deletes = 0$, $substitutions = 0$
2. Create HashTable<k-mer, List<Index» for A, H_A and for B, H_B
3. $count$ = max(#k-mers present in H_A but not in H_B, #k-mers present in H_B but not in H_A)
4. **If** $(count \leq (k * d))$ **then**
5 **Foreach** (k-mer index i in H_A)
6. Find the nearest index j in H_B for the k-mer. if $|i - j| <= d$, remove i from H_A and j from H_B
7 **end Foreach**
8. Create sub-sequences $A_s =< A_1, A_2, \cdots >$ and $B_s =< B_1, B_2, \cdots >$ from indexes in H_A and H_B
9. **Foreach** $((A_i, B_i)$ where $1 \leq i \leq min(|A_s|, |B_s|))$ **do**
10. Compute $Subseq\_Edit\_Distance(A_i, B_i)$ and add $inserts$, $deletes$ and $substitutions$ edit operations of the sub-sequences
11. **end Foreach**
12. $ed_p(A,B) = inserts + deletes + substitutions$
13. **end If**
14. **return** $ed_p(A,B)$, $inserts$, $deletes$, $substitutions$

FUNCTION 2 *Subseq_Edit_Distance(P,Q)*
Input: P, Q
Output: $inserts$, $deletes$, $substitutions$
1. **If** $(|P| \geq |Q|)$ **then**
2. $deletes = |P| - |Q|$
3. **Else**
4. $inserts = |Q| - |P|$
5. **end If**
6. $l = max(|P|, |Q|)$

$dist = abs(|P| - |Q|)$
Foreach $(x \in \Sigma)$ **do**
  $dist+ = |ed(P, x^l) - ed(Q, x^l)|$
end Foreach
$substitutions = (dist/2) - inserts - deletes$
return *inserts*, *deletes*, *substitutions*

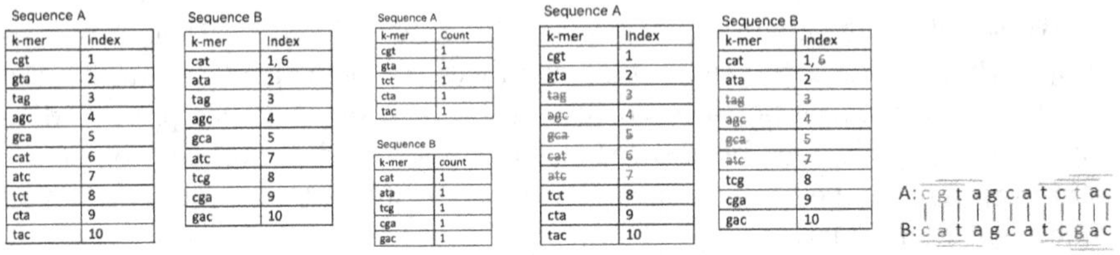

Sequence A

k-mer	Index
cgt	1
gta	2
tag	3
agc	4
gca	5
cat	6
atc	7
tct	8
cta	9
tac	10

Sequence B

k-mer	Index
cat	1, 6
ata	2
tag	3
agc	4
gca	5
atc	7
tcg	8
cga	9
gac	10

(a)

Sequence A

k-mer	Count
cgt	1
gta	1
tct	1
cta	1
tac	1

Sequence B

k-mer	count
cat	1
ata	1
tcg	1
cga	1
gac	1

(b)

Sequence A

k-mer	Index
cgt	1
gta	2
~~tag~~	~~3~~
~~agc~~	~~4~~
~~gca~~	~~5~~
~~cat~~	~~6~~
~~atc~~	~~7~~
tct	8
cta	9
tac	10

Sequence B

k-mer	Index
cat	1, ~~6~~
ata	2
~~tag~~	~~3~~
~~agc~~	~~4~~
~~gca~~	~~5~~
~~atc~~	~~7~~
tcg	8
cga	9
gac	10

(c)

A: c g t a g c a t c t a c
B: c a t a g c a t c g a c

(d)

Fig. 4. Edit distance prediction between sequences $A = cgtagcatctac$ and $B = catagcatcgac$. (a), Each Sequence is converted into a HashMap where key is a k-mer and values are index(es) of the k-mer. (b) Number of different k-mers in each sequences. The product of k-mer length and edit distance threshold should not be greater than the different k-mers count. In this example, for k-mer length 3 and distance threshold 3, the different k-mers count is smaller. This allows to determine whether edit distance is higher than the threshold. (c) For each k-mer, the nearest $|i - j| \leq d$ where $1 < i < |A|$ and $1 < j < |B|$ is discarded, (d) Remaining k-mer positions in each sequence is parsed in linear time to create the subsequences. The sub-sequences are, in A(cg, tac) and in B(ca, gac). Hence, sum of edit distances predicted separately between (cg, ca) and (tac, gac) is the predicted edit distance between A and B.

An example of the proposed method is presented in Fig. 4. For this example, the k-mer length, $k = 3$ and the edit distance threshold, $d = 3$. In Fig. 4 (a), the sequences are converted to HashMaps where key is the k-mers and values are the k-mer indexes. In Fig. 4 (b), k-mers of the sequences are compared to find the k-mers count that are not present in the other sequence. This step is for finding pairs of sequences where the edit distance may be larger than the given thresholds. In case, the number of different k-mers are high enough to surpass the threshold, there is no need to compute the edit distance prediction steps. In Fig. 4 (c), only the non-matching k-mers between the sequences are kept. For a k-mer to be matching, it must satisfy $|i - j| \leq d$ where $1 < i < |A|$ and $1 < j < |B|$. Remaining k-mers in each sequence are sorted based on their indexes. Sub-sequences are generated based on consecutive indexes only. In Fig. 4 (d), corresponding pairs of sub-sequences between$A$ and B are selected from left to right direction. In this example, two such pairs of sub-sequences are selected. For each pair of sub-sequences, we predict the edit distances. Since the sub-sequences are disjoint, sum of edit distances from all pairs of sub-sequences is the predicted edit distance between A and B.

5.1 Differences of Distances with Respect to Pairs of Letters

When the differences of distances between a pair of sub-sequences indicate more than one edit operations, we apply Proposition 2 to compute the edit distances. However, as shown earlier, the differences of distances may not be able to identify multiple edit operations. For example, in Fig. 5, there are two substitution operations between the 2nd pair of sub-sequences, yet the differences of distances predict only one edit operation. Given an alphabet Σ, the probability of correct edit distance prediction for two substitution operations is $\frac{(|\Sigma|-1)+(|\Sigma|-2)^2}{|\Sigma|^2}$. For small alphabet, this probability drops rapidly with multiple edit operations. On the other hand, if the alphabet size is large, the probability of correct prediction is significantly high. In order to take advantage of this property, we consider all possible two consecutive letters as single elements and compare differences of distances with respect to reference sequences based on those elements. As the example in Fig. 5 shows, the differences of distances between the 2nd pair of sub-sequences is six which indicates more than one edit operations between these sub-sequences. It should be noted the for each edit operation, the maximum difference of distances will be four for two letters based reference sequences.

Fig. 5. Edit distance prediction between sub-sequences using differences of distances of pairs of letters

5.2 Computational Complexity

Computation complexity of Function 1 is as follows. In step 2, HashTables for k-mers are created in linear time with respect to the sequence length l. In step 3, differences of k-mers count is computed in linear time. In steps 5–7, shared k-mers are discarded based on their positions. Since the positions are stored in order in the HashTables, the comparison complexity is linear with respect to l. In step 8, sub-sequences are created by grouping consecutive positions of the k-mers in linear time. Computational complexity for steps 9–11 is bounded by length of all the sub-sequences which is bounded by l. Assuming constant alphabet size, computation complexity of steps 9–11 is also linear. Therefore, overall computational complexity of the edit distance prediction is linear with respect to l.

6 Results and Discussions

We conducted extensive experiments to evaluate effectiveness of the proposed methods. The experiments were performed on an Intel(R) Core(TM) i5-4590

CPU @ 3.30 GHz with 24 GB physical memory. Quadratic run time based edit distance calculation was set as the baseline method. In order to find the non-matching sub-sequences between a pair of sequences, we used k-mers of length 8.

We showed superiority of the proposed edit distance prediction method over the existing ones with respect to computation time as well as accuracy. We also evaluated effectiveness of the proposed method on an important NGS dataset application, to find Operational Taxonomic Units (OTU) by creating clusters for high similarity thresholds. Clusters were created based on pairwise distances computed by the proposed method and that of OMH [17]. Then we compared accuracy and creation time of these clusters with those of widely used VSEARCH [15]. The edit distances and the clusters generated from the baseline method are used as ground truth for the experimental results. Supplementary resources of the project are available at https://github.com/edit-distance-prediction/v1.

Table 1. 16S rRNA sequence fragment datasets

Dataset	Mouse	Human	Rat	Soil	Marine Plankton	Zoo Compost	Mice gut 16 s v3
Id	D_m	D_h	D_r	D_s	D_p	D_z	D_{mv3}
#Seq	17993	21575	30079	99742	496314	625178	1985002
Mean Length	252.6	252.9	252.87	253.1	251	250.7	252
St. Dev.	1.23	1.2	0.62	1.12	0	2.45	0

6.1 Datasets

A diverse set of 16 s rRNA NGS sequence datasets obtained from NCBI database are used to assess the performance of our methods. Size of the datasets range from 2×10^4 to 2×10^6. The standard deviation of the sequence length is low because specific regions of the 16 s rRNA are targeted for NGS datasets. But these sequences can still have high edit distances because of insertions, deletions and sequencing errors. The datasets presented in Table 1 are metagenomic samples taken from mouse, human, rat, soil, marine plankton [9], zoo compost [16] and mice gut 16 s v3 region (SRR5634175).

Table 2. Effect of k-mer length on edit distance prediction accuracy

kmer length	3	4	5	6	7	8	9	10
Prediction Accuracy (%)	60	82.1	93.2	97.3	97.6	97.63	97.3	96.9

Table 3. Predicted edit distance accuracy on increasing actual edit distance and dataset size

Data sets	$Ed = 1$	$Ed = 2$	$Ed = 3$	$Ed = 4$	$Ed = 5$
	# Pairs / P_p/OMH	# Pairs / P_p/OMH	# Pairs / P_p/OMH	# Pairs / P_p/ OMH	# Pairs / P_p/OMH
D_m	21837 / 100 / 100	1281230 / 100 / 100	1350954 / 99.9/ 98.8	1140230/99.3/96	745375/98.9/92.7
D_h	22643 / 100 / 100	991935 / 100 / 99.99	1902828 / 99.9/ 99	1541854/99.4/97.2	1465026/98.8/93
D_r	9248 / 100 / 100	58002 / 100 /100	188804 / 99.8 /98.4	877212/99.3/95.7	1372015/98.6/92
D_s	66469 / 100 / 100	618409 / 99.99/ 100	771141 / 99.9 / 98.6	947233/99.4/95.5	1304600/98.8/92.6
D_p	274954/ 100 / 100	4.7×10^6/99.99/100	27.5×10^6/99.9/99.1	54.9×10^6/99.3/96.7	90.5×10^6/98.7/92.2
D_z	479624 / 100 / 100	7.25×10^6/99.99/99.99	34.8×10^6/99.8/98.7	91.3×10^6/99.4/96.3	1.4×10^8/98.63/93.2
D_{mv3}	4.7×10^6/100/100	9.2×10^7/99.99/100	4.8×10^8/99.8/98.8	9.4×10^8/99.3/95.7	1.3×10^9/98.6/92.6
	$Ed = 6$	$Ed = 7$	$Ed = 8$	$Ed = 9$	$Ed = 10$
	# Pairs / P_p/OMH	# Pairs / P_p/OMH	# Pairs / P_p/OMH	# Pairs / P_p/OMH	# Pairs / P_p/OMH
D_m	881950 / 98.3 / 89	847518 / 97.8 / 84	824397 / 98 / 79	943817/97.6/ 73	1063198/97/65
D_h	1866537 / 98.1/ 90.3	1370566 / 97.6/ 85	868773 / 97 / 79	714121/96.5/73.6	622368/95.8/67
D_r	1179662 / 98 / 88	1045195 / 97.3/ 84.5	939344 / 96.4 / 79.6	807129/96/73.1	984443/95.3/67.2
D_s	1837905 / 98.3 / 89.2	2409388 / 97.7 / 84.6	2962163 / 96.8 / 78.5	3437198/96.1/72.3	3730997/95.4/66
D_p	1.02×10^8/98.1/88.9	1.01×10^8/97.5/84.3	90×10^6/97/77.8	84.9×10^6/96.4/73.4	77.3×10^6/95.3/66.2
D_z	1.57×10^8/98.2/89.3	1.66×10^8/97.3/85.3	1.39×10^8/96.8/79.2	1.37×10^8/96.2/73.2	1.26×10^8/95.4/67.4
D_{mv3}	1.35×10^9/98.1/88.3	1.4×10^9/97.1/84.8	1.21×10^9/96.9/78.3	1.31×10^9/96.3/72.8	1.13×10^9/95.6/65.9

P_p = Proposed edit distance prediction method, Ed = Edit distance.

6.2 Selecting K-Mer Length

In Table 2, we present the impact of k-mer length on the proposed edit distance prediction method. For this experiment, we randomly select 10000 pairs of sequences from the mouse, human and soil datasets. Then, we set different k-mer lengths for predicting edit distances between these pairs of sequences. For small k-mer length such as $k = 3$, the resulting sub-sequences correct in approximately 60% cases which cause incorrect edit distance prediction. As k-mer length increases, the resulting sub-sequences become more accurate. However, longer k-mers need more computation time and accuracy does not improve anymore. Based on the observations, we chose k-mer length 8 for our experiments.

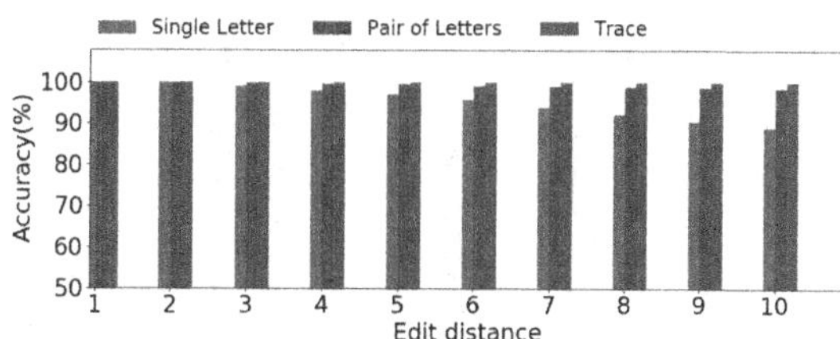

Fig. 6. Effect of proposed strategies on edit distance prediction accuracy

6.3 Comparison of the Proposed Strategies

In Fig. 6, we present comparison of accuracies of the proposed strategies for computing edit distances between the sequences. Edit distance prediction accuracy based on differences of distances with respect to two letters reference sequences over that of single letters improves significantly for higher edit distances. On the other hand, trace based strategy correctly computes the edit distances between the sub-sequences. We will use difference of distances with respect to two letters sequences for edit distance prediction.

6.4 Edit Distance Prediction

We compare accuracy and computation time of the proposed edit distance prediction method with those of Andoni et al. [4] and OMH [17] based edit distance prediction methods. Andoni et al. proposed an edit distance approximation method of near linear time complexity $l^{1+o(1)}$ compared to the proposed method's linear time complexity. The method [4] can approximate up to a factor of $2^{O(\sqrt{(logl)})}$ whereas our proposed method shows more than 97% prediction accuracy for edit distances up to 5 and more than 95% for edit distances up to 10. Since there was no implementation available for Andoni et al., we only compare with respect to theoretical bound.

Edit Distance Prediction Accuracy. We also compare edit distance prediction accuracy between sequences of the experimental datasets in Table 1 using the OMH and the differences of distances with respect to two letters based reference sequences. The comparison is shown in Table 3. Prediction accuracy of the proposed method declines very slowly for larger edit distances. On the other hand, OMH method based pairwise edit similarity prediction accuracy declines more rapidly with higher edit distances. While the OMH based similarity measures are more representative of edit distances compared to that of Weighted Jaccard, they cannot correctly map to the corresponding edit distances.

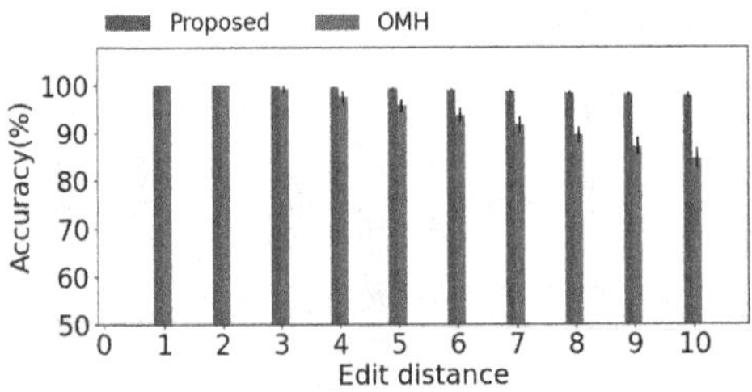

Fig. 7. Cumulative edit distance prediction accuracy on experimental datasets

In Fig. 7, we also present the comparison of cumulative accuracies between the proposed method and the OMH based edit distance predictions. Similar to Table 3, the proposed method shows better accuracy over OMH based predictions. The dashed lines represent the variation in accuracy with the experimental datasets. Superior performance of the proposed method is also demonstrated in

Fig. 8 when compared with respect to both time and accuracy on larger edit distances. For datasets D_h and D_r, the proposed method shows better accuracy with nominal loss in required time.

In the proposed method, edit distances between the corresponding sub-sequences are calculated separately. When there are few edit operations between the sub-sequences, the proposed differences of distances based strategies show very high accuracy. On the other hand, the OMH based method predicts similarity between a pair of sequences based on the ordered k-mer based sketch which is not as accurate as the actual edit distance calculation.

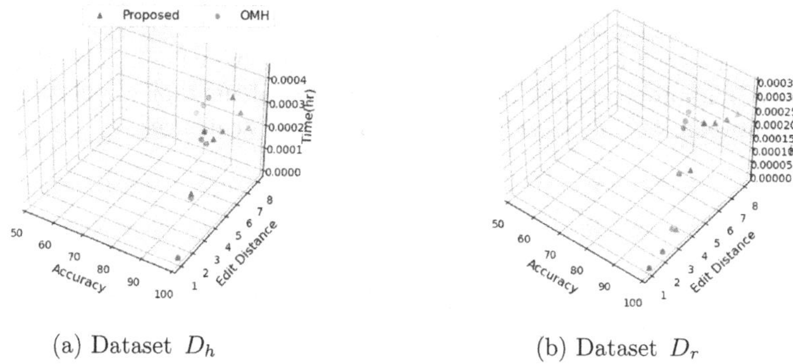

(a) Dataset D_h (b) Dataset D_r

Fig. 8. Comparison of edit distance prediction time and accuracy between the *OMH* and the proposed method

Table 4. Prediction accuracy of edit operations (insertions, deletions and point mutations)

Data sets	$Ed = 1$	$Ed = 2$	$Ed = 3$	$Ed = 4$	$Ed = 5$
	ins / del / mut	ins / del / mut	ins / del / mut	ins / del / mut	ins / del / mut
D_m	100 / 100 / 100	100 / 100 / 100	99.99 / 99.99 / 100	99.4 / 99.5 / 99.35	99.1 / 99.07 / 98.8
D_h	100 / 100 / 100	100 / 100 / 100	99.99 / 99.99 / 100	99.4 / 99.46 / 99.3	99 / 99 / 98.7
D_r	100 / 100 / 100	100 / 100 / 99.97	99.98 / 99.97 / 99.9	99.4 / 99.3 / 99.2	98.9 / 99 / 98.6
D_s	100 / 100 / 100	100 / 99.99 / 100	99.99 / 99.99 / 99.9	99.5/ 99.4 / 99.4	99.05 / 99 / 98.7
D_p	100 / 100 / 100	100 / 100 / 99.99	99.98 / 99.99 / 99.92	99.5 / 99.5 / 99.3	99 / 98.9 / 98.7
	$Ed = 6$	$Ed = 7$	$Ed = 8$	$Ed = 9$	$Ed = 10$
	ins / del / mut	ins / del / mut	ins / del / mut	ins / del / mut	ins / del / mut
D_m	98.32 / 98.2 / 98.3	97.8 / 97.9 / 97.7	96.7 / 96.6 / 96.4	96.5/96.5/96.2	97.1/97.16/95.9
D_h	98.1 / 98 / 97.8	97.5 / 97.8 / 97.5	97 / 96.8 / 96.8	96.4 / 96.6 / 96.5	96 / 95.9 / 95.7
D_r	98 / 97.9 / 97.6	97.6 / 97.6 / 97.2	97.1 / 97.1 / 96.5	96.4 / 96.5 / 96	95.8 / 95.8 / 95.2
D_s	98.3 / 98.2 / 98.2	97.8 / 97.7 / 97.4	96.8 / 96.9 / 96.6	96.2 / 96.4 / 96.1	95.5 / 95.5 / 95.2
D_p	98.1 / 98.1 / 97.8	97.5 / 97.4 / 97.1	96.8 / 96.7 / 96.3	96.1 / 96 / 95.7	95.5 / 95.3 / 95.1

Ed = Edit distance, ins = Insertions, del = Deletions, mut = Point mutations.

Edit Distance Prediction Time. Our proposed method runs in linear time with respect to sequence length l whereas the state-of-the-art method [4] edit distance approximation method has near linear time complexity. On the other hand, OMH requires lower average pairwise distance calculation time because this technique relies on the fixed length ordered k-mer signature for predicting similarity. Figure 9 presents the comparison on average pairwise distance prediction time. For both OMH based method and ours, we include feature vector creation time in the total time. The figure also shows the small variance (dashed line) in computation time signifying nominal impact of varying number of edit operations and their positions in a sequence. It should be noted that the percentage of transformation time becomes smaller with respect to total time with increasing dataset size. The average pairwise distance prediction time is marginally better for the OMH than that of proposed method at the cost of accuracy. The proposed method requires additional time for calculating edit distances between the sub-sequences between pairs of sequences.

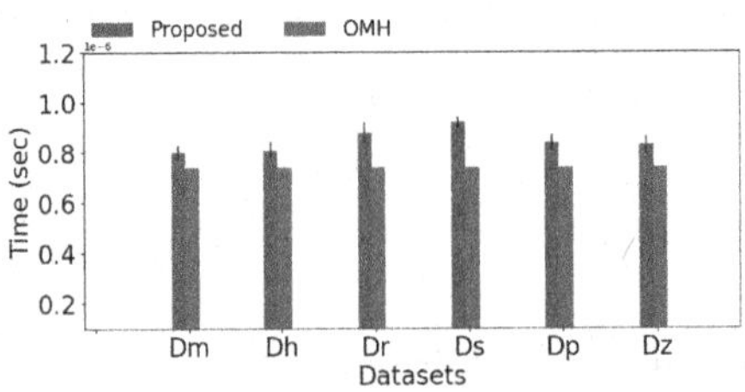

Fig. 9. Average edit distance prediction time of each pair in a dataset

Predicting Types of Edit Operations. The proposed edit distance prediction method also predicts number of each type of edit operations between a pair of sequences. In Table 4, the prediction accuracy are shown for the experiment datasets. Pairs of sequences with up to three edit operations see nearly 100% accuracy in predicting types of edit operations. For larger edit distance, the accuracy degrade gradually. Prediction accuracy of point mutations degrades more rapidly compared to those of insertions and deletions. In the proposed differences of distances based prediction method, when same letter appears in edit operations in the opposite sequences, the point mutations are not captured accurately.

6.5 Hierarchical Clustering

We create average linkage agglomerative hierarchical clusters based on pairwise distances calculated by the proposed predicted edit distance prediction method and OMH [17]. We also create k-mer similarity heuristic based clusters based on VSEARCH [15] for comparison. Lower dendrogram of hierarchical clusters are useful for many NGS sequences analysis techniques such as finding Operational Taxonomic Units (OTUs). Since our focus is large NGS datasets, we choose SparseHC [13], a graph based scalable memory hierarchical clustering method. Performance comparison are based on clustering time as well as NMI and AMI of resulting clusters with respect to those of the ground truth.

Table 5. Comparison of clustering time (in hours) and relative similarity

Similarity Threshold	Data set	Edit distance Calculation Techniques		Heuristics based Techniques	
		Baseline Time	Proposed Method Time/NMI/AMI	VSEARCH Time/NMI/AMI	OMH Time/NMI/AMI
97% Avg dist = 7.5	D_m	0.634	0.09 / 99 / 98	0.003 / 91.5 / 79	0.07 / 95.1 / 90.2
	D_h	0.95	0.11 / 99.1 / 97.8	0.005 / 90.5 / 80.3	0.086 / 95 / 89.6
	D_r	1.796	0.14 / 99 / 97.6	0.008 / 89 / 73	0.11 / 95.6 / 90.3
	D_s	18.01	0.71 / 99.2 / 98	0.043 / 93 / 82	0.56 / 94.7 / 90.1
	D_p	411.5	11.2 / 98.8 / 97.7	0.162 / 90.5 / 79.8	9.1 / 95.3 / 90.4
	D_z	642.3	16.1 / 98.1 / 97.2	0.24 / 90.1 / 79.4	12.8 / 94.9 / 89.7
	D_{mv3}	5974.3	122.5 / 98.4 / 97.9	0.35 / 90.2 / 78.1	98.24 / 95.3 / 90

Clustering Time: Table 5 shows time comparison among the aforementioned edit distance based hierarchical clustering and heuristic based clustering methods. Clustering time using the baseline method takes days with the growth of dataset size. The OMH based clustering is moderately faster than that of proposed predicted edit distance based clustering because of the additional cost of edit distance prediction between sub-sequences in the proposed method. On the other hand, VSEARCH requires the least amount of time among the comparing methods.

Cluster Similarity: In Table 5, we show relative similarity of the clusters generated by the predicted edit distance based methods and OMH. We also compare with the heuristics based method, VSEARCH. We evaluate the NMI and AMI scores of the generated clusters with respect to the ground truth.

The NMI and AMI scores of the VSEARCH method based clusters are significantly lower compared to those of the proposed method. Although the NMI scores are around 0.9, the AMI scores are quite low compared to that of edit distance prediction based methods. OMH based clusters show better NMI and AMI accuracies compared to that of VSEARCH as pairwise distances with this method is more representative of edit distances. On the other hand, the proposed predicted edit distance method based clusters shows nearly 99% relative similarity for 97% similarity threshold.

7 Conclusion

In this paper, we propose a novel edit distance prediction method with linear computation time complexity while giving near exact edit distance prediction. The proposed method uses novel approaches based on non-matching subsequences between the NGS sequences, derived using clustering of k-mers. Our proposed strategies calculate edit distances between the entire sequences based on these non-matching sub-sequences. Extensive experiments have been conducted to show superior performance of our edit distance prediction method with respect to both time and accuracy over those of the existing methods. We have also applied our proposed methods for clustering large sets of NGS sequences and have shown superior performance over those of the existing methods.

Acknowledgement. The authors would like to thank Dr. Shamik Sural of Indian Institute of Technology - Kharagpur. The research work is partially funded by the US National Science Foundation (grants #IIS-0414576 and #IIS-0414594) and Collaborative Project with Scientists & Technologists of Indian Origin Abroad Program (CP-STIO), Department of Science and Technology, Govt. of India.

References

1. Pramanik, S., Islam, A.T., Sural, S.: Predicted edit distance based clustering of gene sequences. In: 2018 IEEE International Conference on Data Mining (ICDM), pp. 1206–1211. IEEE (2018)
2. Wagner, R.A., Fischer, M.J.: The string-to-string correction problem. J. ACM (JACM) **21**(1), 168–173 (1974)
3. Abboud, A., Backurs, A.: Towards hardness of approximation for polynomial time problems. In: 8th Innovations in Theoretical Computer Science Conference (ITCS 2017). Schloss Dagstuhl-Leibniz-Zentrum fuer Informatik (2017)
4. Andoni, A., Onak, K.: Approximating edit distance in near-linear time. SIAM J. Comput. **41**(6), 1635–1648 (2012)
5. Andoni, A., Krauthgamer, R., Onak, K.: Polylogarithmic approximation for edit distance and the asymmetric query complexity. In: 2010 IEEE 51st Annual Symposium on Foundations of Computer Science, pp. 377–386. IEEE (2010)
6. Backurs, A., Indyk, P.: Edit distance cannot be computed in strongly subquadratic time (unless SETH is false). In: Proceedings of the Forty-Seventh Annual ACM Symposium on Theory of Computing, pp. 51–58. ACM (2015)
7. Chakraborty, D., Das, D., Goldenberg, E., Koucky, M., Saks, M.: Approximating edit distance within constant factor in truly sub-quadratic time. In: 2018 IEEE 59th Annual Symposium on Foundations of Computer Science (FOCS), pp. 979–990. IEEE (2018)
8. Edgar, R.C.: Search and clustering orders of magnitude faster than blast. Bioinformatics **26**(19), 2460–2461 (2010)
9. ERX2155923: Inter-comparison of marine plankton metagenome analysis methods (2017)
10. Islam, A.T., Pramanik, S., Mirjalili, V., Sural, S.: RESTRAC: reference sequence based space transformation for clustering. In: 2017 IEEE International Conference on Data Mining Workshops (ICDMW), pp. 462–469. IEEE (2017)
11. Landau, G.M., Myers, E.W., Schmidt, J.P.: Incremental string comparison. SIAM J. Comput. **27**(2), 557–582 (1998)
12. Masek, W.J., Paterson, M.S.: A faster algorithm computing string edit distances. J. Comput. Syst. Sci. **20**(1), 18–31 (2013)
13. Nguyen, T.-D., Schmidt, B., Kwoh, C.-K.: SparseHC: a memory-efficient online hierarchical clustering algorithm. Procedia Comput. Sci. **29**, 8–19 (2014)

14. Rasheed, Z., Rangwala, H., Barbará, D.: 16S rRNA metagenome clustering and diversity estimation using locality sensitive hashing. BMC Syst. Biol. **7**(4), S11 (2013)
15. Rognes, T., Flouri, T., Nichols, B., Quince, C., Mahé, F.: VSEARCH: a versatile open source tool for metagenomics. PeerJ **4**, e2584 (2016)
16. SRX1537393: 16S rRNA sequencing of Sao Paulo zoo compost (2016)
17. Marçais, G., DeBlasio, D., Pandey, P., Kingsford, C.: Locality-sensitive hashing for the edit distance. Bioinformatics **35**(14), i127–i135 (2019)

MixCL: Mixed Contrastive Learning for Relation Extraction

Jinglei Zhang[1,2], Bo Li[1,2], Xixin Cao[2(✉)], Minghui Zhang[3], and Wen Zhao[1]

[1] National Engineering Research Center for Software Engineering, Peking University, Beijing, China
{jinglei.zhang,deepblue.lb}@stu.pku.edu.cn, zhaowen@pku.edu.cn

[2] School of Software and Microelectronics, Peking University, Beijing, China
cxx@stu.pku.edu.cn

[3] Handan Institute of Innovation, Peking University, Handan, Hebei, China
zhangmh@pkuhd.cn

Abstract. Entity representation plays a fundamental role in modern relation extraction models. Previous efforts usually explicitly distinguish entities from contextual words, e.g., by introducing position embedding w.r.t. entities or surrounding entities with special tokens. Inspired by this observation, we propose improving relation extraction via a novel entity-level contrastive learning, which contrasts an entity with both other ones and its contextual words in a mini-batch. To generate high-quality negatives for contrast, we equip our entity-level contrastive learning with an innovative Mixup strategy, which interpolates feature representations of negative entities and contextual words to create new diversified negative examples. Extensive experiments on TACRED, TACRED-revisited, and SemEval2010 show that our method delivers robust performance improvements base on a strong relation extraction baseline. Furthermore, we propose a new metric to measure the overall hardness of the negative examples by considering their dissimilarities with the anchor instance as well as their diversities, explaining the superiority of our method in-depth.

Keywords: Relation Extraction · Contrastive learning · data augmentation

1 Introduction

Relation Extraction (RE) is the task of extracting relations between given entities under their corresponding context. It can be applied to various natural language processing tasks, such as knowledge graph construction [13] and question answering [28].

Existing neural relation extraction models usually utilize representations of the two given entities to classify relation types [32]. Many large language models [27] with entity-aware pre-training objectives also achieve impressive RE performance by generating better entity representations. These efforts indicates

Z. Jinglei and L. Bo—Equal Contribution.

D.-N. Yang et al. (Eds.): PAKDD 2024, LNAI 14648, pp. 92–103, 2024.
https://doi.org/10.1007/978-981-97-2238-9_7

the importance of entity representations for relation extraction, inspiring us to explore better methods to learn entity representations for RE.

Contrastive learning (CL) is a popular representation learning method, which guides the embeddings of the same input from different views to be close (positive examples), while the embeddings of different inputs to be apart from each other (negative examples). CL achieves remarkable performance in both computer vision [6] and natural language processing [16].

In our RE scenario, we can implement CL as a dictionary lookup process [6] to learn better entity representations. Specifically, given an anchor entity, we use the representation of one view of the anchor entity as the key and then retrieve the anchor entity from a batch of representations, which contains the representation of the anchor entity from another view.

The construction of positive representations for anchor entities is straightforward, while constructing negative examples of anchor entities is non-trivial. One possible method is simply using all the other entities (except the anchor entity) within a batch as negative examples. In this setting, the CL objective aims to distinguish the representation of the anchor entity from representations of other entities, implicitly characterizing more accurate semantics contributing to a specific relation.

Alternatively, we can also use contextual words (i.e., all words other than entities) within a batch as negative examples, considering that distinguishing representations of entities and contextual words has been proven useful for relation extraction in previous work. For example, [29] propose to leverage relative position embeddings w.r.t. *entities* in relation extraction, achieving promising results. A common wisdom from most recent relation extraction models [12,21,23] is to surround entities with two special tokens (e.g., '@' for head entities and '#' for tail entities), which also shows improved performance in relation extraction. By explicitly highlighting entities, these two classes of methods distinguish entity representations from representations of other words. From a similar perspective, incorporating contextual words as negative examples in contrastive learning potentially benefit relation extraction.

Indeed, we observed in our experiments that using either entities or contextual words as our negative examples in contrastive learning can improve upon a strong relation extraction system based on large pre-trained transformers. However, there is not a further improvement when using *both* entities and contextual words. One possible reason for this phenomenon maybe the quality of negative examples are not good enough. [10,20] demonstrate that easy negative examples contribute small to the optimization process of contrastive learning. Intuitively, for an anchor entity in a batch, contextual words are easier to distinguish compared to other entities. Thus it is a non-trivial task to design a reasonable strategy to generate proper negative examples.

To this end, we introduce Mixup [8,30]—a technique linearly interpolating feature representations of different examples to generate a new one—into our entity-level contrastive learning. Specifically, for an anchor entity in our **Mix**ed **C**ontrastive **L**earning (MixCL), we treat all the other entities in a batch as the

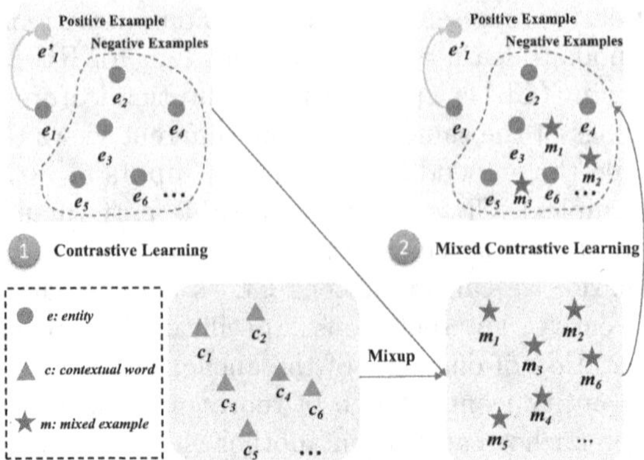

Fig. 1. The process of MixCL in a mini-batch. We use entities and contextual words to generate more mixed examples for contrastive learning.

original negative examples, and then exploit mixup of entities and contextual words to generate more diversified negative ones. We experiment with MixCL on three popular relation extraction datasets. The empirical results show that MixCL achieves competitive results compared to recent models, while being conceptually simpler.

To verify the conjecture that mixing entity embeddings with contextual word embeddings leads to enhanced hardness of negative examples, we further propose Q-index (shorthand for Quality Index) to measure the overall quality (or hardness) of negative examples. Q-index comprehensively considers the similarity between an anchor example and its negatives as well as the diversity among all the negatives, and evaluation on this metric roughly reveals the main reason behind our MixCL's superiority. To sum up, our contributions are as follows:

- We propose improving relation extraction from a novel perspective of entity-level contrast, and further incorporating Mixup into contrastive learning. By innovatively mixing entities and contextual words, our method generates diversified high-quality negative examples that facilitate contrastive effects.
- Our experimental results show that MixCL yields remarkable performance improvement over a strong relation extraction baseline in three popular relation extraction dataset.
- We propose a systematic metric to measure the overall hardness of the negative examples, revealing why MixCL works more in-depth.

2 Mixed Contrastive Learning

In this section, we introduce **Mix**ed **C**ontrastive **L**earning (MixCL), which leverages contrastive learning and mixup for RE task. We work with a batch of training examples of size N, denoted as $\{s_i, e_i^h, e_i^t\}_{i=1,...,N}$. Where s_i is the i-th input sentence, e_i^h and e_i^t are two entities belong to s_i. $f_{enc}(\cdot) \in \mathbb{R}^d$ represents an encoder that outputs the final encoder hidden representation of each token in sentence s_i. For example, the encoder could be a BERT-base model [2], thus $d = 768$. Without loss generality, we denote entity and contextual word as e and w, and their corresponding low-dimensional vectors after encoding are denoted as $\mathbf{e}$ and $\mathbf{w}$. For the contrastive learning, we use $\{e, e^{'}\}$ as positive pair, where e and $e^{'}$ are semantically related.

Our methods intent to leverage contrastive learning for better entity representations. For a given entity positive pair $\{e, e^{'}\}$, we first use the rest entities in a mini-batch as negative examples. To obtain better contrastive learning performance, we then mixing entities with contextual words to generate more high-quality negative examples. Compared with directly using entities and contextual words as negatives at the same time, our method can make full use of the entity and contextual words in a mini-batch for better contrastive learning performance, and finally guide model to learn more powerful entity representations.

Before describing ours, we first briefly introduce a neural relation extraction baseline model.

2.1 Neural Relation Extraction Baseline Model

We now introduce a strong baseline for relation extraction, it predicts the relation under the given input sentence s and two named entities e^h, e^t, this baseline contains three components:

Input Module. To highlight named entities in the input sentence, the input module encloses entity spans using special tokens, thus modifying s to the format of "...@e^h@...#e^t#...". The modified input sentence $\hat{s}$ is then passed into the encoder. **Encoder.** The encoder $f_{enc}(\cdot) \in \mathbb{R}^d$ could be a BERT or other pre-trained language models that are suitable for relation extraction task. Since the head entity and tail entity contain a few subtokens, we take the embeddings of the first subtokens in both spans of e^h and e^t as head entity embedding $\mathbf{e}^h$ and as tail entity embedding $\mathbf{e}^t$. The resulting relation representation $\mathbf{r}$ is a concatenation of the two embeddings $\mathbf{r} = [\mathbf{e}^h; \mathbf{e}^t]$. **Classifier.** We then employ a single hidden layer f with relu activation function to resize the relation representation $\mathbf{r}$. Finally, another single hidden layer g with softmax activation function is employed to predict the final relation class y.

$$p(y|s, e^h, e^t) = g((f(\mathbf{r})), \tag{1}$$

The loss function used in relation extraction is *cross-entropy loss*, denoted as $\mathcal{L}_{CE}$, where q is the one-hot label for each training instance.

$$\mathcal{L}_{CE} = -logp(y|s, e^h, e^t), \tag{2}$$

It is worth to notice that the above baseline is similar to [32] proposed recently.

2.2 Entity Centralized Contrastive Learning for Relation Extraction

We first incorporate CL in the relation extraction scenario by contrasting different entity in a mini-batch, as shown in Fig. 1. Intuitively, the CL could generate more discriminative entity representations, thus benefits for relation extraction task. Now we introduce contrasting entities for relation extraction in detail.

Consider a mini-batch with N training examples, there will be $2N$ entity embeddings after encoding and l_2 normalized, denoted as $\mathbf{e}_1, \mathbf{e}_2, ..., \mathbf{e}_{2N}$, where $\mathbf{e}_i \in \mathbb{R}^d$. Similar to [3], we feed the same input sentence to the encoder twice with different dropout masks, and obtain two embeddings of the same entity e, denoted as $\mathbf{e}_i$ and $\mathbf{e}_i^{'}$. For a given anchor entity embedding $\mathbf{e}_i$, we consider $\mathbf{e}_i^{'}$ as the positive sample, and the rest $2N - 1$ entity embeddings $[\mathbf{e}_1^{'}, ..., \mathbf{e}_{i-1}^{'}, \mathbf{e}_{i+1}^{'}, ..., \mathbf{e}_{2N}^{'}]$ as negative examples. Thus the contrastive loss among entities would be

$$\mathcal{L}_{CL} = -log \frac{exp(\mathbf{e}_i^T \mathbf{e}_i^{'})/\tau}{\sum_{j=1}^{2N} exp(\mathbf{e}_i^T \mathbf{e}_j^{'})/\tau}. \tag{3}$$

τ is a temperature parameter and all gradients are scaled by $1/\tau$. The above contrastive loss guides the learned representation to map positive pair to be close in the semantic space, and negative pairs farther apart. Combined with cross-entropy, the overall loss for RE model is:

$$\mathcal{L} = \mathcal{L}_{CE} + \alpha \mathcal{L}_{CL}, \tag{4}$$

where α is a scalar weighting factor.

2.3 Mixed Contrastive Learning for Relation Extraction

Although contrasting entities could gain remarkable improvements in relation extraction task, we believe the performance can be improved if we use more negative examples. However, we observe that the RE model can not obtain further improvements when we use the entities and contextual words as negative examples at the same time. This may because directly contrasting entities with contextual words is too easy, this loss contributes less or even useless in relation extraction task. We therefore further propose MixCL, which leverages mixup for generating more hard negatives, as shown in Fig. 1. Specifically, we treat the embeddings of contextual word as reasonable noise for entity embeddings. Then we consider linearly interpolate entity embeddings and contextual embeddings to obtain mixed negative examples. We intend the above mixed negative examples to force the entity embeddings far apart from the semantic space of contextual words. Besides, compared with entity embeddings, the mixed negative examples carry more complicated semantic meanings, which can be seen as hard negative examples to some extend. MixCL can generate harder and useful negatives for contrastive learning, finally improves the RE performance by a large margin.

Formally, let $\mathbf{m}$ be a mixed example to be generated, the mixing procedure is shown below:

$$\mathbf{m} = \lambda \mathbf{e} + (1 - \lambda)\mathbf{w}, \tag{5}$$

$$\lambda = max(\lambda, 1 - \lambda), \tag{6}$$

where λ is a random coefficient and $\lambda \in (0, 1)$, Eq. (6) guarantees that most parts of the mixed examples come from entity embeddings. $\mathbf{e}$ and $\mathbf{w}$ are randomly sampled from entity embeddings and contextual embeddings within the same mini-batch. Based on Eq. (5), we can obtain M mixed examples as additional negative examples.

Then, for a given anchor entity e_i, we use both the other entities and the mixed examples as negative examples. Thus the loss of MixCL is computed as follows:

$$\mathcal{L}_{MixCL} = -log \frac{exp(\mathbf{e}_i^T \mathbf{e}_i^{'})/\tau}{\sum_{j=1}^{2N} exp(\mathbf{e}_i^T \mathbf{e}_j^{'})/\tau + \sum_{k=1}^{M} exp(\mathbf{e}_i^T \mathbf{m}_k^{'})/\tau}. \tag{7}$$

Note that the M mixed examples are used as the additional negative examples for all anchor entities. The MixCL loss can help RE model to learn more powerful entity embeddings for relation extraction task.

Finally, the overall training loss is shown bellow:

$$\mathcal{L} = \mathcal{L}_{CE} + \alpha \mathcal{L}_{MixCL}. \tag{8}$$

3 Experiments

3.1 Datasets and Comparison Models

Datasets. We use three English relation extraction datasets for evaluation, which are TACRED [31], TACRED-revisited [1] and SemEval2010 [7].

Comparison Models. We use the following models for comparison. 1) **MTB** [21] built task agnostic relation representations solely from entity-linked text; 2) **KnowBERT** [17] jointly trained the masked language model and entity linker and achieved better performance than BERT on RE; 3) The pre-train objective of **SpanBERT** [9] is to predict a contiguous random spans rather than random tokens; 4) **LUKE** [27], designed with the pre-trained objectives that predicting randomly masked words and entities in a large entity-annotated corpus retrieved from Wikipedia, which is the strong model; 5) **ERICA** [18] further trained RoBERTa using contrastive learning on relation level and distantly supervised dataset from Wikipedia; 6) **GDPNet** [26] based on SpanBERT and proposed to construct a latent multi-view graph to capture various possible relationships among tokens; 7) **A-CGN** [23] is a dependency-driven approach for relation extraction with attentive graph-based networks. Note that the first five methods leverage external datasets or entity-aware pre-training objectives upon powerful language models, while our method only use the RE dataset.

3.2 Main Results

Table 1. The test set F1-scores on various relation extraction datasets. Underlined results are the previous highest score, and the result better than the previous score show in **bold**. Model with * used external resources or entity-aware pre-training objectives. All the results are directly retrieved from published papers.

Model	TACRED	TACRED-revisited	SemEval2010
MTB*[21]	71.5	-	89.5
KnowBERT* [17]	71.5	79.3	-
SpanBERT* [9]	70.8	78.0	-
LUKE* [27]	<u>72.7</u>	<u>80.6</u>	-
ERICA* [18]	69.8	-	89.2
GDPNet$_{SpanBERT}$ [26]	70.5	80.2	-
A-CGN$_{BERT\text{-}large}$ [23]	-	-	<u>89.8</u>
BERT-RE$_{base}$	68.5	77.1	88.6
BERT-RE$_{base}$ + **CL (Contextual word)**	69.0(+0.5)	78.2(+1.1)	89.0(+0.4)
BERT-RE$_{base}$ + **CL (Entity)**	69.3(+0.8)	78.6(+1.5)	89.1(+0.5)
BERT-RE$_{base}$ + **MixCL**	70.1(+1.6)	79.3(+2.2)	89.8(+1.2)
BERT-RE$_{large}$	69.6	77.5	89.6
BERT-RE$_{large}$ + **MixCL**	70.8(+2.3)	80.1(+3.0)	**90.4**(+1.8)
RoBERTa-RE$_{large}$	70.3	80.4	90.1
RoBERTa-RE$_{large}$ + **MixCL**	71.6(+3.1)	**81.4**(+4.3)	**90.8**(+2.2)

Our main results on English datasets are shown in Table 1. The first block summarizes results of previously proposed models (see Sect. 3.1).

We experiment our model variants with $\text{BERT}_{\text{base}}$ and results are in the second block. When using contextual words as negative examples (**BERT-RE**$_{base}$+ **CL (Word)**), we improve our base model by 0.5 to 1.1 F1 across the three datasets. When negative examples are entities (**BERT-RE**$_{base}$+**CL (Entity)**), we obtain larger improvement over the base model (+0.8 to + 1.5 F1). It is as expected since entities are highly related to resulting relations, leading to a relatively better contrast effect. When we employ MixCL (**BERT-RE**$_{base}$ + **MixCL**), which uses entities and mixtures of entities and contextual words as negative examples, the gains over our base RE model become even larger (+1.6 to +2.2 F1). The above results demonstrate the CL can generate more powerful entity representations and achieve better RE performances, and MixCL could further improve the results by a large margin.

In the third block, we implement our model variants with larger and stronger pre-trained models (i.e., $\text{BERT}_{\text{large}}$ and $\text{RoBERTa}_{\text{large}}$). The trends are similar with base sized models. Our MixCL models (**BERT-RE**$_{large}$ + **MixCL** and **RoBERTa-RE**$_{large}$ + **MixCL**) improves over their baselines by 1.8 to 4.3 F1.

Compared with previous models (the first block), even the models used various external resources (marked with *) and task-specific entity-aware pre-training objectives, our approach still achieves very competitive or even better results across the three datasets. **RoBERTa-RE**$_{large}$ + **MixCL** obtains 81.4 on TACRED-revisited and 90.8 on SemEval2010, showing a significant improvements. We also achieve competitive results on TACRED compared to LUKE (71.6 v.s. 72.7). Note that LUKE used more external resources (a large entity-annotated corpus retrieved from Wikipedia) and designed entity-aware pre-training objectives. In addition, the number of parameters of LUKE is 560M, which is nearly twice of our large size model (340M). Lastly, compared to LUKE, MixCL is much easier to implement and conceptually simpler.

4 Analysis

Without loss of generality, in this section, for an anchor entity, we name the rest entities as *original* negative examples, and the contextual words or the mixed examples as *additional* negative examples.

Table 2. The Q-index of different additional negative examples, higher Q-index means better quality. The mean value of Gaussian noise we use is 0.1, and the variance is 0. The final results are the average of ten runs.

Methods	Additional Negative Examples	Mix Target	TACRED		SemEval2010	
			F1-score	Q-index	F1-score	Q-index
Contrastive Learning	-	-	69.3	-	89.1	-
	Contextual Words	-	69.4(+0.1)	1.0016	89.1(+0.0)	1.0039
Mixed Contrastive Learning	Mixed Examples	Entity	69.4(+0.1)	1.0022	89.1(+0.0)	1.0037
		Gaussian Noise	69.5(+0.2)	1.0027	89.2(+0.1)	1.0050
		Contextual Word	**70.1(+0.8)**	**1.0101**	**89.8(+0.7)**	**1.1143**

4.1 Metric of Quality of Negative Examples

[20] measures the hardness of a negative example (x^-) by computing the similarity between the anchor example x and x^-. For example, a large inner product (similarity) between x and x^- generally means the x^- is a hard negative example.

The indicator proposed by [20], however, only considers the similarity between the anchor example and the negative examples, ignoring the diversity between different negative examples. We believe that the diversity of negative samples are also crucial for contrastive effects. Considering an extreme case, if we repeat the entities as the additional negative examples, these negatives certainly do not provide any helpful information for CL.

Based on the above analysis, ideal high-quality negative examples should keep certain dissimilarities with the anchor without losing their own diversities.

In this paper, we propose **Quality Index**, termed as **Q-index**, to measure the hardness of negative examples. Q-index systematically considers the similarity between an anchor example and its negatives as well as the diversity among all the negatives. Taking MixCL as an example, suppose a mini-batch has N instances with $2N$ entities and M mixed examples. For an anchor entity, suppose that the rest $2N-1$ entities are the original negative examples, and M mixed examples are the additional negatives. Then the mathematical expressions of the Q-index for a given anchor entity are shown as following:

$$\mathbf{value}_s = \frac{1}{M}\sum_{k=1}^{M} d(\mathbf{e}_i, \mathbf{m}_k^{'}), \tag{9}$$

$$\mathbf{value}_d = \frac{1}{2N-1}\frac{1}{M}\sum_{\substack{j=1 \\ j\neq i}}^{2N}\sum_{k=1}^{M} d(\mathbf{e}_j^{'}, \mathbf{m}_k^{'}), \tag{10}$$

$$\textbf{Q-index} = \frac{\mathbf{value}_s}{\mathbf{value}_d}. \tag{11}$$

where $d(\mathbf{x},\mathbf{y})$ is the function that computes the similarity between $\mathbf{x}$ and $\mathbf{y}$. We simply use inner product here. Equation (9) computes the similarity between the anchor entity and the additional negative examples, which indicates the hardness of the additional negative examples, and the higher, the harder. Equation (10) represents the diversity between the original and additional negative examples, low $value_d$ indicates small similarity and large diversity. Finally, we use Eq. (11) to measure the quality of the additional negative examples. A high-quality mixed example should have higher $value_s$ and lower $value_d$.

4.2 Why Does MixCL Work?

Followed by Previous studies [3,6,20], we believe that MixCL provides better negative samples for CL. To verify our assumption, we first evaluation the quality of negative examples based on our proposed Q-index metric.

The Quality of Negative Examples. We use Q-index to measure the quality of various additional negative examples. Table 2 shows the performances and the Q-index under different settings. All the experiments using entities in a mini-batch as the original negative examples. From Table 2 we can see that using contextual words as additional negative samples has a very low Q-index. Since the $value_s$ is too small, distinguishing contextual words are relatively easy compared with the original negative examples. As for MixCL, mixing original negatives with entities or Gaussian noise still have low Q-index, as those mixed examples are very similar to original negative examples. That is why there is no significant performance improvements. When mixing entities with contextual words, we observe that the Q-index and performance both increased by a large margin, which indicates that our approach could generate high-quality additional negatives examples and thus benefit relation extraction performance.

The Number of Negative Examples. We also investigate that how the number of negative examples influences the model's performance. Figure 2 shows that as the number of examples increases, the improvements for both TACRED and TACRED-revisited rise at first, and then reduce quickly when the number larger than 48. This indicates that too few mixed examples are not sufficient to leverage the information from contextual words, and too many mixed examples may bring noises for contrastive learning, hence hurting the performances. In general, we find the 48 is the best choice for all datasets.

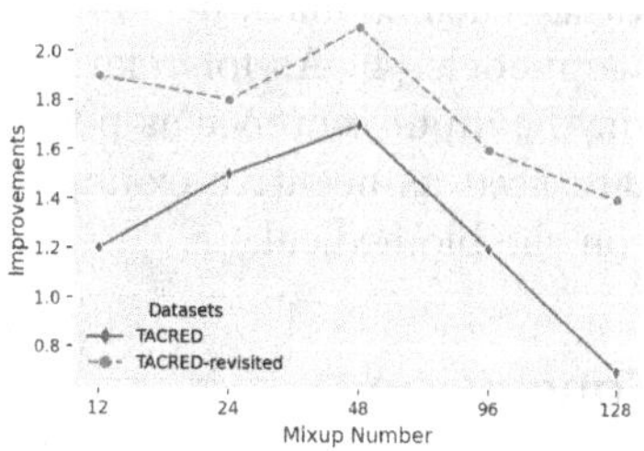

Fig. 2. The effectiveness of mixup numbers.

Based on the above analysis, we conclude that by providing a suitable number of negative examples for contrast learning, MixCL generates powerful entity representations, and finally gain remarkable improvements on various RE datasets.

5 Related Work

5.1 Neural Relation Extraction

The target entity embeddings play a crucial role in relation extraction. To generate high-quality entity representations, some researchers designing powerful neural network architectures. For example, [15] use convolutional neural networks or recurrent neural networks for RE. More recently, graph-based models [5,14] are popular in RE community, as graph-based methods can provide long-distance connections between informative words and target entities.

Moreover, some researchers use different knowledge graphs to incorporate structured knowledge for better entity representations, such as [25]. Others [12, 19] leverage the Wikipedia to refine entity representations. Recently, researchers try to build more powerful language models with various external resources and entity-aware pre-training objectives, such as MTB [21], KnowBERT [17] and LUKE [27].

5.2 Contrastive Learning

Early works are mainly focused on computer vision community with unsupervised settings. For example, [11] proposed to use labels as supervised singles and designed a supervised contrastive learning framework. Some works also explored from various aspects of contrastive learning. For example, [10,20] explored the effectiveness of hard negative examples.

Contrastive learning also achieved great success in natural language processing community. [3] proposed to use dropout as data augmentation methods for semantic textual similarity task. Contrastive learning also has been applied in other tasks, such as machine translation [16], text retrieval [24] and pre-trained language models [4]. As for relation extraction task, [22] simply changed some words in the input sentence as positive example, and other sentences in a mini-batch are used as negative examples, this approach achieved slightly improvements on the biomedical.

6 Conclusion

In this paper, we propose MixCL, a contrastive learning framework for the relation extraction task. MixCL applies contrastive learning at entity-level, and generates high-quality negative examples for contrastive learning by mix entities and tokens. By enabling models to better distinguish the semantic spaces of entities and contextual words, our MixCL yields more powerful entity representations for relation extraction. We also propose Quality Index, a metric to measure the quality of mixed examples, by which we reveal why our MixCL works. By incorporating MixCL into a neural RE baseline, we achieve significant improvements on two out of three relation extraction datasets. MixCL is easy to implement and does not need external datasets or pre-training procedures, serving as a solid baseline for future works of relation extraction.

Acknowledgment. This work is supported by the Research and Application of Intelligent Regional Industrial Brain Platform.

References

1. Alt, C., Gabryszak, A., Hennig, L.: TACRED revisited: a thorough evaluation of the TACRED relation extraction task. In: ACL 2020 (2020)
2. Devlin, J., Chang, M., Lee, K., Toutanova, K.: BERT: pre-training of deep bidirectional transformers for language understanding. In: NAACL-HLT 2019 (2019)
3. Gao, T., Yao, X., Chen, D.: SimCSE: simple contrastive learning of sentence embeddings. CoRR abs/2104.08821 (2021). https://arxiv.org/abs/2104.08821
4. Gunel, B., Du, J., Conneau, A., Stoyanov, V.: Supervised contrastive learning for pre-trained language model fine-tuning. In: ICLR 2021 (2021)
5. Guo, Z., Zhang, Y., Lu, W.: Attention guided graph convolutional networks for relation extraction. In: Korhonen, A., Traum, D.R., Màrquez, L. (eds.) ACL 2019 (2009)
6. He, K., Fan, H., Wu, Y., Xie, S., Girshick, R.B.: Momentum contrast for unsupervised visual representation learning. In: CVPR 2020 (2020)
7. Hendrickx, I., et al.: SemEval-2010 task 8: multi-way classification of semantic relations between pairs of nominals. In: SemEval@ACL 2010 (2010)
8. Inoue, H.: Data augmentation by pairing samples for images classification. CoRR abs/1801.02929 (2018). http://arxiv.org/abs/1801.02929
9. Joshi, M., Chen, D., Liu, Y., Weld, D.S., Zettlemoyer, L., Levy, O.: SpanBERT: improving pre-training by representing and predicting spans. Trans. Assoc. Comput. Linguist. (2020)
10. Kalantidis, Y., Sariyildiz, M.B., Pion, N., Weinzaepfel, P., Larlus, D.: Hard negative mixing for contrastive learning. In: NeurIPS 2020 (2020)

11. Khosla, P., et al.: Supervised contrastive learning. In: NeurIPS 2020 (2020)
12. Li, B., Ye, W., Huang, C., Zhang, S.: Multi-view inference for relation extraction with uncertain knowledge. In: AAAI 2021 (2021)
13. MacDonald, E., Barbosa, D.: Neural relation extraction on Wikipedia tables for augmenting knowledge graphs. In: CIKM 2020 (2020)
14. Mandya, A., Bollegala, D., Coenen, F.: Graph convolution over multiple dependency sub-graphs for relation extraction. In: COLING 2020 (2020)
15. Miwa, M., Bansal, M.: End-to-end relation extraction using LSTMs on sequences and tree structures. In: ACL 2016 (2016)
16. Pan, X., Wang, M., Wu, L., Li, L.: Contrastive learning for many-to-many multilingual neural machine translation. In: ACL/IJCNLP 2021 (2021)
17. Peters, M.E., et al.: Knowledge enhanced contextual word representations. In: EMNLP-IJCNLP 2019 (2019)
18. Qin, Y., et al.: ERICA: improving entity and relation understanding for pre-trained language models via contrastive learning. In: ACL/IJCNLP 2021 (2021)
19. Ren, F., et al.: Neural relation classification with text descriptions. In: COLING 2018 (2018)
20. Robinson, J.D., Chuang, C., Sra, S., Jegelka, S.: Contrastive learning with hard negative samples. In: ICLR 2021 (2021)
21. Soares, L.B., FitzGerald, N., Ling, J., Kwiatkowski, T.: Matching the blanks: distributional similarity for relation learning. In: ACL 2019 (2019)
22. Su, P., Peng, Y., Vijay-Shanker, K.: Improving BERT model using contrastive learning for biomedical relation extraction. In: BioNLP@NAACL-HLT 2021 (2021)
23. Tian, Y., Chen, G., Song, Y., Wan, X.: Dependency-driven relation extraction with attentive graph convolutional networks. In: ACL/IJCNLP 2021 (2021)
24. Xiong, L., et al.: Approximate nearest neighbor negative contrastive learning for dense text retrieval. In: ICLR 2021 (2021)
25. Xu, P., Barbosa, D.: Connecting language and knowledge with heterogeneous representations for neural relation extraction. In: NAACL-HLT 2019 (2019)
26. Xue, F., Sun, A., Zhang, H., Chng, E.S.: GDPNet: refining latent multi-view graph for relation extraction. In: AAAI 2021 (2021)
27. Yamada, I., Asai, A., Shindo, H., Takeda, H., Matsumoto, Y.: LUKE: deep contextualized entity representations with entity-aware self-attention. In: EMNLP 2020 (2020)
28. Yu, M., Yin, W., Hasan, K.S., dos Santos, C.N., Xiang, B., Zhou, B.: Improved neural relation detection for knowledge base question answering. In: ACL 2017 (2017)
29. Zeng, D., Liu, K., Lai, S., Zhou, G., Zhao, J.: Relation classification via convolutional deep neural network. In: COLING 2014 (2014)
30. Zhang, H., Cissé, M., Dauphin, Y.N., Lopez-Paz, D.: mixup: beyond empirical risk minimization. In: ICLR 2018 (2018)
31. Zhang, Y., Zhong, V., Chen, D., Angeli, G., Manning, C.D.: Position-aware attention and supervised data improve slot filling. In: EMNLP 2017 (2017)
32. Zhou, W., Chen, M.: An improved baseline for sentence-level relation extraction. CoRR abs/2102.01373 (2021). https://arxiv.org/abs/2102.01373

Decomposing Relational Triple Extraction with Large Language Models for Better Generalization on Unseen Data

Boyu Meng[1], Tianhe Lin[1], and Deqing Yang[1,2](✉)

[1] School of Data Science, Fudan University, Shanghai, China
bymeng21@m.fudan.edu.cn, {thlin20,yangdeqing}@fudan.edu.cn
[2] Shanghai Key Laboratory of Data Science, Shanghai, China

Abstract. Despite the significant achievements of existing methods for relational triple extraction (RTE), most of them only have weak generalization for the data unseen or partially seen in the training set. Recently, large language models (LLMs) have attracted more interests for their powers in various natural language processing tasks, but they still suffer from some issues, including less sensitivity to the order of subjects and objects and unsatisfactory output formats. To harness LLMs to achieve more accurate RTE especially on the unseen data, we propose a novel framework *LRTE* in which the RTE task is decomposed into three subtasks of a pipeline: relation extraction, entity extraction, and triple filtering. To evaluate all models' RTE performance more truthfully, we also refined two RTE benchmarks through removing noisy triples and complementing the missing triples. Our extensive experiments upon the refined datasets demonstrate our framework's superior performance over the previous competitors.

Keywords: relational triple extraction · large language model · small language model

1 Introduction

Relational triple extraction (RTE) aims to extract the entity pair and their relation in the form of (subject, relation, object), (s, r, o) in short, which is a representative task significant to many downstream tasks including large-scale knowledge base construction [18], question answering [4], and text mining [24]. Although many effective RTE models [12,15,23] have been proposed and most of them perform well for the entities they have seen, they are not generalized well to those entities unseen in the training data [7,26].

Recently, large language models (LLMs) have exhibited their potential on various natural language processing (NLP) tasks [13,14], including named entity

This work was supported by Chinese NSF Major Research Plan (No. 92270121), Shanghai Science and Technology Innovation Action Plan (No. 21511100401).

D.-N. Yang et al. (Eds.): PAKDD 2024, LNAI 14648, pp. 104–115, 2024.
https://doi.org/10.1007/978-981-97-2238-9_8

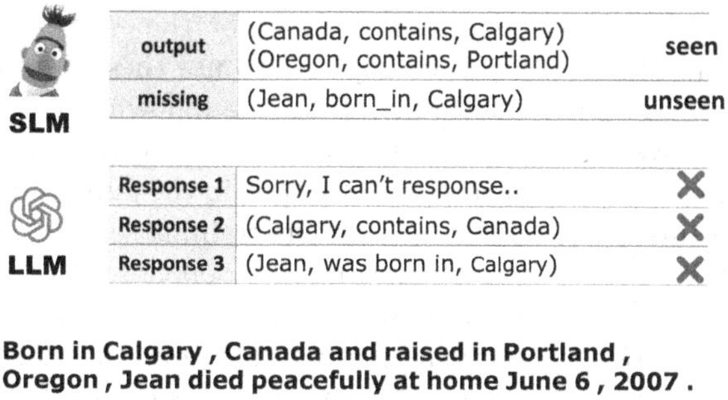

SLM

output	(Canada, contains, Calgary) (Oregon, contains, Portland)	seen
missing	(Jean, born_in, Calgary)	unseen

LLM

Response 1	Sorry, I can't response..	✗
Response 2	(Calgary, contains, Canada)	✗
Response 3	(Jean, was born in, Calgary)	✗

Born in Calgary , Canada and raised in Portland , Oregon , Jean died peacefully at home June 6 , 2007 .

Fig. 1. A traditional small language model (SLM) can easily memorize and recall the seen data but is not generalized well on the unseen data of RTE. While the LLM struggles to provide satisfactory responses due to its less sensitivity to the subject/object order and less following output format.

recognition (NER) [10,19] and relation extraction (RE) [17]. It inspires us to accomplish RTE with LLMs as they can effectively identify entities not observed in the training data, due to their solid comprehension of the sentence meaning. Moreover, they can also detect the noisy triples not reflected in the sentence.

However, LLMs still yield unsatisfactory performance on RTE [8] as shown in Fig. 1, mainly due to the following reasons. *1)* The limitation of input token length hinders incorporating more demonstration instances (demos) into LLMs' prompt for RTE achievement. Also, the bound tokens of the prompt barely assist LLMs in comprehending the extensive relation (type) list and defining the format of desired outputs. *2)* Designing a perfect prompt to aid LLMs in obtaining satisfactory RTE, encompassing task understanding, selection of suitable demonstration instances, and response formatting, is a difficult task. *3)* The limited RTE performance results from LLMs' lower sensitivity to the order of subject and object entities [3].

To address the aforementioned issues when harnessing LLMs to achieve RTE, we propose a pipeline framework of RTE based on LLMs, denoted as LRTE, that decomposes an RTE task into three independent sub-tasks as follows:

1) *Relation Extraction*: At first, given LLMs' weakness of extracting the relation(s) from the sentence directly, we adopt a BERT-based model to extract the relation(s) and retain top K potential relations for higher recall.
2) *Entity Extraction*: Next, we instruct an LLM to extract the corresponding entities (subjects and objects) for the preserved K relations. To prevent the LLM from confusing the order of subjects and objects, we issue a basic query for an accurate response rather than guiding the LLM with a carefully crafted prompt. As a result, the LLM is more likely to produce a satisfactory response.
3) *Triple Filtering*: We use the LLM's explanations for every triple generated in the previous stages to efficiently filter out the incorrect triples, including those that are not communicated in the sentence, thereby enhancing the final accuracy.

In addition, through our in-depth analysis for the existing benchmarks of RTE, we also found that it is not appropriate to use these datasets to evaluate the LLM-based RTE models truthfully, given the following two issues. 1) Most of these datasets have few unseen triples, but these unseen triples are frequently encountered in real-world scenarios. For example, in the famous dataset NYT, its test set only contains 1.75% triples unseen in its training set. Thus, the high recall of traditional RTE models' performance mainly depends on the fact that most triples have been seen in the training set. 2) Many datasets were constructed by distant supervision (DS), introducing many noisy samples where the sentence's semantics fail to convey the annotated relation(s). Xie et al. [22] have found that about 30% of samples in NYT dataset are noisy. It implies that many previous RTE models' high performance scores are resulted from their overfitting rather than truly good performance. Therefore, to evaluate RTE models' performance more truthfully, we have revised the test sets of two benchmarks, *i.e.*, NYT11-HRL and NYT10-HRL, to reduce the noisy samples and focus more on unseen triples in our experiments.

In summary, we make the following contributions in this paper.

1. In order to harness LLMs' power to achieve RTE with good generalization on the unseen data, we propose a novel RTE framework LRTE, where the RTE task is decomposed into three sub-tasks of a pipeline.
2. We design a collaboration framework encompassing a small language model (SLM) and an LLM, in which the SLM is used extract the relation at first, and the LLM is then used to extract entities with a two-step QA strategy.
3. We conduct extensive experiments upon our revised datasets to truthfully demonstrate that the special designs in our LRTE are indeed helpful to achieve the RTE of high recall and precision, especially for the unseen triples.

2 Methodology

2.1 Problem Formalization

Given a sentence $X = \{x_1, x_2, ..., x_N\}$ with N tokens, RTE's goal is to identify all possible triples $T = [(s_i, r_i, o_i)|s_i, o_i \in E, r_i \in R]_{i=1}^{L}$ in X, where s_i, r_i, o_i represent the subject, relation and object of the i-th triple respectively, and L is the number of triples in X. E represents the entity set, and the relation is from a pre-defined relation list R with M types. Note that entities and relations might be shared among different triples in X. In order to make our proposed LRTE better achieve RTE, we decompose the RTE task into three sub-tasks of a pipeline as follows.

2.2 Sub-task 1: Relation Extraction

For the given sentence X, this sub-task aims to predict the top K relations most likely to appear in the sentence from a set of M predefined relations,

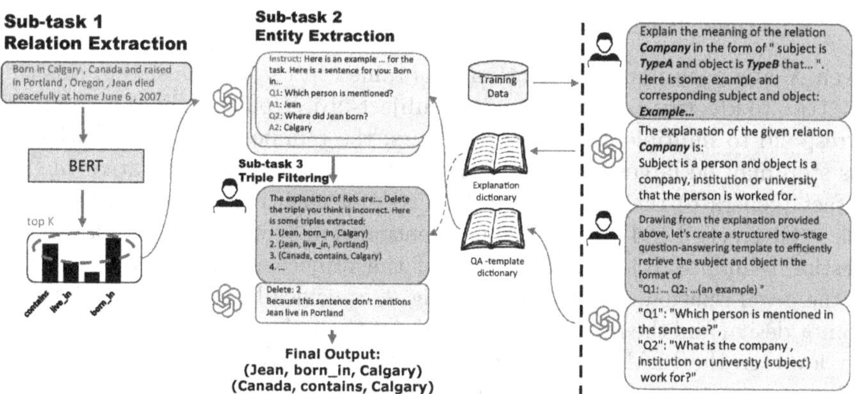

Fig. 2. The pipeline of LRTE is decomposed into three sub-tasks. In sub-task 1, a BERT-based model is employed to select top K relations from the input sentence, such as 'contains', 'live_in' and 'born_in'. In sub-task 2, a sentence from the training set is first retrieved as an example in the prompt by cosine similarity. Then, we leverage the questions stored in the QA-template dictionary to prompt an LLM to generate responses for the subject and the object associated with the identified relation obtained in the preceding sub-task. In sub-task 3, the LLM eliminates the incorrect triples with the help of relation explanations from the explanation dictionary.

which is denoted as $Y_r(X) = \{r_i | r_i \in R\}$ $(1 \leq i \leq K$, and $K \leq M)$. We have found that LLMs are inferior to small language models (SLMs) in extracting the relation(s) directly from the sentence. So we use an SLM rather than an LLM to achieve this sub-task. Specifically, we adopt a relation extractor based on BERT [1] in this sub-task. Formally, the input for this model is $X' = \{[CLS], x_1, x_2, ..., x_N, [SEP]\}$, and we obtain its representation $H_{rc}(X') = \{\mathbf{h}_0, \mathbf{h}_1, \mathbf{h}_2, ...\mathbf{h}_N, \mathbf{h}_{N+1}\}$, where $\mathbf{h}_i \in \mathbb{R}^d$ $(0 \leq i \leq N+1)$ is the output embedding from the BERT encoder. Then, the relation extractor's output is a probability vector of all relation types, denoted as $\hat{\mathbf{y}}_{rc}(X') = \{\hat{y}_{rc}^1, \hat{y}_{rc}^2, ..., \hat{y}_{rc}^M\}$, where $\hat{y}_{rc}^i \in (0, 1)$ $(1 \leq i \leq M)$ is the probability that input sentence X expresses relation $R_i \in R$. $\hat{\mathbf{y}}_{rc}$ is computed as

$$\begin{aligned} \mathbf{h}_{avg} &= \mathrm{Avg}\left(H_{rc}(X')\right), \\ \hat{\mathbf{y}}_{rc}(X') &= \sigma(W_{rc}\mathbf{h}_{avg} + \mathbf{b}_{rc}), \end{aligned} \tag{1}$$

where $\mathrm{Avg}(\cdot)$ represents the operation of average pooling, $W_{rc} \in \mathbb{R}^{M \times d}$ is the trainable weight matrix, $\mathbf{b}_{rc} \in \mathbb{R}^M$ is the bias and σ is Sigmoid activation function. We use binary cross entropy (BCE) loss to train this relation extractor.

We rank all relations according to $\hat{\mathbf{y}}_{rc}(X')$, and select the top K relations as the output of this sub-task. In the subsequent experiment section, we will demonstrate the impact of using different values of K.

2.3 Sub-task 2: Entity Extraction

Given X and each potential relation r_i obtained in the previous step, this sub-task tries to identify r_i's corresponding subject-object pair(s). Since each r_i may correspond to multiple subject-object pairs, the number of triples identified in this step may be larger than K. In this sub-task, we use a generative LLM to extract (generate) the subject-object pair(s) for each relation obtained in sub-task 1. Given the LLM's shortcomings mentioned in Sect. 1, we design a two-step question template to let the LLM answer the subject and object entity in turn for the given relation, as illustrated in Fig. 2. Specifically, to avoid sophisticated prompt design and the LLM's confusion about the order of subject and object, we first leverage the LLM to generate all relations' question templates in advance with some examples from the training set and store them in a QA-template dictionary for subsequent usage.

We use the LLM to collect the question templates for all relations efficiently. The logic behind this approach is that each relation is associated with certain specific types of subject and object entities. Therefore, when given a relation, we merely need to pose a straightforward question to the LLM to identify the corresponding entity types. We select ten distinct sentences with a specific relation from the training set and then prompt LLM to interpret the meaning of the relation using these sentences in the format of 'subject is ... object is ...'. Subsequently, using the entity types relevant to the given relation, LLM can autonomously create its question templates. We preserve the question templates and the interpretations of the relation in two separate dictionaries. For instance, given the relation 'place_lived', the LLM can infer from its contexts that the subject should be a person and the object should be a place. Then, the first sub-question in the template could be 'Which people are mentioned in the sentence?', which is used to extract the subject. For object extraction, the second sub-question of the template is generated by combining a relation-specific template with the extracted subject. For example, we can use 'Which place does subject live or from?' to extract the object of relation 'place_lived'.

Most Similar Sentence Retrieval. In addition, we also need a demonstration example in the prompt to help the LLM better achieve entity extraction, as shown in Fig. 2. To this end, we first store the sentence-level representations for all samples (sentences) in the training set. Then, we find the sample that are most semantically close to each test sample through computing the cosine similarity between the test sentence's representation and these stored representations.

2.4 Sub-task 3: Triple Filtering

Given X and each predicted $t_i = (s_i, r_i, o_i)$ from the previous steps, this sub-task judges t_i's correctness, based on which the incorrect or not-conveyed triples are filtered out from the final results. The previous sub-tasks may introduce some incorrect triples, including those that are not conveyed by the sentence's semantics. Thus, in this sub-task, we still leverage an LLM to filter out the incorrect triples obtained before, based on the LLM's understanding of the semantics of each extracted triple and the sentence.

We also observe that some relations such as 'company' (company, institution or university) and 'place_lived' (live or from) encompass broader semantics across different contexts than the original semantics of the relation terms (names). Thus, it is challenging for the LLM to understand the relations' semantics sufficiently and correctly if only the relation names are provided in the prompt. In order to help the LLM better understand each relation's semantics and then successfully remove the incorrect triples, we also adopt the in-context learning method with only a few examples for the LLM to automatically generate the explanation for each given relation. Then, the explanations of all relations constitute an explanation dictionary.

Specifically, the inputs of this sub-task include the sentence X, all triples obtained before, and the explanations of all relations in these triples that are fetched from the explanation dictionary, as illustrated in Fig. 2. Then, the LLM carefully evaluates each inputted triple and eliminates the incorrect triples and the triples that are not conveyed by the sentence. Through this sub-task, LRTE improves the precision of final outputs.

3 Experiments

3.1 Datasets

We considered a widely used public dataset in our experiments, i.e., New York Times corpus (NYT), which was developed based on distant supervision without sufficient labeling. Thus, NYT contains relations with inherent noises and can not evaluate models' RTE performance truthfully. To evaluate LRTE's performance on the dataset with different numbers of relation types, we focused on two versions of NYT, i.e., NYT10-HRL and NYT11-HRL [16][1], which have 29 relations and 12 relations respectively. For more truthful evaluation, we recruited three volunteers to annotate the labels (triples) of their test samples. When two annotators provided different results for one sample, the third one was consulted to make the final judgment.

NYT11-New: NYT11-HRL contains 53,395 training samples and 368 test samples in total, of which the labels were manually annotated. However, there are still some missing triples. Thus, our volunteers re-annotated all test samples and 278 missing triples were complemented. This revised dataset is denoted as NYT11-New.

NYT10-New: NYT10-HRL has 69,988 training samples and 4,006 test samples in total. Different to NYT11-HRL, the labels of NYT10-HRL were obtained by distant supervision which introduced some noises. Thus the volunteers randomly sampled 180 sentences from NYT10-HRL's test set, and carefully removed the noisy triples (117) and complemented the missing triples (161) for them. This revised dataset is denoted as NYT10-New.

[1] https://github.com/redreamality/-RERE-data.

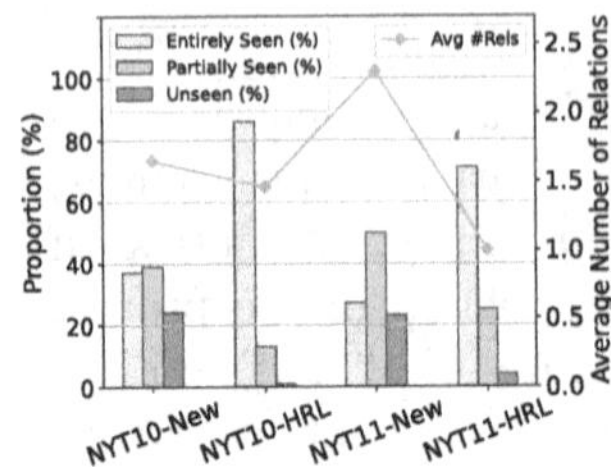

Fig. 3. The average number of relations per sentence and the proportion of three triple types in the four datasets.

Table 1. All models' RTE performance on the two refined datasets. `k-shot` denotes using an LLM to generate final results directly with K demonstration examples.

Method	NYT11-New			NYT10-New		
	Prec.	Rec.	F1	Prec.	Rec.	F1
CasRel	56.31	34.52	42.80	55.66	42.03	47.90
RERE	51.91	37.93	43.83	51.82	43.73	47.43
PRGC	48.33	31.27	37.97	48.26	37.97	42.50
ChatGLM3-6B	<u>56.39</u>	35.83	43.82	48.84	38.53	43.08
Mixtral-8x7B	56.29	37.67	45.14	<u>56.18</u>	43.12	48.79
GPT-3.5$_{\texttt{5-shot}}$	10.98	21.52	14.54	7.66	20.02	11.08
GPT-3.5$_{\texttt{10-shot}}$	11.37	27.19	16.03	11.50	27.61	16.24
GPT-4$_{\texttt{5-shot}}$	31.62	34.08	32.80	30.66	36.43	33.29
GPT-4$_{\texttt{10-shot}}$	33.04	32.26	32.65	37.41	<u>44.30</u>	40.56
LRTE$_{\texttt{GPT-3.5}}$	48.36	<u>43.50</u>	<u>45.80</u>	55.36	43.73	<u>48.86</u>
LRTE$_{\texttt{GPT-4}}$	**66.37**	**80.65**	**72.82**	**59.12**	**72.88**	**65.28**

To specifically evaluate LRTE's performance on the unseen data, we follow Lee et al. [7] to divide all triples in the test set into three types - *entirely seen, partially seen* and *unseen* as follows. Suppose T is a triple set of training data. Then, given a triple $t = (s, r, o)$ in the test set, t is an *entirely seen* triple if $(s, r, o) \in T$. Otherwise, t is a *partially seen* triple if $(s, r, \cdot) \in T \vee (\cdot, r, o) \in T$. Other triples are categorized as *unseen* triples. Figure 3 depicts the average number of relations per sentence and the proportion of three types of triples in the four datasets, showing that the *entirely seen* triples account for the majority of the original datasets, while our annotations significantly involve more *partially seen* and *unseen* triples.

3.2 Compared Methods and Evaluation Metrics

We evaluate our LRTE with five strong baselines consisting of classic SLM baseline CasRel [21], RERE [22] and PRGC [27], and QLora-based fine-tuning LLM baseline ChatGLM3-6B [2] and Mixtral-8x7B [6]. We train three SLM baseline on a single RTX3090, and fine-tune two LLM baseline on a workstation with eight A100-80G. We report precision (Prec.), recall (Rec.) and the standard micro F1 score for all results. We use *Exact Matching* in our evaluations, i.e., a triple is regarded as a hit case only when both its relation and the whole span of the subject and object entities are extracted correctly.

3.3 Overall Performance

Besides comparing five fine-tuning LM baselines, we also compare LRTE with a variant which achieves RTE by an LLM with K demonstration examples in one-stage rather than the three sub-tasks, denoted as GPT-3.5/4$_{\texttt{K-shot}}$. All compared models' overall RTE performance scores are listed in Table 1 where the

Table 2. All compared models' performance on the three types of test triples in NYT10-New and NYT11-New. GPT_{QA} denotes the ablated variant of LRTE without triple filtering.

	Method	Entirely Seen			Partially Seen			Unseen		
		Prec.	Rec.	F1	Prec.	Rec.	F1	Prec.	Rec.	F1
NYT10-New	CasRel	56.44	85.32	67.94	50.00	20.87	29.45	70.00	9.86	17.28
	RERE	54.55	**88.99**	67.64	40.68	20.87	27.59	66.67	11.27	19.28
	PRGC	51.79	80.73	63.10	32.08	14.78	20.24	**77.78**	9.86	17.58
	ChatGLM3-6B	54.32	72.73	62.19	42.25	23.62	30.30	32.00	10.13	15.38
	Mixtral-8x7B	54.26	84.30	66.02	**60.00**	23.62	33.90	69.23	11.39	19.57
	GPT-3.5$_{QA}$	70.59	55.05	61.86	34.29	52.17	41.38	31.76	38.03	34.62
	LRTE$_{GPT-3.5}$	**70.73**	53.21	60.73	48.51	42.61	45.37	44.00	30.99	36.36
	GPT-4$_{QA}$	58.33	84.40	**68.99**	43.37	**73.91**	54.66	31.43	**61.97**	41.71
	LRTE$_{GPT-4}$	58.44	83.49	68.75	59.56	70.43	**64.54**	59.72	60.56	**60.14**
NYT11-New	CasRel	57.02	76.57	65.37	53.96	23.15	32.40	63.64	9.52	16.57
	RERE	54.98	85.14	66.82	45.03	23.77	31.11	63.33	12.93	21.47
	PRGC	53.45	70.86	60.93	50.82	19.14	27.80	25.00	10.88	15.17
	ChatGLM3-6B	**60.00**	77.14	67.50	53.19	23.01	32.12	48.98	15.79	23.88
	Mixtral-8x7B	56.98	**86.29**	**68.64**	54.48	24.23	33.55	59.26	10.53	17.88
	GPT-3.5$_{QA}$	58.55	64.57	61.41	35.38	55.25	43.13	26.69	45.58	33.67
	LRTE$_{GPT-3.5}$	59.02	61.71	60.34	48.64	38.58	43.03	34.04	32.65	33.33
	GPT-4$_{QA}$	57.14	82.29	67.45	55.74	**82.41**	66.50	37.01	**77.55**	50.11
	LRTE$_{GPT-4}$	58.06	82.29	68.09	**70.70**	81.17	**75.57**	**69.09**	**77.55**	**73.08**

best scores are **bold** and the second best scores are underlined. It shows that LRTE$_{GPT-4}$ outperforms all baselines and achieves new SOTA RTE performance on the two revised datasets. In addition, GPT-4$_{10\text{-}shot}$ is still inferior to the SLM baselines, although it employs the SOTA LLM. This deficiency is possibly due to the issues of LLMs mentioned before, including imperfect prompt design, sample selection, and the tendency to invalid responses.

3.4 Model Generalization Evaluation

Furthermore, Table 2 lists the performance of all compared models on the three types of test triples. From the table, we find that:

1. LRTE$_{GPT-4}$'s relative performance improvements over the baselines on the partially seen and unseen triples are much more obvious than that on the entirely seen triples. It demonstrates that our designs in LRTE's pipeline sufficiently leverage the LLM's advantage of generalization.
2. To compare LLMs' generalization capability more fairly, we propose an ablated variant of LRTE, denoted as GPT_{QA}, in which the sub-task 3 of triple filtering is removed. It shows that with the help of the LLM, GPT-3.5$_{QA}$ is also much better than the SLM baselines on the partially seen and unseen triples, although it is inferior to the baselines on the entirely seen triples.

Table 3. Comparisons of relation extraction performance between BERT-based models and LLMs. In fact, $\text{BERT}_{\text{topk}}$ is the ablated variant of LRTE only having sub-task 1, and $_{\text{topK}}$ represents top K relations are retained for the given sentence.

Method	NYT11-New			NYT10-New		
	Prec.	Rec.	F1	Prec.	Rec.	F1
$\text{BERT}_{\text{top1}}$	**70.48**	52.22	**59.99**	**58.27**	70.93	**63.98**
$\text{BERT}_{\text{top2}}$	46.82	66.29	54.88	43.46	77.18	55.61
$\text{BERT}_{\text{top3}}$	36.83	80.74	50.58	28.65	85.87	42.97
$\text{BERT}_{\text{top4}}$	27.05	**85.87**	41.14	24.18	**86.57**	37.80
GPT-3.5	35.00	27.81	31.00	32.42	26.01	28.86
GPT-4	58.76	28.38	38.27	64.50	33.61	44.19

3.5 Ablation Studies

Rationality of Using an SLM for Relation Extraction. To justify the rationale of adopting a BERT-based small model instead of an LLM to achieve the sub-task 1 of relation extraction in LRTE, we focused on the ablated variant of LRTE without sub-task 2 and sub-task 3, denoted as $\text{BERT}_{\text{topK}}$, where $_{\text{topK}}$ represents top K relations are retained for the given sentence. For comparisons, we also directly asked GPT-3.5 and GPT-4 to output the relations for the given sentence through the prompt of providing all candidate relation types. The relation extraction performance scores of these compared models are listed in Table 3, obviously showing that $\text{BERT}_{\text{topK}}$ performs much better than GPT-3.5/4 in terms of F1. In general, we tend to select a big K in LRTE's sub-task 1 for keeping a high recall. Then, the precision is improved mainly through LRTE's sub-task 3. Given that $\text{BERT}_{\text{top4}}$'s F1 score is too low although it has the highest recall, we selected $\text{BERT}_{\text{top3}}$ in LRTE's sub-task 1, i.e., retaining top 3 ($K = 3$) predicted relations in sub-task 1.

Rationality of Two-Step QA in Entity Extraction. To justify adopting the two-step question strategy in LRTE's sub-task 2 to achieve entity extraction given the potential rations obtained in sub-task 1, we provided new compared variants GPT-3.5/4_{K}. For example, in GPT-3.5_5, with the prompt containing a potential relation accompanied by 5 demonstration examples having this relation that are randomly selected from the training set, we ask GPT-3.5 to directly output the triple(s) involving this relation for the given sentence. Furthermore, to justify the impact of the only one example in the prompt of sub-task 2 that is retrieved from the training set, we also provided other ablated variants of LRTE, denoted as GPT-3.5/$4_{\text{QA}-none/random}$, where *none* and *random* represent the two-step QA strategy not using demonstration examples and using a randomly selected example, respectively. From the Table 4 we have the following observations and analysis.

Table 4. RTE performance comparisons between GPT-3.5/4_{QA} and the variants (denoted as GPT-3.5/4_K) using the LLM to extract entities directly with K demonstration examples in sub-task 2. *none* and *random* represent the two-step QA strategy not using demonstration examples and using a randomly selected example in sub-task 2, respectively.

Method	NYT11-New			NYT10-New		
	Prec.	Rec.	F1	Prec.	Rec.	F1
GPT-3.5_5	20.55	28.35	23.83	18.10	29.64	22.48
GPT-3.5_{10}	20.98	30.80	24.96	19.38	35.01	24.95
GPT-3.5_{15}	22.51	32.99	26.76	19.89	35.18	25.41
GPT-4_5	26.40	40.96	32.11	27.68	44.54	34.15
GPT-4_{10}	29.59	45.10	35.74	34.62	53.19	41.94
GPT-4_{15}	32.87	44.49	37.81	38.55	49.55	43.36
GPT-3.5_{QA}-*none*	33.07	58.50	42.25	34.72	55.96	42.86
GPT-3.5_{QA}-*random*	33.80	59.72	43.17	37.30	48.77	42.27
GPT-3.5_{QA}	37.79	55.57	44.99	42.61	49.83	45.94
GPT-4_{QA}-*none*	42.33	63.40	50.77	35.01	55.35	42.89
GPT-4_{QA}-*random*	46.01	68.91	55.18	36.92	72.48	48.92
GPT-4_{QA}	**50.53**	**81.27**	**62.31**	**44.72**	**74.92**	**56.00**

1. Both GPT-3.5_K and GPT-4_K improve their performance as K increases.
2. Any variant of GPT-3.5/4_{QA} is better than GPT-3.5/4_K, justifying the two-step QA strategy in sub-task 2 can help the LLM identify the subject and object entity more accurately. Even compared with GPT-4_{15}, GPT-3.5_{QA} is still superior.
3. GPT-4_{QA}'s performance improvement over GPT-$4_{QA-none}$ is more remarkable than GPT-3.5_{QA}'s performance improvement over GPT-3.$5_{QA-none}$. It implies that an optimal demonstrate example in LRTE's sub-task 2 can inspire GPT-4's advantage more sufficiently.

Rationality of Triple Filtering. Recall Table 2, we find that $\text{LRTE}_{\text{GPT-4}}$ improves F1 apparently compared with GPT-4_{QA}, as it detects and filters out the incorrect triples through sub-task 3. Thus, the rationality of triple filtering in LRTE is also justified.

4 Related Work

Relational Triple Extraction. Most of RTE methods can be categorized into pipeline-based methods and joint extraction methods. *1) Pipeline-based methods* aim to extract entities and relations from the given sentence through two separate phases. Wei et al. [21] proposed a cascade framework in which all possible subjects in the sentence are identified at first, resulting in redundant relations for judgment. In general, the pipeline-based methods neglect the potential interactions of the sub-tasks and may suffer from error propagation [11]. *2) Joint extraction approaches* perform entity recognition and relation classification simultaneously [15,20]. These methods still have some problems such as entity nesting, overlapping relationships, and insufficient data extraction.

Large Language Models. Large language models (LLMs) obtain impressive zero and few-shot learning performance by the pre-training of self-supervised next-token prediction with massive scales. More recently, supervised fine-tuning on a large number of downstream tasks has been proven helpful to improve LLMs' accuracy, robustness, fairness, and generalization to unseen tasks [5,25]. [8] proposed a comprehensive and systematic framework to evaluate LLMs' performance, but found that LLMs perform poorly in the RTE task. [9] introduced a unified text-to-structure generation framework capable of modeling different IE tasks. In addition, [10] proposed an adaptive filter-then-re-rank paradigm. However, this paradigm does not fully explore LLMs' extraction ability. In contrast, our proposed three-sub-task pipeline framework in LRTE fully harnesses the strengths of LLMs, and thus achieves the RTE of high precision and recall.

5 Conclusion

In this paper, we propose a novel LLM-based RTE framework, namely LRTE, in which the RTE task is decomposed into three sequential sub-tasks: relation extraction, entity extraction, and triple filtering. Through these sub-tasks, LRTE maximizes the strengths of LLMs to achieve the RTE of high recall and precision especially on the unseen data. We have conducted extensive experiments to demonstrate that our LRTE outperforms the SOTA RTE models more remarkably on the partially seen and unseen triples, and the rationality of our designs in LRTE were also justified.

References

1. Devlin, J., Chang, M.W., Lee, K., Toutanova, K.: BERT: pre-training of deep bidirectional transformers for language understanding. In: Proceedings of NAACL (2019)
2. Du, Z., et al.: GLM: general language model pretraining with autoregressive blank infilling. In: Proceedings of ACL (2022)
3. Han, R., Peng, T., Yang, C., Wang, B., Liu, L., Wan, X.: Is information extraction solved by ChatGPT? An analysis of performance, evaluation criteria, robustness and errors. CoRR abs/2305.14450 (2023)
4. Ho, X., Nguyen, A.K.D., et al.: Analyzing the effectiveness of the underlying reasoning tasks in multi-hop question answering. In: Findings of EACL (2023)
5. Iyer, S., et al.: OPT-IML: scaling language model instruction meta learning through the lens of generalization. CoRR abs/2212.12017 (2022)
6. Jiang, A.Q., et al.: Mixtral of experts. arXiv preprint arXiv:2401.04088 (2024)

7. Lee, J., Lee, M.J., et al.: Does it really generalize well on unseen data? Systematic evaluation of relational triple extraction methods. In: Proceedings of NAACL (2022)
8. Li, B., et al.: Evaluating ChatGPT's information extraction capabilities: an assessment of performance, explainability, calibration, and faithfulness. arXiv preprint arXiv:2304.11633 (2023)
9. Lu, Y., Liu, Q., et al.: Unified structure generation for universal information extraction. In: Proceedings of ACL (2022)
10. Ma, Y., et al.: Large language model is not a good few-shot information extractor, but a good reranker for hard samples! CoRR abs/2303.08559 (2023)
11. Nayak, T., et al.: Deep neural approaches to relation triplets extraction: a comprehensive survey. Cogn. Comput. **13**, 1215–1232 (2021)
12. Ning, J., Yang, Z., et al.: OD-RTE: a one-stage object detection framework for relational triple extraction. In: Proceedings of ACL (2023)
13. OpenAI: ChatGPT (2022). https://openai.com/blog/chatgpt
14. OpenAI: GPT-4 technical report (2023)
15. Shang, Y.M., Huang, H., Mao, X.L.: OneRel: joint entity and relation extraction with one module in one step. In: Proceedings of the AAAI (2022)
16. Takanobu, R., Zhang, T., Liu, J., Huang, M.: A hierarchical framework for relation extraction with reinforcement learning. In: Proceedings of AAAI (2019)
17. Wan, Z., Cheng, F., et al.: GPT-RE: in-context learning for relation extraction using large language models. CoRR abs/2305.02105 (2023)
18. Wan, Z., Cheng, F., Liu, Q., Mao, Z., Song, H., Kurohashi, S.: Relation extraction with weighted contrastive pre-training on distant supervision. In: Findings of EACL (2022)
19. Wang, S., et al.: GPT-NER: named entity recognition via large language models. CoRR abs/2304.10428 (2023)
20. Wang, Y., Yu, B., et al.: TPLinker: single-stage joint extraction of entities and relations through token pair linking. In: Proceedings of COLING (2020)
21. Wei, Z., Su, J., Wang, Y., Tian, Y., Chang, Y.: A novel cascade binary tagging framework for relational triple extraction. In: Proceedings of ACL (2019)
22. Xie, C., Liang, J., et al.: Revisiting the negative data of distantly supervised relation extraction. In: Proceedings of ACL (2021)
23. Xu, B., Wang, Q., et al.: EmRel: joint representation of entities and embedded relations for multi-triple extraction. In: Proceedings of NAACL (2022)
24. Yang, Z., Huang, Y., Feng, J.: Learning to leverage high-order medical knowledge graph for joint entity and relation extraction. In: Findings of ACL (2023)
25. Yuan, S., et al.: Distilling script knowledge from large language models for constrained language planning. In: Proceedings of ACL (2023)
26. Yuan, S., et al.: Generative entity typing with curriculum learning. In: Proceedings of ACL (2022)
27. Zheng, H., et al.: PRGC: potential relation and global correspondence based joint relational triple extraction. In: Proceedings of ACL/IJCNLP (2021)

Multi-Query Person Search with Transformers

Ying Chen[1], Zhihui Li[2], and Andy Song[1(✉)]

[1] RMIT University, Melbourne, Australia
Andy.Song@rmit.edu.au
[2] University of Science and Technology of China, Hefei, China

Abstract. We propose a transformer-based multi-query person search (MQPS) method that jointly performs person detection and person re-identification (re-id) in an end-to-end framework. Most existing person search methods employ hand-crafted components and involve multiple steps and stages to detect and identify the target person, which are computationally inefficient and brutal to generalise to different datasets. The recent advance in end-to-end object detection with transformers, mainly the DETR family, employ object queries to learn objects and directly predict a set of bounding boxes and object classes. However, this approach uses one object query per object so that the detected object is centred around the object spatial location, which is not ideal for small and occluded objects during feature representation learning. Therefore, we propose a multi-query method for person detection and person feature representation learning. Specifically, MQPS utilises multiple adjacent object queries to learn a target person object with multi-scale features. Moreover, to improve the feature representation learning of intra-identity objects, we employ a margin ranking loss to bring closer the intra-identity person instances in the feature space. Experiments on CUHK-SYSU and PRW datasets demonstrate the effectiveness of the proposed method.

Keywords: Multi-Query Person Search · Transformers

1 Introduction

Person search [20] aims to detect and identify a target person in a gallery of scene images. Person search addresses the challenges of two sub-tasks of person detection and person re-identification and jointly optimizing them in a unified framework. [21] pointed out the conflict between person detection and re-id sub-tasks because of the different objectives [14]. Person detection deals with common human appearance, while re- id focuses on a person's uniqueness. Person detection strives to perform object-level recognition and localization by differentiating the person category from the background. On the other hand, re-id aims to match a person instance to an identity, thereby requiring to discriminate among instances of different persons within the same identity captured on multiple cameras (Fig. 1).

D.-N. Yang et al. (Eds.): PAKDD 2024, LNAI 14648, pp. 116–128, 2024.
https://doi.org/10.1007/978-981-97-2238-9_9

Fig. 1. Multi-query method employs multiple spatial object queries to detect the most relevant parts of the target object with less noises for re-id. This is particularly useful for small and occluded person objects. Single-query method such as DETR tends to detect objects around the object center location, which may achieve higher Intersection over Union (IoU). However, higher IoU also indicates higher noise for occluded object. The proposed multi-query method strives to find most relevant person body, leading to better re-id performance.

Typical person search solutions tackle the problem in two-step or one step. Two-step person search tackles the two sub-tasks in two steps, respectively, where person detection and re-id are performed separately. Firstly, a detection network is utilised to detect and crop out person objects into a fixed resolution. Secondly, a re-id network is used to identify cropped person instances. While achieving promising performance, the two-step person search is computationally expensive. In contrast, one-step approaches simultaneously detect and identify the target person using a single network, in which a shared network extracts person features. Then two branches within the same network perform person detection and re-id.

Despite the promising performance in recent person search works, two-step and one-step approaches employ hand-designed components, such as non-maximum suppression (NMS), to filter out duplicate region predictions for each person object. Recently, transformer-based object detectors have alleviated the need to employ different hand-designed components, leading to a simpler end-to-end trainable architecture. Further, the transformer architecture can be easily extended to a multi-task learning framework. Despite their success in general object detection tasks, transformers are yet to be the dominant framework for person search due to the challenges of detecting and matching small and occluded person objects. In this work, we propose an end-to-end transformer-based person

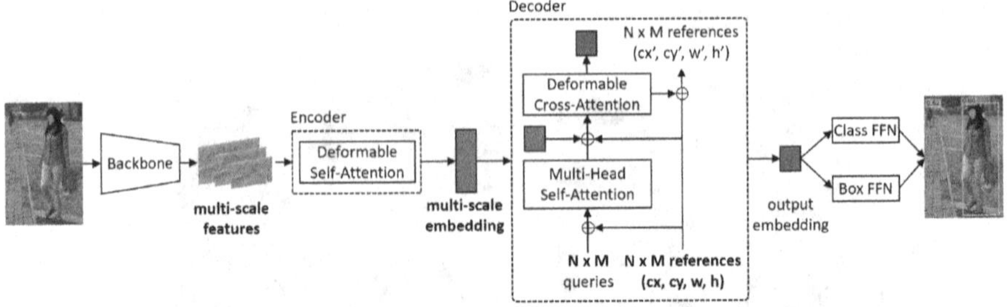

Fig. 2. Multi-query person search with transformer workflow. We use N × M queries in the proposed method to learn N target objects using multi-scale features. We employ M spatially adjacent object queries for each target object to locate and learn the target object, which effectively locates small and occluded objects.

search framework with a specialised multi-query decoder for improved performance in detecting and learning small and occluded objects.

The proposed method adapts the Deformable-DETR [28] architecture and introduces a specialised multi-query decoder to locate suitable person objects for feature learning. DETR family of object detection methods use one query per target object when learning object representations which is ineffective for occluded objects and spatially adjacent objects, in which significant objects dominate the cross-attention learning. Therefore we propose a multi-query per object learning strategy so that a single target object is learnt using multiple object queries. The best matching query is selected with minimal matching cost and is used for classification and bounding box regression. Moreover, inter-camera person instances that belong to the same identity are difficult to match due to different poses and lighting conditions. Most existing person search methods inherit from traditional object detection methods, omitting that person search is a ranking problem at inference time, that the inference is based on the ranking of the visual similarity of detected person objects. Therefore, we propose to rank re-id probability during training with a margin ranking loss. The architecture of the proposed framework is illustrated in Fig. 2. The main contributions of this paper are:

- We propose a specialised multi-query transformer decoder for joint person detection and feature representation learning that utilises adjacent objects with multi-scale features to locate and learn target person object characteristics effectively.
- We employ a margin ranking loss to improve identity matching of intra-identity person objects, which is handy to associate intra-identity person instances across multiple cameras.
- Extensive experiments and model analysis demonstrate the effectiveness of the proposed method for person search.

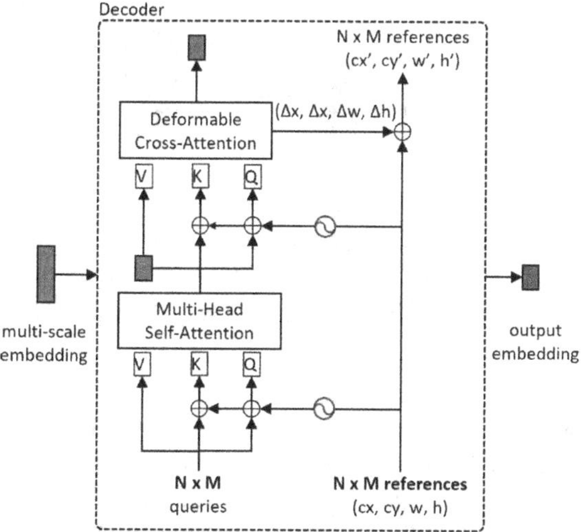

Fig. 3. Multi-Query Decoder. N × M object queries are utilised in the proposed method to learn N objects. M spatially adjacent object queries are used for each target object, effectively locating small and occluded objects. The encoder performs self-attention on the multi-scale features and outputs the multi-scale embedding as input to the decoder. The decoder employs N × M object queries to learn object representations with the multi-scale embeddings in the self-attention and cross-attention steps. The decoder output embedding are used by the classification and regression header to predict the classes and bounding boxes. These decoder embeddings are also used for a person matching during inference.

2 Method

2.1 Multi-Query Decoder

In the proposed MQPS method, the detection and re-id encoder-decoder is built upon Deformable-DETR. The encoder-decoder consists of a six-layer encoder and a six-layer decoder, utilizing the multi-scale features as input. An encoder layer has a deformable self-attention module and a MLP module. Consequently, the first decoder layer inputs the multi-scale features and the N object queries. As illustrated in Fig. 3, a decoder layer contains a multi-head self-attention module, a deformable cross-attention layer, and a MLP layer. Following Deformable-DETR, we use a feature length of 256 for the encoders and decoders. The decoder output features are utilized in a prediction head for classification and bounding box regression. These output features are further used for person instance matching during inference.

Scale variation poses a significant challenge in person matching. We employ multi-scale features as the input to address this issue and perform re-id feature generation. Multi-scale features are constructed at various scales using backbone-extracted feature maps. As shown in Fig. 2, the multi-scale features first go

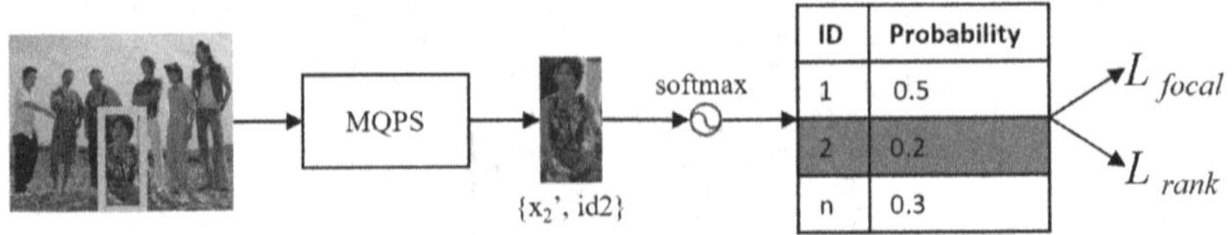

Fig. 4. Focal loss and margin ranking loss. Person feature representation are activated with Softmax function. The output probabilities are evaluated using focal loss and ranking loss so that intra-identity instances are separated from inter-identity instances.

through the deformable encoder self-attention. Then the multi-scale encoder embedding passes through the decoder self-attention and cross-attention to get the final query features. During training, these discriminative decoder features are supervised by independent re-id losses. During inference, these multi-scale decoder embeddings are concatenated for person matching.

DETR family of object detection methods use one query to learn a target object which is ineffective for small objects and spatially adjacent objects because the significant objects dominate cross-attention learning. Therefore we propose a multi-query per target object learning strategy so that a single target object is learnt using multiple object queries. Specifically, we use one central reference point plus a few vertically adjacent points to locate an object. The reason for using vertically adjacent points is that the pedestrians are standing or walking in person search datasets. Therefore, adjacent vertical points effectively locate the different body parts. This is especially useful for small and occluded objects where a centre point doesn't contain discriminative information. Still, the other part of the body, such as the head, shoulder or leg, can be used to locate and identify the target person. During training, only centre queries reference points are learnable parameters. The rest of the query reference points are vertically adjacent to the centre points. For example, we use 300 queries for most of out models. Among them, 60 query reference points are learnable parameters. The rest 240 queries are from adjacent locations. There are 60×5 reference points R, which are R_{0-59}, R_{60-119}, $R_{120-179}$, $R_{180-239}$, $R_{240-199}$ Only $R_{120-179}$ are learnable parameters. The rest are vertically adjacent points to $R_{120-179}$. We constrain the adjacent queries to be within 50% of the vertical distance. The reason is that objects with an IoU match of 0.5 are considered good detection during inference. The multi-query location can be constructed as:

$$
\begin{aligned}
r &= (c_x, c_y, w, h), \\
r_{adj} &= (c_x, c_y + m, w, h), m < 0.5
\end{aligned}
$$

where r is a learnable parameter representing a spatial location. c_x and c_y are the central x and y coordinates, w and h are the width and height of the spatial location respectively. r_{adj} is an adjacent location to r and is m distance away from r.

2.2 Detection and Identification Losses

Detection Loss. Following DETR, the detection loss is a linear combination of the L1 bounding box loss and the generalized IoU loss. The box loss is defined as:

$$\mathcal{L}_{box}(b_i, \hat{b}_{\sigma(i)}) = \lambda_{iou} L_{iou}(bi, \hat{b}_{\sigma(i)}) + \lambda_{L1}||bi, \hat{b}_{\sigma(i)}|| \tag{1}$$

where λ_{iou}, λ_{L1} are hyperparameters. b_i is the ground-truch bounding box, $\hat{b}_{\sigma(i)}$ is the predicted bounding box from query embedding $\sigma(i)$.

To learn discriminative person features for person matching, we employ the OIM loss [19] and focal loss [13] to guide person feature representation learning. We further utilise a margin ranking loss to force intra-identity features a marginal distance away from the inter-identity features in feature space.

OIM Loss. OIM [19] effectively closes the query-person gap utilising labelled and unlabeled identities from training data. The probability of detected person features x being recognised as the identity with class-id i by a Softmax function:

$$p_i = \frac{\exp(v_i^T x/\tau)}{\sum_{j=1}^{L} \exp(v_j^T x/\tau) + \sum_{k=1}^{Q} \exp(u_k^T x/\tau)}. \tag{2}$$

Where v_i^T is the labelled person features for the i_{th} identity in the lookup table (LUT). v_j^T is the j_{th} labelled person features in the LUT. u_k^T is the k_{th} unlabelled person features in the LUT. τ regulates probability distribution. OIM objective is to maximize the expected log-likelihood of the target t.

$$\mathcal{L}_{oim} = \mathrm{E}_{\mathrm{x}}\left[\log p_t\right]. \tag{3}$$

Following [19], during training, we employ a lookup table V and a circular queue U to guide re-id feature learning. We store re-id features of all L labelled identities in V and re-id features of Q unlabeled identities from recent mini-batches in U. Labeled features are calculated as a running average of all intra-identity features. At each iteration, we first compute similarities between re-id features (e.g., F1) in the current mini-batch and all features in V and U. We then calculate online instance matching (OIM) loss based on similarities measured as dot products of the re-id features and online instances.

Focal Loss. We utilise focal loss [13] to learn person features for imbalanced training samples. Focal loss is expressed as following:

$$\mathcal{L}_{focal}(p_t) = -\alpha(1 - p_t)^{\gamma} \log(p_t) \tag{4}$$

where p_t is the softmax output probability. Following [13], we set α to 0.5, γ to 2 during training. The focal loss is depicted in Fig. 4.

Margin Ranking Loss. Besides the re-id losses, We employ a margin ranking loss to force intra-identity features a marginal distance away from the inter-identity features. OIM and focal loss close the gap between intra-identity person instances in feature space. The intuition of the margin ranking loss is that intra-identity person objects should be more intimate than inter-identity in feature space, even though those intra-identify instances are visually dissimilar. Using a suitable margin to separate intra-identity features from inter-identity features further. Therefore, it improves search accuracy during inference. This is especially helpful for visually dissimilar person instance matching. The margin ranking loss is defined as:

$$\mathcal{L}_{rank} = \sum_{i=1}^{N} [max(0, -y * (s_i - s_i') + m)]]. \tag{5}$$

where s_i and s_i' are scores to be ranked, $y = \{1| - 1\}$ indicating whether the first or second value should be ranked higher, m is the objective margin between the scores.

3 Experiments

3.1 Datasets

CUHK-SYSU [19] and PRW [27] are the most commonly used datasets for image-based person search. CUHK-SYSU contains 18,184 images, 8,432 person identities, and 99,809 annotated bounding boxes. The training set contains 11,206 images and 5,532 query identities. The test set has 6,978 images and 2,900 query identities. PRW dataset has 11,816 frames and 43,110 person bounding boxes. 34,304 people have identities ranging from 1 to 932, and the rest are assigned identities of −2. The PRW training set has 5,704 images and 482 identities, and the test set has 6112 pictures and 450 identities.

3.2 Evaluation Metrics

Cumulative matching characteristics (CMC top-K) and mean averaged precision (mAP) are the primary evaluation metrics for person search. In CMC, the top-K predicted bounding boxes are ranked according to the Intersection over Union (IoU) overlap with the ground-truth equal to or greater than 0.5. The mAP is a popular evaluation metric in object detection, in which an averaged precision (AP) is calculated for each query person. Then the final mAP is computed as an average of all APs.

3.3 Implementation Details

We conduct experiments with a pre-trained ResNet50 backbone. Our models are trained on a single Tesla V100 GPU using AdamW optimizer. We employ multi-scale feature and focal loss for the PRW dataset and OIM loss for the

Table 1. Performance of image-based person search methods on CUHK-SYSU, PRW datasets.

Method	CUHK-SYSU		PRW	
	mAP	top-1	mAP	top-1
Two-step person search				
MGTS [3]	83.00	83.70	32.60	72.10
CLSA [11]	87.20	88.50	38.70	65.00
IGPN + PCB [6]	90.30	91.40	47.20	87.00
RDLR [8]	93.00	94.20	42.90	70.20
TCTS [17]	93.90	**95.10**	46.80	87.50
One-step multi-stage person search				
OIM [19]	75.50	78.70	21.30	49.90
IAN [18]	76.30	80.10	23.00	61.90
OIAM [7]	76.98	77.86	51.02	69.85
IOIM [15]	79.78	79.90	21.00	63.10
RCAA [2]	–	81.30	–	–
I-NET [9]	79.50	81.50	–	–
KD-OIM [16]	83.80	84.20	–	–
CGPS [23]	84.10	86.50	33.40	73.60
PFFN [10]	84.50	89.80	34.30	73.90
SMG [26]	86.30	86.50	–	–
FPSP [12]	86.99	89.87	44.45	70.58
BINet [5]	90.80	91.60	47.20	83.40
NAE+ [4]	92.10	92.90	44.00	81.10
DKD [25]	93.09	94.24	50.51	87.07
COAT [24]	94.20	94.70	53.30	87.40
One-step one-stage person search				
AlignPS [22]	93.10	93.40	45.90	81.90
COAT [24]	–	–	43.30	78.80
PSTR [1]	93.50	**95.00**	49.50	87.80
MQPS (ours)	**94.23**	94.07	**52.20**	**87.86**

CUHK-SYSU dataset respectively. The model is trained for 59 epochs and uses a mini-batch size of 2. The initial learning rate is set to 0.0001, and we decrease the learning rate by 0.10 every 10 epochs.

For testing, we rescale the test images to a fixed size of 1500 × 900 pixels during inference. Testing of the MQPS model first needs to detect the query probes from the probe dataset, then use the same model to detect target objects from the image gallery, and finally calculate the mAP and Top-1 results based on the dot products of the probes and target objects.

Table 2. Effect of the OIM loss and Focal loss on the CUHK-SYSU and PRW datasets. Experiment with multi-query, ranking loss and 6-scale features.

		CUHK-SYSU		PRW	
OIM	Focal loss	mAP	top-1	mAP	top-1
✓		**94.23**	**94.07**	46.69	82.33
	✓	87.52	92.66	**52.20**	**87.86**

Table 3. Ablation of multi-query on the CUHK-SYSU and PRW datasets. Experiment with 6-scale features, OIM loss for CUHK-SYSU dataset, Focal loss for PRW dataset and ranking loss

	CUHK-SYSU		PRW	
Multi-Query	mAP	top-1	mAP	top-1
	93.22	93.16	49.22	85.36
✓	**94.23**	**94.07**	**52.20**	**87.86**

3.4 Comparison to the State-of-the-Art Approaches

The proposed MQPS method achieves promising results compared to the state-of-the-art person search methods. For a fair comparison, we only compare the MQPS with the most recent one-step, one-stage methods. As reported in Table 1, the MQPS method outperforms most recent person search methods. MQPS outperforms the sophisticated transformer-based COAT and PSRT methods. MQPS also achieves better result than the FCOS-based AlignPS method. The MQPS method ranks first in mAP for the CUHK-SYSU dataset, the mAP and top-1 results for the PRW dataset also leading the state of the art methods. It is worthwhile mention that CNN-based two-step methods RDLR and TCTS achieve very high evaluation results, because the search probes are cropped images and are available for direct comparison with detected person objects. While in the transformer-based methods, the search probes need to be detected first which may results in some detection error (Table 3).

3.5 Model Analysis

OIM Loss and Focal Loss. OIM loss was initially proposed in [19] and had been adopted by several person search methods. OIM loss is effective for the CUHK dataset with many identities since the OIM maintains the running average of the instance features. CUHK-SYSU dataset is collected from video clips so that person instances in different scenes are visually similar, therefore the running average of person instances is a good person representation. However, OIM doesn't work well for the PRW dataset because the PRW dataset is collected from six cameras, and inter-camera person instances of the same identity vary significantly. Therefore, the running average of person instances does not represent

Table 4. Effect of the multi-scale features on CUHK-SYSU, PRW datasets. Experiment with multi-query, ranking loss, OIM loss for CUHK-SYSU dataset, Focal loss for PRW dataset.

					CUHK-SYSU		PRW	
4	5	6	7	8	mAP	top-1	mAP	top-1
✓					89.71	91.83	45.79	83.33
	✓				93.52	93.76	46.22	87.56
		✓			**94.23**	**94.07**	**52.20**	**87.86**
			✓		93.43	93.65	45.32	84.36
				✓	92.30	92.50	45.00	82.10

Table 5. Effect of margin ranking loss on the CUHK-SYSU and PRW datasets. Experiment with 6-scale features, OIM loss for CUHK-SYSU dataset, Focal loss for PRW dataset and margin 0.2 in the ranking loss.

	CUHK-SYSU		PRW	
Ranking loss	mAP	top-1	mAP	top-1
	92.82	92.15	48.23	83.26
✓	**94.23**	**94.07**	**52.20**	**87.86**

the person well. Consequently, we employ the focal loss for the PRW dataset to address the complex intra-identity samples and imbalanced data. Figure 2 shows that OIM works for CUHK-SYSU dataset, and focal loss works better on PRW dataset.

Effect of Multi-Scale Features. We analyse the effect of the multi-scale feature ranging from scale 4 to 8. The multi-scale re-id decoder takes the last N feature maps as the inputs for different re-id decoder branches. Six-scale features achieves the best performance as shown in Table 4. Multi-scale features are essential to detect objects at various sizes. Large objects can be easily detected using their shape, while small objects rely more on critical features such as head and foot (Table 5).

Effect of the Margin Ranking Loss. Person search is eventually a ranking problem at the inference time. Evaluation performance is ranked based on the visual similarity of the person objects. We demonstrate the benefits of the Margin Ranking Loss in Table 4. Without the ranking loss, the person feature representation learning is equivalent to a standard object detection task. Therefore, by introducing the ranking loss in training, the evaluation result is notably improved, especially for hard intra-identity samples which are visually different but belong to the same identity.

4 Conclusion

We propose a transformer-based multi-query person search (MQPS) method that jointly performs person detection and person re-id in an end-to-end manner. The proposed MQPS method effectively improves detection and re-id accuracy for small and occluded instances, utilising multiple adjacent queries to locate the target object for re-id, considering most relevant body parts and salient features. Furthermore, to improve the matching of intra-identity person instances, we employ a margin ranking loss to discriminate intra-identity person features during training so that intra-identity feature representations are forced to be a marginal distance away from inter-identity feature representation. Experiments on CUHK-SYSU and PRW datasets demonstrate the effectiveness of the proposed MQPS method. We aim to design efficient person search models using fewer computational resources in future work.

References

1. Cao, J., et al.: PSTR: end-to-end one-step person search with transformers (2022)
2. Chang, X., Huang, P.-Y., Shen, Y.-D., Liang, X., Yang, Y., Hauptmann, A.G.: RCAA: relational context-aware agents for person search. In: Ferrari, V., Hebert, M., Sminchisescu, C., Weiss, Y. (eds.) ECCV 2018. LNCS, vol. 11213, pp. 86–102. Springer, Cham (2018). https://doi.org/10.1007/978-3-030-01240-3_6
3. Chen, D., Zhang, S., Ouyang, W., Yang, J., Tai, Y.: Person search via a mask-guided two-stream CNN model. In: Ferrari, V., Hebert, M., Sminchisescu, C., Weiss, Y. (eds.) ECCV 2018. LNCS, vol. 11211, pp. 764–781. Springer, Cham (2018). https://doi.org/10.1007/978-3-030-01234-2_45
4. Chen, D., Zhang, S., Yang, J., Schiele, B.: Norm-aware embedding for efficient person search. In: CVPR (2020)
5. Dong, W., Zhang, Z., Song, C., Tan, T.: Bi-directional interaction network for person search. In: Proceedings of the IEEE/CVF Conference on Computer Vision and Pattern Recognition (CVPR), June 2020
6. Dong, W., Zhang, Z., Song, C., Tan, T.: Instance guided proposal network for person search. In: 2020 IEEE/CVF Conference on Computer Vision and Pattern Recognition (CVPR), pp. 2582–2591 (2020). https://doi.org/10.1109/CVPR42600.2020.00266
7. Gao, C., Yao, R., Zhao, J., Zhou, Y., Hu, F., Li, L.: Structure-aware person search with self-attention and online instance aggregation matching. Neurocomputing **369** (2019). https://doi.org/10.1016/j.neucom.2019.08.038. https://www.sciencedirect.com/science/article/pii/S0925231219311762
8. Han, C., et al.: Re-ID driven localization refinement for person search. In: 2019 IEEE/CVF International Conference on Computer Vision (ICCV), pp. 9813–9822 (2019). https://doi.org/10.1109/ICCV.2019.00991
9. He, Z., Zhang, L.: End-to-end detection and re-identification integrated net for person search. In: Jawahar, C.V., Li, H., Mori, G., Schindler, K. (eds.) ACCV 2018. LNCS, vol. 11362, pp. 349–364. Springer, Cham (2019). https://doi.org/10.1007/978-3-030-20890-5_23
10. Hong, Z., Liu, B., Lu, Y., Yin, G., Yu, N.: Scale voting with pyramidal feature fusion network for person search. IEEE Access **7** (2019). https://doi.org/10.1109/ACCESS.2019.2943112

11. Lan, X., Zhu, X., Gong, S.: Person search by multi-scale matching. In: Ferrari, V., Hebert, M., Sminchisescu, C., Weiss, Y. (eds.) ECCV 2018. LNCS, vol. 11205, pp. 553–569. Springer, Cham (2018). https://doi.org/10.1007/978-3-030-01246-5_33
12. Li, J., Liang, F., Li, Y., Zheng, W.S.: Fast person search pipeline. In: 2019 IEEE International Conference on Multimedia and Expo (ICME), pp. 1114–1119, July 2019. https://doi.org/10.1109/ICME.2019.00195. ISSN 1945-788X
13. Lin, T., Goyal, P., Girshick, R.B., He, K., Dollár, P.: Focal loss for dense object detection. In: IEEE International Conference on Computer Vision, ICCV 2017, Venice, Italy, 22–29 October 2017, pp. 2999–3007. IEEE Computer Society (2017). https://doi.org/10.1109/ICCV.2017.324
14. Lin, X., Ren, P., Xiao, Y., Chang, X., Hauptmann, A.: Person search challenges and solutions: a survey. In: Zhou, Z.H. (ed.) Proceedings of the Thirtieth International Joint Conference on Artificial Intelligence, IJCAI 2021, pp. 4500–4507, August 2021. https://doi.org/10.24963/ijcai.2021/613, survey Track
15. Liu, H., Shi, W., Huang, W., Guan, Q.: A discriminatively learned feature embedding based on multi-loss fusion for person search. In: 2018 IEEE International Conference on Acoustics, Speech and Signal Processing (ICASSP), pp. 1668–1672, April 2018. https://doi.org/10.1109/ICASSP.2018.8462484. ISSN 2379-190X
16. Munjal, B., Galasso, F., Amin, S.: Knowledge distillation for end-to-end person search. In: 30th British Machine Vision Conference 2019, BMVC 2019, Cardiff, UK, 9–12 September 2019, p. 216. BMVA Press (2019)
17. Wang, C., Ma, B., Chang, H., Shan, S., Chen, X.: TCTS: a task-consistent two-stage framework for person search. In: Proceedings of the IEEE/CVF Conference on Computer Vision and Pattern Recognition (CVPR), June 2020
18. Xiao, J., Xie, Y., Tillo, T., Huang, K., Wei, Y., Feng, J.: IAN: the individual aggregation network for person search. Pattern Recognit. **87**, 332–340 (2019)
19. Xiao, T., Li, S., Wang, B., Lin, L., Wang, X.: Joint detection and identification feature learning for person search. In: CVPR (2017)
20. Xu, Y., Ma, B., Huang, R., Lin, L.: Person search in a scene by jointly modeling people commonness and person uniqueness. In: Hua, K.A., Rui, Y., Steinmetz, R., Hanjalic, A., Natsev, A., Zhu, W. (eds.) Proceedings of the ACM International Conference on Multimedia, MM 2014, Orlando, FL, USA, 03–07 November 2014, pp. 937–940. ACM (2014)
21. Xu, Y., Ma, B., Huang, R., Lin, L.: Person search in a scene by jointly modeling people commonness and person uniqueness. In: Proceedings of the 22nd ACM international conference on Multimedia, MM 2014, pp. 937–940. Association for Computing Machinery, New York, November 2014. https://doi.org/10.1145/2647868.2654965
22. Yan, Y., et al.: Anchor-free person search. arXiv:2103.11617 [cs], March 2021
23. Yan, Y., Zhang, Q., Ni, B., Zhang, W., Xu, M., Yang, X.: Learning context graph for person search. In: Proceedings of the IEEE/CVF Conference on Computer Vision and Pattern Recognition (CVPR), June 2019
24. Yu, R., et al.: Cascade transformers for end-to-end person search, pp. 7267–7276 (2022)
25. Zhang, X., Wang, X., Bian, J.W., Shen, C., You, M.: Diverse knowledge distillation for end-to-end person search. arXiv:2012.11187 [cs], December 2020
26. Zheng, D., Xiao, J., Huang, K., Zhao, Y.: Segmentation mask guided end-to-end person search. Signal Process. Image Commun. **86**, 115876 (2020). https://doi.org/10.1016/j.image.2020.115876. http://www.sciencedirect.com/science/article/pii/S0923596520300849

27. Zheng, L., Zhang, H., Sun, S., Chandraker, M., Yang, Y., Tian, Q.: Person re-identification in the wild. In: Proceedings of the IEEE Conference on Computer Vision and Pattern Recognition (CVPR), July 2017
28. Zhu, X., Su, W., Lu, L., Li, B., Wang, X., Dai, J.: Deformable DETR: deformable transformers for end-to-end object detection. In: 9th International Conference on Learning Representations, ICLR 2021, Virtual Event, Austria, 3–7 May 2021. OpenReview.net (2021). https://openreview.net/forum?id=gZ9hCDWe6ke

BioReX: Biomarker Information Extraction Inspired by Aspect-Based Sentiment Analysis

Weiting Gao[1], Xiangyu Gao[1], Wenjin Chen[2], David J. Foran[2], and Yi Chen[1(✉)]

[1] New Jersey Institute of Technology, Newark, NJ 07102, USA
{wg97,xg77,yi.chen}@njit.edu

[2] Rutgers Cancer Institute of New Jersey, Rutgers, The State University of New Jersey, New Brunswick, NJ 08901, USA
{chenwe,foran}@rutgers.edu

Abstract. Biomarkers are critical in cancer diagnosis, prognosis, and treatment planning. However, this information is often buried in unstructured text form. In this paper, we make an analogy between Biomarker Information Extraction and Aspect-Based Sentiment Analysis. We propose a system, Biomarker and Result Extraction Model (BioReX). BioReX employs BERT post-training methods to augment the BioBERT model with domain-specific and task-specific knowledge for biomarker extraction. It uses syntactic-based and semantic-based attention to associate results to corresponding biomarkers. Evaluation demonstrates the effectiveness of the proposed approach.

Keywords: Biomarker Extraction · Information Extraction · Aspect-Based Sentiment Analysis · Large Language Model · Attention Mechanism · Natural Language Processing

1 Introduction

Biomarkers are molecules like genes, DNA, proteins, and metabolites that signify whether the processes in the body are regular or irregular [1]. Cancer markers are critical in cancer diagnosis, prognosis, and treatment planning. However, cancer biomarker information buried in free-text pathology reports is difficult to identify, index, or query. Extracting biomarker information from pathology reports is a critical step to the development of a comprehensive clinic data warehouse [3].

Biomarker information encompasses biomarker mentions and the associated results. For example, sentence "the tumor is negative for ER" contains a biomarker mention "ER" and its result "negative". A biomarker result can be either *positive*, *negative*, or *unknown* (indicating no results associated, as results are uncertain, pending, or not mentioned). In this paper, we also refer biomarker results as *polarity*. Limited studies [4,5,7,16] have been conducted on Biomarker Information Extraction (BIE) from pathology reports, with the majority focusing on the extraction of biomarkers while overlooking their associated results.

D.-N. Yang et al. (Eds.): PAKDD 2024, LNAI 14648, pp. 129–141, 2024.
https://doi.org/10.1007/978-981-97-2238-9_10

BIE imposes several technical challenges. First, extracting biomarker mentions is challenging since the vocabularies of biomarkers overlap with the vocabularies of non-biomarkers, referred to as *mention overlapping*. For example, the occurrences of term "ER" in Table 1 have different meanings. The "ER" in Case 2 refers to a biomarker (i.e., Estrogen Receptor), while the "ERs" in Case 1&3 do not. To differentiate the same mentions with different meanings, the identification of biomarkers needs to depend on the context of their occurrences.

Second, it is challenging to correctly associate a biomarker with its result since one sentence usually contains multiple biomarkers with different results, referred to as *mix-polarity*. Proximity may not always indicate the correct association. For example, consider Case 4 in Table 1 the closest result to biomarker "CK20" is "positive", however, the correct one is "negative".

Furthermore, associating a biomarker with its result is challenging since sometimes results are uncertain or unavailable in the report, which is referred to as *no-result*. For example, "HER2" in Case 5 does not have a result. However, since it appears alongside biomarkers that have associated results, a model often tends to erroneously assign other biomarkers' results to it.

To tackle these challenges, we propose to make an analogy of BIE with Aspect-Based Sentiment Analysis (ABSA) [6,9,10,15], as illustrated in Table 2. ABSA mines people's opinions towards aspects of a product from unstructured reviews. Given a sentence in the review (e.g., The pizza was delicious, but the waiter was rude.), ABSA aims to extract aspects of products (e.g., "pizza" and "waiter"), and associated polarity (e.g., positive for "pizza", negative for "waiter"). We analogize aspects with biomarker mentions and analogize polarity with biomarker results. We propose to leverage the technical advances in ABSA to the relatively unexplored domain of BIE.

We have developed a Biomarker and Result Extraction Model (BioReX), which adapts the state-of-the-art ABSA methods [10,15] for BIE. Other ABSA methods, including but not limited to, [6,9], can also be applied for BIE.

Adapting from existing techniques in ABSA, BioReX addresses the unique challenges in BIE compared with ABSA. To address the first challenge *mention overlapping* in the medical domain, especially in the situation when only limited labeled data is available, we post-train a Large Language Model (LLM) in medical domain, BioBERT [8], which was trained by biomedical article abstracts, for sequence labeling to extract cancer biomarkers given context information.

The second challenge *mix-polarity* on BIE is more intensive than it is on ABSA, since the number of biomarkers and results in a sentence in BIE is typically larger than the number of aspects and sentiments in a sentence in ABSA. Existing ABSA methods (e.g., [6,10,15]) handle the mix-polarity challenge by capturing the sequential interactions between words and the proximity with the aspect. However, when the occurrences of biomarker and result are intense, the closest result is not necessarily to be the correct one (e.g., Case 4 in Table 1).

The third challenge *no-result* is unique on BIE. Unlike product reviews where customers almost always express their sentiment along with each aspect, it is not uncommon that a biomarker does not have an associated result in the pathology report as the results are unknown.

Table 1. Biomarker Information Extraction Cases. **Bold**: Biomarkers. *Italic*: Results

Categories	Cases
Mention Overlapping	1. Part E is labeled ... right posterior X, EO left anterior X, EP EQ left posterior X, ER (Specimen Label) right anterior Z, ES right mid Z ...
	2. The tumor ... is negative for **ER** (Estrogen Receptor), **PR**, **RCC**, **CA19.9**, **PAX8**, and **Vimentin**.
	3. ... patient who presented to the ER (Emergency Room) at [location] on [date] with abdominal pain and bloating, immunostain for **P53** is strongly and diffusely *positive*.
Mix-polarity	4. Immunohistochemical stains as follows: **CK7** is *positive*, **CK20** is *negative*, **WT1** is *positive*, **PAX8** is *positive*, **S100P** is *negative*, **GATA3** is *negative*.
No-result	5. Note: **ER** = *Negative*, **PR** = *Negative*, **HER2** by fish is pending and the result will be given in an addendum report

Table 2. Analogy of BIE and ABSA

Tasks	Sentences	Biomarkers/Aspects	Results/Polarities
BIE	Immunostain for MART-1 antigen is positive, immunostain for CK8 is negative.	MART-1	positive
		CK8	negative
ABSA	The pizza was delicious, but the waiter was rude.	pizza	positive
		waiter	negative

To address the unique challenges associated with intensified *mix-polarity* and *no-result*, we propose two attention modules in BioReX, syntactic-based and semantic-based attentions, to adjust the weights of the associations between the biomarkers and the context words. Syntactic features are considered since the correct corresponding results are syntactic-dependent words of the target biomarker. Semantic features are integrated as they capture the interaction between words based on their semantic meanings.

To summarize, the contributions of our work are as follows:

- This is the first study that extracts both cancer biomarkers and their results in a limited data setting.
- We propose to make an analogy between BIE and ABSA. Our research opens pathways and shows opportunities for the less-explored BIE domain to leverage insights from the well-studied ABSA field.
- The proposed method, BioReX, adapts techniques in ABSA and addresses the unique challenges in BIE. BioReX post-trains BioBERT, a medical LLM, to enhance it with domain and task knowledge to address the *mention over-*

lapping challenge on limited data. It uses syntactic-based and semantic-based attention modules to address the *mix-polarity* and *no-result* challenges.
- Experimental results demonstrate the effectiveness of BioReX.
- The source code and models are publically available[1].

2 Related Work

While there are many studies on biomarker extraction from images [2] or multi-omics [1], limited works have been conducted on extracting biomarker information from textual pathology reports. They can be divided into three categories: clinical NLP Softwares [13,14], machine learning (ML) methods [5,7], and deep learning (DL) methods [4,16].

CLAMP [14] and Apache cTAKES [13] are clinical NLP software that extracts biomarker mentions from unstructured electronic health records, mostly relying on pre-defined dictionaries. Later studies adopt ML methods, such as support vector machines [5] and tree-based method [7], to consider contextual features so that the model can recognize biomarkers more comprehensively. However, they cannot differentiate instances where the same mention carries different meanings, i.e., *mention overlapping* cases. Besides, ML methods require complex feature engineering.

Recent works leverage DL methods to address the *mention overlapping* cases for biomarker extraction. The authors in [4] propose a hybrid method that first applies Long Short-Time Memory Neural Network (LSTM) to filter out the sentences without biomarkers then uses a dictionary-based module to perform exact-match and fuzzy-match on the sentences to identify biomarkers. LSTM can filter out some cases where the sentence doesn't contain any biomarkers (Case 1 in Table 1). However, the dictionary-based method restricts biomarker extraction to predefined terms, while the fuzzy-match strategy introduces noises. [16] pre-trains BERT model on 3 GB of clinical corpus in Chinese and fine-tunes it on 8,473 sentences from six types of clinical notes with annotation to extract various entities and attributes for breast cancer, including biomarkers and their results. Pre-training on large datasets enhances the model's ability to capture the contextual information of the biomarkers, making it possible to handle *mention overlapping* cases. However, this approach is not applicable to applications where only limited data is available. In this work, we leverage the BERT post-training technique to handle such situations.

Note that [16] is the only existing work that extracts results besides biomarkers. It formulates biomarker result extraction as the entity-relation extraction problem. It fine-tunes the pre-trained models to classify relations between all potential entity pairs in a sentence. Thus, biomarkers are linked with multiple entities, including those not indicating results. This approach has a high chance of associating multiple results to a biomarker in *mix-polarity* cases, and considering other medical entities as the results of a biomarker in *no-result* cases.

[1] https://github.com/NJIT-AI-in-Healthcare/Pathology-Biomarker-Information-Extraction.

In this paper, we analogize BIE with ABSA, a research topic that identifies aspects and their sentiments from product reviews [6,9,10,15]. Existing ABSA models use BERT models with fine-tuning [6,15] and Attention mechanism [10]. As discussed earlier, we adapt existing ABSA techniques for BIE, handling its unique challenges by post-training a LLM, the BioBERT model for limited data in medical domain, and using two attention models to handle intensive *mix-polarity* and *no-result* cases.

3 Methodology

Figure 1 gives an overview of the proposed system. It consists of two parts: Biomarker Extraction (BE) and Result Association (RA). Given a pathology report, the goal is to extract all biomarkers and the corresponding results from the report. A pathology report is composed of D sentences $[s_1, s_2, ...s_D]$, where a sentence s_i contains N tokens, $s_i = [w_{i1}, w_{i2}, ..., w_{iN}]$. In the BE module, we first post-train the BioBERT model to enhance the domain and task knowledge, then the post-trained model, BioX, is used for sequence labeling to extract cancer biomarkers $A_i = [a_{i1}, a_{i2}, ..., a_{iP}]$ from s_i, where P is the number of biomarkers in s_i. In the RA module, for each extracted biomarker a_{ip}, we identify the result c_{ip} associated with a_{ip} from s_i, where $c_{ip} \in \{Positive, Negative, None\}$. The associations between biomarkers and contextual words are learned by syntactic-based and semantic-based attention models. The result is predicted through the reweighted embeddings. As a result, for sentence s_i, we get the list of biomarkers and the corresponding results $[(a_{i1}, c_{i1}), (a_{i2}, c_{i2}), ..., (a_{iP}, c_{iP})]$.

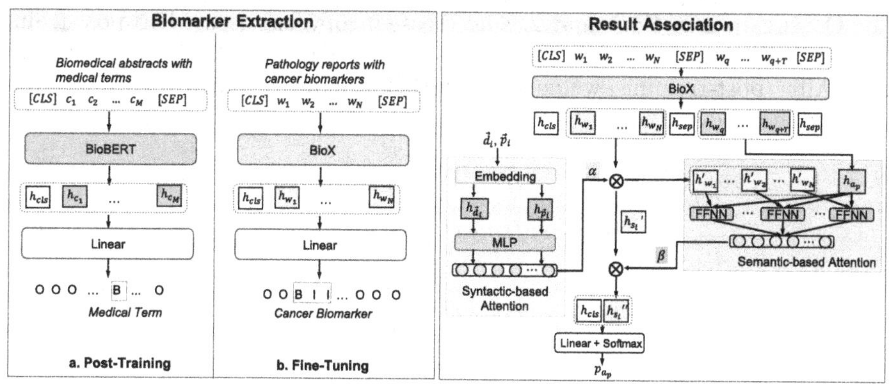

Fig. 1. An Overview of BioReX

3.1 Biomarker Extraction (BE)

Post-Training. In the face of the limited data which is insufficient to train the model for cancer biomarkers extraction, we leverage a large language model in medical domain, BioBERT [8]. BioBERT is trained using medical research papers and abstracts, with two objectives, masked language model and next sentence prediction. To enrich the model with domain and task knowledge, the BERT-PT [15] technique post-trains the BERT model using external data sources. Specifically, apart from pre-training and fine-tuning steps, a post-training step is added in between. The domain knowledge is learned from unsupervised in-domain data, while the task knowledge is learned from supervised but out-of-domain data.

In our work, we adopt the idea of post-training to improve the model's ability to capture the context of biomarkers in order to address the mention overlapping and limited data challenges. However, instead of enhancing the domain and task knowledge separately in different tasks, we post-train the BioBERT model through the task of medical term extraction using supervised in-domain data, so that both domain and task knowledge are enhanced in one task. Medical terms refer to medical concepts which include biomarkers.

The medical term extraction problem is formulated as a sequence labeling task, where each token in a sentence is labeled with {Begin, Inside, or Outside}. The continuous tokens labeled as B and followed by zero or more Is are considered as a medical term. For example, for input "dual targeting of HER2 and IGF-IR", the model generates output {O, O, O, B, O, B, I, I} where the first B corresponds to biomarker "HER2" and {B, I, I} corresponds to biomarker "IGF - IR".

Specifically, we feed sentences $C_j = [c_{j1}, c_{j2}, ..., c_{jM}]$ into BioBERT followed by a dense layer to get p_{C_j}, the probability of the tokens belonging to labels B, I, or O, as shown in Eq. 1 and 2. The cross-entropy loss is used to post-train BioBERT (Eq. 3), where y_c is the label of the token, $\hat{y}_c$ is the predicted token label. After post-training, we get the BioBERT model enhanced with domain and task knowledge, BioX.

$$z_j = [[CLS], c_{i1}, c_{i2}, ..., c_{iM}, [SEP]]; h_j = BioBERT(z_j) \tag{1}$$

$$p_{z_j} = softmax(W_1 \cdot h_j + b_1) \tag{2}$$

$$\mathcal{L}_{post-train}(y_c, \hat{y}_c) = -\sum_{j=1}^{M} y_{c_j} \log(\hat{y}_{c_j}) \tag{3}$$

Fine-Tuning for Biomarker Extraction. The biomarker extraction is conducted at the sentence level. Similar to medical term extraction, biomarker extraction is modeled as a sequence labeling task. We segment each pathology report into sentences using the clinical data processing toolkit, CLAMP [14]. Given the input sentence $s_i = [w_{i1}, w_{i2}, ..., w_{iN}]$, we convert it to the BERT format, then apply the post-trained BioX model to get the hidden representation

h_i, followed by a dense layer to get p_{x_i}, the probability of the tokens belonging to labels B, I, or O, as shown in Eq. 4 and 5.

$$x_i = [[CLS], w_{i1}, w_{i2}, ..., w_{iN}, [SEP]]; h_i = BioX(x_i) \tag{4}$$

$$p_{x_i} = softmax(W_2 \cdot h_i + b_2) \tag{5}$$

The cross-entropy loss is used to fine-tune the model on biomarker extraction. A sequence of continuous tokens labeled as B and followed by zero or more Is are considered as a biomarker.

3.2 Result Association (RA)

This module associates each extracted biomarker with a result (if any), handling the *mix-polarity* and *no-result* challenges discussed in Sect. 1. Let a_{ip} be one of the extracted biomarkers in sentence s_i, where the index span of the biomarker a_{ip} is $[q, ..., q+T]$. The result association is formulated as a classification problem. Given the sentence $s_i = [w_{i1}, ...w_{iq}, ..., w_{i(q+T)} ..., w_{iN}]$ and the biomarker $a_{ip} = [w_{iq}, ..., w_{i(q+T)}]$, we aim to classify a_{ip} into one of the three classes: Positive, Negative or None.

The model aims to adjust the contribution of context words to the target biomarker based on their syntactic and semantic features. We use syntactic dependency features since words denoting results are typically dependent on the corresponding biomarker in the grammar structure. The semantic features help the model identify the words indicating the results.

To achieve that, we build two attention modules, syntactic-based and semantic-based attention modules. The idea of using two attentions to adjust the contribution is inspired by an aspect sentiment classification model, Cabasc [10]. Cabasc adjusts word contributions based on sequential interactions and relative word position. In contrast, our proposed attention module learns the weights through syntactic and semantic features. Since true corresponding results are syntactic-dependent words of the target biomarker, the model leverages syntactic features to assign higher weights to the true results. Semantic features capture the interaction between words based on their semantic meanings. Integrating semantic features strengthens the connection between a biomarker and the words indicating results.

The architecture is shown in Fig. 1(Right). Given the sentence s_i and target biomarker a_{ip}, we format the input as Eq. 6. We use BioX to get word embeddings h_x and pooled embedding h_{cls}, and extract the word embeddings for sentence h_{s_i} and biomarker $h_{a_{ip}}$ based on their indices in the formatted input, as shown in Eq. 7 and 8. Then, syntactic-based and semantic-based attention modules are built to learn two attention weights to adjust the contributions of context words to the target biomarker.

$$x = [[CLS], w_{i1}, ..., w_{iN}, [SEP], w_{iq}, ..., w_{i(q+T)}, [SEP]] \tag{6}$$

$$h_{cls}, h_x = BioX(x) \tag{7}$$

$$h_{s_i} = h_x[1 : 1 + N]; h_{a_{ip}} = h_x[2 + N : 2 + N + T] \tag{8}$$

Syntactic-Based Attention Module. For the words in s_i, we incorporate both syntactic dependency types and POS (Part Of Speech) tags of the words as the input features, which are extracted by a dependency parser, CoreNLP [11]. To mitigate the impact of unrelated words, we focus solely on syntactic features associated with biomarkers. These features include words located within two hops in the syntactic dependency graph generated by CoreNLP.

Specifically, we build a dependency type vector $\boldsymbol{d_i} = \{d_{in}\}_{1<=n<=N}$ and POS tag vector $\boldsymbol{p_i} = \{p_{in}\}_{1<=n<=N}$. For the words that are not dependent on the biomarker, d_{in} and p_{in} are set to *<pad>*. While for the words that are dependent on the biomarker, d_{in} is the dependency type (e.g., *dep*, *amod*, *nusbj*, etc.) and p_{in} is the POS tag (e.g., *noun*, *adj*, *verb*, etc.). The dependency type embedding $\boldsymbol{h_{d_i}}$ and POS tag embedding $\boldsymbol{h_{p_i}}$ are obtained by a lookup table $f(*)$.

$$\boldsymbol{h_{d_i}} = f(\boldsymbol{d_i}); \boldsymbol{h_{p_i}} = f(\boldsymbol{p_i}) \tag{9}$$

To combine the impact of dependency types and POS tags, we concatenate $\boldsymbol{h_{d_i}}$ and $\boldsymbol{h_{p_i}}$ and feed it into an MLP layer followed by a Softmax function to get the syntactic-based attention weights $\boldsymbol{\alpha_i}$, as shown in Eq. 10 and 11. Finally, the word embeddings $\boldsymbol{h_{s_i}}$ are re-weighted by $\boldsymbol{\alpha_i}$ to get $\boldsymbol{h'_{s_i}}$ as shown in Eq. 12.

$$\boldsymbol{h_{dep}} = W_3(\boldsymbol{h_{d_i}}||\boldsymbol{h_{p_i}}) + b_3 \tag{10}$$

$$\boldsymbol{\alpha_i} = \frac{exp(\boldsymbol{h_{dep}})}{\sum_{n=1}^{N} exp(\boldsymbol{h_{dep}})} \tag{11}$$

$$\boldsymbol{h_{s_i}}' = \boldsymbol{\alpha_i} \odot \boldsymbol{h_{s_i}}; \boldsymbol{h_{s_i}}' = \{h'_{w_{i1}}, h'_{w_{i2}}, ..., h'_{w_{iN}}\} \tag{12}$$

Semantic-Based Attention Module. We adjust the word contributions based on the semantic features in this module. Specifically, we use the feed-forward neural network (FFNN) to score how important a word w_{in} is to the biomarker a_{ip}. The score c_{in} of w_{in} is calculated based on the context word embedding h'_{w_in} and the biomarker embedding $h_{a_{ip}}$, as shown in Eq. 13, where $h_{a_{ip}}$ is the averaged word embeddings of the words belonging to the target biomarker a_{ip}. Then the semantic-based attention weight β_{ij} of w_{ij} is calculated in Eq. 14.

$$c_{in} = W_4 tanh(W_5 h'_{w_in} + W_6 h_{a_{ip}} + b_4) \tag{13}$$

$$\beta_{ij} = \frac{exp(c_{ij})}{\sum_{n=1}^{N} exp(c_{in})} \tag{14}$$

The semantic-based attention weight for all words in the sentence is computed to obtain the attention weight vector $\boldsymbol{\beta_i}$. The weighted word embeddings $\boldsymbol{h_{s_i}}''$ is calculated by (Eq. 15). Finally, $\boldsymbol{h_{s_i}}''$ are concatenated with h_{cls} and fed into a dense layer to get the probability of the biomarker belonging to each result (Eq. 16).

$$\boldsymbol{h_{s_i}}'' = \boldsymbol{\beta_i} \odot \boldsymbol{h_{s_i}}' \tag{15}$$

$$\boldsymbol{p_{a_{ip}}} = softmax(W_7(\boldsymbol{h_{cls}}||\boldsymbol{h_{s_i}}'') + b_7) \tag{16}$$

We aim to minimize the cross-entropy loss between the true biomarker result $y_{a_{ip}}$ and the model's prediction $\hat{y}_{a_{ip}}$ in model training.

$$\mathcal{L}_{RA}(y_a, \hat{y}_a) = -\sum_{i=1}^{P} y_{a_{ip}} \log(\hat{y}_{a_{ip}}) \tag{17}$$

4 Experiments

4.1 Experiment Setting

Datasets. Rutgers Cancer Institute of New Jersey provides 995 de-identified pathology reports with 43,423 sentences in total, with an average of 15 words per sentence. The biomarkers and the corresponding results are annotated by medical professionals. There are 15% of sentences containing more than 3 biomarkers, and 41.8% biomarkers aren't associated with results. For post-training, we use the MedMentions dataset [12] for medical term extraction. MedMentions dataset contains about 4,000 abstracts manually annotated for UMLS entities by human experts. It has 47,722 sentences and 321,899 entities. Typically biomarker results are either stated explicitly as *positive* or *negative* in the report. Sometimes other indicators are used, for instance, *active*, *amplified*, *no loss of expression* indicate *positive*. We used a dictionary to identify *positive* or *negative* results and focus on the technical challenge of correctly associating results with corresponding biomarkers.

Setting. We adopt the hyper-parameter settings recommended by [8] for BioBERT. The maximum sequence length is set to 128 and the batch size for training and testing is 16. We exclude stopwords from the syntactic-dependent words of the biomarkers. The dimensions of the dependency type and POS tag embeddings are set to 300. We train the model for 2 epochs and use the Adam optimizer with a learning rate of 2e−5. We use five-fold cross-validation, compute the precision, recall, and F1 score for each fold, and present the result as the average over all the folds. All experiments are run on the computer with NVIDIA GeForce RTX 3090 Ti. The post-training, biomarker extraction, and result association take 64, 10, 7 min respectively.

Proposed Model and Its Variations. BioX denotes the model we proposed in Sect. 3.1 for biomarker extraction, which is also evaluated for RA. **BioX+SynAtt** and **BioX+SemAtt**, denote BioX with a syntactic-based attention or a semantic-based attention, respectively. **BioReX** refers to BioX with the syntactic and then semantic attention. We also tested the reverse order of combining the syntactic and semantic attention, which shows slightly inferior results and is omitted here.

Comparison Models. We compare the proposed model with the following baselines. **CBEx** [4] is a hybrid method that combines LSTM and a dictionary-based module. **cTAKES** [13] and **CLAMP** [14] are two clinical NLP software. **BioBERT** [8] is a pre-trained language model for the biomedical domain. We only evaluate CBEx, cTAKES, and CLAMP for BE as they don't have the function to identify results. BioBERT is evaluated for both BE and RA. The only existing work that extracts biomarker results [16] is a pre-trained model trained using non-English datasets and is not publicly available, and thus can't be compared with.

Table 3. Model Performance for BE and RA

Methods		Biomarker Extraction			Result Association		
		Precision	Recall	F1	Precision	Recall	F1
Baselines	**cTAKES**	0.622	0.144	0.224	–	–	–
	CLAMP	0.283	0.374	0.323	–	–	–
	CBEx	0.615	0.884	0.725	–	–	–
	BioBERT	0.916	0.922	0.918	0.927	0.919	0.920
Proposed Models	**BioX**	0.934	0.937	0.935	0.929	0.922	0.923
	BioX+SynAtt	–	–	–	0.945	0.940	0.946
	BioX+SemAtt	–	–	–	0.939	0.932	0.934
	BioReX	–	–	–	0.954	0.950	0.952

4.2 Results and Analysis

Table 3 shows the superior performance of the proposed method compared to the baselines. For BE, we can see clinical NLP tools, cTAKES and CLAMP have inferior performance, as they cannot recognize the non-standard formats of biomarkers. The hybrid model, CBEx, is much better than Clinical NLP tools as it is trained by LSTM to capture the context of biomarkers, but inferior to BioBERT and BioX. BioX outperforms BioBERT as the post-trained model is enhanced with the medical terms knowledge and the token classification tasks.

For RA, BioBERT and BioX have similar performances, indicating that post-training itself cannot improve result association performance. Syntactic-based or semantic-based attention module each introduces performance gain, and their combination further boosts the performance.

Case Studies. We demonstrate the proposed method's ability to handle the complex cases mentioned in Table 1 in this section.

Mention Overlapping Cases. Table 4 provides the extracted biomarkers for Case 1–3 in Table 1 by different models. Note that among three cases, only the "ER" in Case 2 is a biomarker. As we can see, cTAKES and CLAMP can't extract "ER" as a biomarker since they cannot identify the acronyms of biomarkers. CBEx correctly identifies "ER" as not a biomarker in Case 1. However, when non-biomarker "ER" occurs with other biomarkers (e.g., Case 3), CBEx still incorrectly considers it as a biomarker. BioBERT makes the correct extraction for the first two cases, but it tends to include other medical terms as biomarkers in Case 3. Only BioX makes the correct extraction for all three cases.

Mix-polarity and No-result Cases. Table 5 provides the identified biomarkers' results for Case 4–5 in Table 1. Case 4 is a complex mix-polarity case that contains six biomarkers where three biomarkers (e.g., CK20, WT1, and S100P) are adjacent to words indicating different results. The impacts of different results offset with each other, making BioBERT and BioX hard to make the correct association. For Case 5, HER-2 is not associated with any results, but both BioBERT and BioX assign other biomarkers' results to it. In contrast, BioReX made the correct association in both cases.

Table 4. Biomarker Extraction Case Study

	Ground Truth	**Extracted Biomarkers**				
		cTAKES	**CLAMP**	**CBEx**	**BioBERT**	**BioX**
Case 1	None	None	None	None	None	None
Case 2	ER, PR, RCC, CA19.9, PAX8, Vimentin	Vimentin	RCC, PAX8, Vimentin	ER, PR, RCC, CA19.9, PAX8, Vimentin	ER, PR, RCC, CA19.9, PAX8, Vimentin	ER, PR, RCC, CA19.9, PAX8, Vimentin
Case 3	P53	None	P53	ER, P53	ER, P53, pain, bloating	P53

Table 5. Mix-polarity and No-result Cases. Biomarkers with positive, negative or no results are shown in red, blue, brown, respectively.

	Case 4 (Mix-polarity)						Case 5 (No-result)		
	CK7	CK20	WT1	PAX8	S100P	GATA3	ER	PR	HER2
BioBERT	✓	✗	✗	✗	✗	✓	✓	✓	✗
BioX	✓	✗	✗	✓	✗	✓	✓	✓	✗
BioReX	✓	✓	✓	✓	✓	✓	✓	✓	✓

5 Conclusions

In this paper, we introduce a system BioReX for Biomarker Information Extraction (BIE) that extracts cancer biomarkers and their associated results. We draw parallels between Aspect-Based Sentiment Analysis (ABSA) and BIE, and propose to leverage the advances in ABSA to the relatively unexplored domain of BIE. We identify the challenges in BIE and adapt existing ABSA models to address these unique challenges. To tackle challenges of limited data and overlapping mentions, we use BERT post-training techniques to enhance BioBERT, a LLM model, with domain-specific and task-specific knowledge. To address challenges of *mix-polarity* and *no-result cases*, we introduce syntactic-based attention and semantic-based attention to adjust word contributions to the target biomarkers. Evaluation shows the effectiveness of BioReX. We plan to deploy BioReX in clinical data warehouses.

Acknowledgment. The work is partially supported by a grant from the National Institutes of Health (UL1TR003017), the Martin Tuchman'62 Chair Endowment, the Leir Foundation, and the National Science Foundation (CNS 2237328). We gratefully acknowledge Nancy Sazo, Huiqi Chu, and Evita Sadimin for medical knowledge support.

References

1. Dhillon, A., Singh, A., Bhalla, V.K.: A systematic review on biomarker identification for cancer diagnosis and prognosis in multi-omics: from computational needs to machine learning and deep learning. Arch. Comput. Meth. Eng. **30**(2), 917–949 (2023)
2. Echle, A., Rindtorff, N.T., Brinker, T.J., Luedde, T., Pearson, A.T., Kather, J.N.: Deep learning in cancer pathology: a new generation of clinical biomarkers. Br. J. Cancer **124**(4), 686–696 (2021)
3. Foran, D.J., et al.: Roadmap to a comprehensive clinical data warehouse for precision medicine applications in oncology. Cancer Inform. **16**, 1176935117694349 (2017)
4. Gao, X., et al.: CBEx: a hybrid approach for cancer biomarker extraction. In: BIBM, pp. 2958–2958. IEEE (2020)
5. Islam, M.T., Shaikh, M., Nayak, A., Ranganathan, S.: Extracting biomarker information applying natural language processing and machine learning. In: ICBBE, pp. 1–4. IEEE (2010)
6. Karimi, A., Rossi, L., Prati, A.: Adversarial training for aspect-based sentiment analysis with BERT. In: 2020 25th International Conference on Pattern Recognition (ICPR), pp. 8797–8803. IEEE (2021)
7. Lee, J., et al.: Automated extraction of biomarker information from pathology reports. BMC Med. Inform. Decis. Mak. **18**(1), 1–11 (2018)
8. Lee, J., et al.: BioBERT: a pre-trained biomedical language representation model for biomedical text mining. Bioinformatics **36**(4), 1234–1240 (2020)
9. Liu, H., Chatterjee, I., Zhou, M., Lu, X.S., Abusorrah, A.: Aspect-based sentiment analysis: a survey of deep learning methods. IEEE Trans. Comput. Soc. Syst. **7**(6), 1358–1375 (2020)

10. Liu, Q., Zhang, H., Zeng, Y., Huang, Z., Wu, Z.: Content attention model for aspect based sentiment analysis. In: WWW, pp. 1023–1032 (2018)
11. Manning, C.D., Surdeanu, M., Bauer, J., Finkel, J.R., Bethard, S., McClosky, D.: The stanford CoreNLP natural language processing toolkit. In: Proceedings of 52nd Annual Meeting of the Association for Computational Linguistics: System Demonstrations, pp. 55–60 (2014)
12. Mohan, S., Li, D.: MedMentions: a large biomedical corpus annotated with UMLS concepts. In: 1st Conference on Automated Knowledge Base Construction, AKBC 2019, Amherst, MA, USA, 20–22 May 2019 (2019). https://doi.org/10.24432/C5G59C
13. Savova, G.K., et al.: Mayo clinical text analysis and knowledge extraction system (cTAKES): architecture, component evaluation and applications. JAMIA **17**(5), 507–513 (2010)
14. Soysal, E., et al.: CLAMP-a toolkit for efficiently building customized clinical natural language processing pipelines. JAMIA **25**(3), 331–336 (2018)
15. Xu, H., Liu, B., Shu, L., Philip, S.Y.: BERT post-training for review reading comprehension and aspect-based sentiment analysis. In: ACL, pp. 2324–2335 (2019)
16. Zhang, X., et al.: Extracting comprehensive clinical information for breast cancer using deep learning methods. Int. J. Med. Informatics **132**, 103985 (2019)

IR Embedding Fairness Inspection via Contrastive Learning and Human-AI Collaborative Intelligence

Heng Huang[1], Yunhan Bai[1], Hongwei Liang[1], and Xiaozhong Liu[2](✉)

[1] Xiaohongshu, Shanghai, China
[2] Worcester Polytechnic Institute, Worcester, USA
xliu14@wpi.edu

Abstract. Xiaohongshu's search daily serves tens of millions active users in social networks, that presents a challenge to existing log-based embedding based retrieval (EBR) system: how to endorse individual document exposure fairness to diversify the search results. Conventional EBR models optimize relevance between query and document by leveraging massive user behavior data, e.g. clicks, purchase, etc., however, search log derived retrieval outcomes can deviate from true relevance distribution, that may result in less opportunity to retrieve for low-popularity or long-tailed documents. To address this problem, in this study, we propose a novel semi-supervised model, Gaussian process based contrastive learning (GPCL), which minimizes the discrepancy between model prediction distribution and true relevance distribution via taking advantage of contrastive samples adaptively generated from small human-labeled data. We validated the effectiveness of the proposed methodology by comparing with a set of baselines and observed significant metrics gains via online A/B testing. We discuss the entire system including model deployment and parameter-tuning. Also the new dataset is publicly available, which associates manually labeled relevance samples and massive click-logs.

Keywords: embedding based retrieval · collaborative contrastive learning · deep representative learning

1 Introduction

As a social networking platform, tens of millions users share their life stories on Xiaohongshu and our search engine helps people to access huge amount of information from billions of documents that Xiaohongshu users composed. Embedding-based retrieval (EBR) is a common technique in IR field and also developed in Xiaohongshu to offer relevant and diversified search results. In this paper we discuss the recent development and invention of our EBR model which improved metrics significantly.

D.-N. Yang et al. (Eds.): PAKDD 2024, LNAI 14648, pp. 142–153, 2024.
https://doi.org/10.1007/978-981-97-2238-9_11

Conventionally, general EBR methods utilize large scale random samples from implicit user feedback, such like clickthrough data, to fit 'user tastes', which is expected to approximate query-document relevance. However, user behavior data collected from long-tailed or low-popularity documents in social networks can be extremely sparse, that easily prejudice model's relevance judgements due to presentation bias as [13] explored. For instance, when users search *"COVID-19 prevention policy"*, because of user preference and topic popularity observed through search log, the EBR system may return popular documents focusing on *"nucleic acid testing"* and *"contact precaution"* topics, while we can expect 'mutual reinforcement' of topic exposure bias and ranking prejudice. Then, long-tailed topics, e.g., *"control measures for primary schools"* and *"supplies in quarantine area"* may take longer to popup in the search result. Such an excessive log-dependency EBR model can hardly satisfy individual exposure fairness for all relevant documents.

The erratic discrepancy between the distributions of biased document clicks and true relevance triggers EBR fairness concerns. To address this problem, we propose a **Gaussian process based contrastive learning model (GPCL)** via human-AI collaborative intelligence. GPCL can estimate the learning bias comparing with true relevance distribution by leveraging a small set of human-rated data, and produce Gaussian process modeled contrastive samples aiming at model weakness. The hypothetical intuition is that small human-labeled data can provide essential complements to the log-based EBR system to endorse retrieval trustworthiness, result diversification, and individual exposure fairness.

The contributions of our work can be concluded by following three-fold:

Application: To the best of our knowledge, this is a pioneer investigation to address individual fairness of embedding generation for EBR models with human-AI collaboration

Methodology: We propose a novel Gaussian Process based Contrastive Learning (GPCL) framework that can minimize the EBR-generated embedding bias leveraging non-parametric Gaussian method. Meanwhile, GPCL exhausts small human-labeled data utilization and adaptively produces corresponding contrastive samples via query reformulation.

Evaluation/Dataset: In order to validate the proposed model, we collect an innovative dataset from Xiaohongshu's search, with massive user click logs plus a small set of human labeled samples. Results show that the proposed model outperforms series of baselines. In order to motivate scholars further investigate this novel but important problem, we make dataset publicly available.

2 Related Works

With thriving of approximated nearest neighbor (ANN) algorithms [1,14,17], EBR system has witnessed significant advancement and are increasingly being deployed in practice. DSSM [11], as a learning-to-match paradigm, employs a siamese neural network with shared parameters to predict the click through rate

(CTR) for any given query-document pair, and takes the outputs of last DNN layer as the queries/documents embeddings. Many subsequent studies improve this framework via replace the fundamental text encoders with other NLP tools, such like CNN [21], LSTM [19], Transformer [26,27], etc. Based on the principle method of word2vec [16,18], list embedding [7] or graph embedding [8,22] are also alternative models for EBR system.

For these log-dependent approaches, negative sampling plays a big part towards data sparsity problem. Facebook's EBR model [10] deploys a strategy of negative sampling that produces numerous negatives via random sampling (from all the indexed documents) to enhance the embedding semantics. However these easy negatives merely carry informativeness to enhance the model regarding to relevance. In [5], authors presented MOBIUS that introduced a batch of external samples judged via an extra relevance discriminator. This approach offered important potential to surge negative samples from user behavior log, but it is vulnerable to the instability of the discriminator for judging the relevance of a sample. Enlightened by the success of contrastive learning in fields of CV and NLP, ANCE [28] provides global negatives constrained by being-optimized dense retrieval model rather than random or in-batch local negatives to sampling strategy, which leads to lower variance of the stochastic gradient estimation and faster learning convergence. However, relevance criteria are missed in this framework, which are vital to IR matching phase. All of the excessive log-dependent methods are easily dominated by ad-hoc user preferences, which often leads to document retrieval unfairness.

Prior efforts in fairness-aware IR methods can be generally classified into two categories: optimizing rank list for group fairness or individual fairness. The former methods [6,9,15,20,29] investigate fairness trustworthiness at the level of subject groups, like age, gender, race, etc., and preserve a certain proportion of impressions for the protected groups. FAIR-PG-RANK [24], for instance, formulated the fair-exposure policy into learning-to-rank (LTR) model with a trade-off between relevance and group fairness. Similarly, [2] offered a set of pair-wise metrics to measure the group unfairness for algorithmic rankings, and DELTR [30] was proposed as another LTR method to address potential group discrimination and inequality in rankings at training time. On the other side, eliminating individual unfairness is proved to improve group fairness in the work from [3], which proposed equity-of-attention ranking mechanism to achieve amortized fairness. [23] explored the individual fairness and estimated the position bias from probabilistic perspective. However, all those approaches make trade-off between utility and fairness in ranking based on relevance judgements, which fundamentally differentiate log-based EBR ecosystems.

In this paper, we present an innovative framework named as Gaussian process based contrastive learning (GPCL) that involves both large-scale unlabeled data from search log and human knowledge derived from a few manually labeled samples to eliminate the embedding unfairness. GPCL estimates the erratic discrepancy between two distributions from EBR model (biased document clicks) and true relevance (human assessment) with non-parametric Gaussian method and contributes informative contrastive samples to the model for reducing the deviation. To the best of our knowledge, this is a pioneer investigation to deploy Gaussian process to augment a finite set of human rated data as contrastive samples via query reformulation, which can enhance the log-dependent embedding model.

3 Methodology

3.1 Model Overview

In an EBR system, a query q^k carries a candidate set of documents $d^k = \{d_1^k, d_2^k, \cdots, d_n^k\}$ with a corresponding set of real-valued matching score as $s^k = \{s_1^k, s_2^k, \cdots, s_n^k\}$ and the matching phase can be considered as a ranking problem based on the scores. Our model is agnostic to how the matching score s is calculated for retrieving n-best items which are relevant to the query. The model is a two-towered neural network as Fig. 1, which contains three major components: feature encoder, fusion module and matching layer. The feature encoder has two parts: discrete feature encoder which is a 4-layer DNN and textual feature encoder which is a 4-layer BERT. In order to provide rich information containing both of textual and discrete features to the top matching layer, the outputs of each feature encoder are concatenated together and passed into a 4-layer DNN for reconciliation. And then the top matching layer calculates the similarity between the user-query Q and document D using inner-product. This typical two towered neural network can be readily trained via search click log with binary cross entropy and we demonstrate how GPCL works based on this fundamental architecture in next section.

3.2 GP Based Contrastive Learning

As aforementioned, click-based EBR can trigger individual exposure fairness and result diversity concerns. An intuitive solution is conducting human assessment over the search results with these queries, however, the cost of large scale annotation is prohibitive. To yield productive samples beyond human assessment, we introduce a novel semi-supervised contrastive learning framework based on a small set of human-rated query-document pairs. The key point of the approach is to estimate the model prediction deviation to the relevance distribution inferred from human-rated data and calculate MAP of query reformulation via Gaussian process which can adaptively produce contrastive samples. Let $< q, d >$ denote a labeled pair, and GPCL extends q to $\tilde{q}$ for exploiting more relevant documents $\tilde{d}$ and contributes massive contrastive pairs $\tilde{x}$ (like $(q, \tilde{d})$ or $(\tilde{q}, d)$) with soft label $\tilde{y}$ to the model. In other words, it maximizes the posterior probability to seek $\tilde{x}_m$ may cause maximum prediction deviation $\Delta\tilde{y}$:

$$\tilde{q}^* = argmax\, P(\tilde{q}|\tilde{x}_m, q) \tag{1}$$

Note that $\tilde{x}_m$ is only corresponding to $\tilde{q}$, hence Eq. 1 can be equal to:

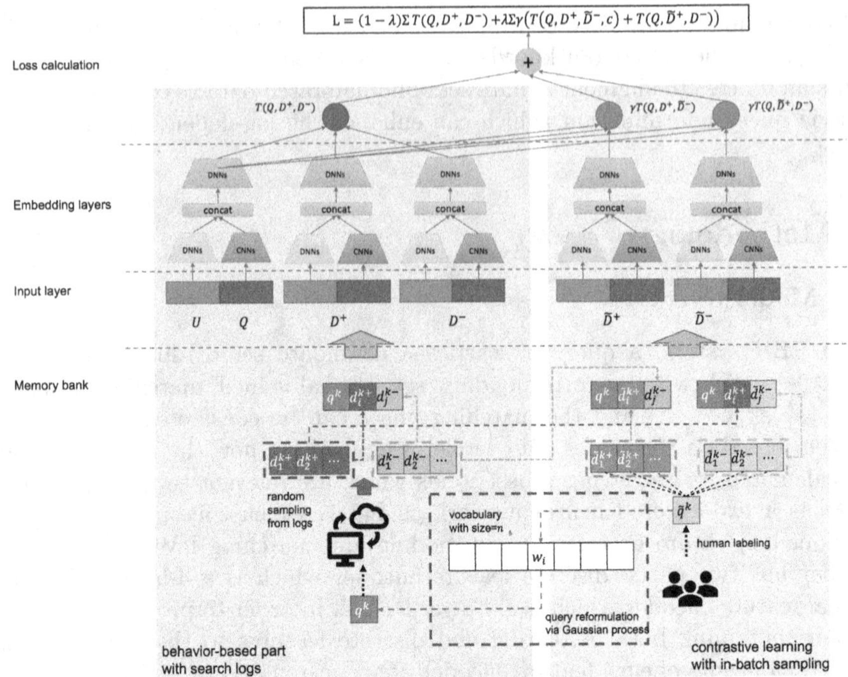

Fig. 1. The fundamental model is a typical two-towered neural network.

$$
\begin{aligned}
\tilde{q}^* &= argmax\, \frac{P(\tilde{x}_m|\tilde{q})P(\tilde{q}|q)P(q)}{P(\tilde{x}_m)} \\
&\propto argmax\, P(\tilde{x_m}|\tilde{q})P(\tilde{q}|q),
\end{aligned}
\tag{2}
$$

where $P(\tilde{x}_m)$ can be considered as a uniform distribution to omit and $P(q)$ is the observation of human-labeled data that can be omit in calculation too. $P(\tilde{q}|q)$ is the conditional distribution of query reformulation that can be generated via Gibbs sampling. Considering a query as multi-hot representation of tokens $q = \{w_0, w_1, \cdots, w_i\}$, $\tilde{q}$ can be generated via random sampling with the conditional probability $P(w_t|q)$ if a series of seed queries are given. Then relevant and irrelevant documents $\tilde{d}^{+/-}$ are sampled from click logs with the reformulated query $\tilde{q}$, and $\tilde{x}_m$ can be checked to obtain $P(\tilde{x}_m|\tilde{q})$ by statistics of these sampled contrastive pairs $(q, \tilde{d}^{+/-})$ or $(\tilde{q}, d^{+/-})$. Hence, searching for $\tilde{q}$ amounts to such a sampling process as follows:

$$
\tilde{q}^* = argmax(\frac{1}{N}\sum(f(\tilde{q}^k, d_i^k)) + \frac{1}{M}\sum(f(q^k, \tilde{d}_i^k))). \tag{3}
$$

Let $\Delta y = |y - y^*|$ denote the deviation between ground-truth y and model prediction y^*. To produce informative contrastive data, the augmentation strategy is required to sample more data if Δy becomes larger. Considering two implicit functions $y = r(x)$ and $y^* = m(x)$ to estimate the relevance label of any pair $x = (q, d)$ as human-assessment and model prediction respectively, for a new sampled pair $\tilde{x}$ we can infer $\tilde{y} = r(\tilde{x})$ and $\tilde{y}^* = m(\tilde{x})$ via Gaussian process. Moreover, $r(\tilde{x})$ and $m(\tilde{x})$ are sampled individually from different Gaussian distributions, hence, $\Delta\tilde{y} = f(\tilde{x}) = r(\tilde{x}) - m(\tilde{x})$ follows Gaussian distribution too. So, for a given data set $X \in \mathcal{R}^{m*n}$ and corresponding bias set $\Delta Y \in [0,1]^{m*1}$, GP samples a distribution function $\tilde{f} \sim \mathcal{N}(\mu, \Sigma)$ to generate the estimated bias $\Delta\tilde{y} = \tilde{f}(\tilde{x})$ for a given new pair $\tilde{x}$:

$$\begin{bmatrix} f(X) \\ \tilde{f}(\tilde{x}) \end{bmatrix} \sim \mathcal{N}\left(\begin{matrix} \mu(x) \\ \mu(\tilde{x}) \end{matrix}, \begin{bmatrix} k(X,X) & k(X,\tilde{x}) \\ k(\tilde{x},X) & k(\tilde{x},\tilde{x}) \end{bmatrix} \right) \tag{4}$$

where $k(\cdot)$ is kernel function and we choose squared exponential kernel as follows:

$$k(x_i, x_j) = \exp(-\frac{1}{2\theta^2} \left\| x_i - x_j \right\|^2) \tag{5}$$

Note that θ is a smoothing hyper-parameter and set as 1.0. A new sample $\tilde{x}$ is selected according to its Upper Confidence Bound (UCB) based on the distribution $\tilde{f}(\tilde{x})$, which can be calculated as follows:

$$\tilde{x} = argmax(GP\text{-}UCB(\tilde{x})) = argmax(\mu(\tilde{x}) + \kappa\delta(\tilde{x})), \tag{6}$$

where μ and δ are *mean* and *std* which are both sampled from GP. Note that κ is generally set as 1.0. Then, the model minimize the loss with soft label $\tilde{y}$ as follows:

$$L(S(\tilde{x})) = -(\tilde{y} \cdot \log S(\tilde{x}) + (1 - \tilde{y}) \cdot \log (1 - S(\tilde{x}))) \tag{7}$$

If $\Delta\tilde{y} > max(\Delta Y)$, the augmented data $\tilde{x}$ and observation $\Delta\tilde{y}$ can be preserved to update the existing X and ΔY for next iterations.

In order to reduce computation complexity, we propose two heuristic strategies when Gibbs sampling is performed rather than search entire query distribution. The two rules are simply effective to make contrastive pairs $\tilde{X}$ as follows:

Query Expansion (QE) is applied for yielding contrastive negative pairs. QE operation expands query q to q_e via adding tokens, so that q_e is more specific than q. Then we sample negative documents $\tilde{d}^-$ from search logs and make contrastive negative pairs $(q, \tilde{d}^-)$ or (q_e, d^-).

Query Reduction (QR) is to produce contrastive positive pairs. QR operation reduces tokens from original query q as q_r that leads to less informativeness for q_r. And the contrastive positive pairs can be formed as $(q, \tilde{d}^+)$ or (q_r, d^+).

The entire sampling algorithm is showed as Algorithm 1. Noteworthy, GPCL estimates the model prediction deviation towards real distribution as the accumulation of Δy, which can exactly adjust the coefficient λ to control rectification intensity learnt from true relevance labeled data. Hence λ can be calculated as follows:

$$\lambda = \frac{\frac{1}{N}\sum \Delta y}{(\frac{1}{N}\sum \Delta y + \frac{1}{M}\sum \Delta y_u)}, \tag{8}$$

Algorithm 1. Gaussian Process to Produce Contrastive Samples

Input: human-labeled data X
Output: $\tilde{X}$ as contrastive samples
set $\tilde{X} = \emptyset$, $\Delta Y = F(X)$, $t = 0$, $N = 4$, $K = 8$
initialize query seeds $\{q_i\}$ in descending order by Δy_i
while $t < t_{max}$ && $size(\tilde{X}) < batch\_size$ **do**
    sample N $\tilde{q}$ based on q_t via constrained Gibbs sampling
    sample K $\tilde{x}$ based on $\{\tilde{q}_i\}$ in descending order by GP-UCB
    update: $\tilde{x} \rightarrow \tilde{X}$
    if $\Delta\tilde{y} > max(\Delta Y)$ **then**
        update: $\tilde{x} \rightarrow X$, $\Delta\tilde{y} \rightarrow \Delta Y$
        update: $\mu(x)$, $\Sigma(x)$
    end if
    $t = t + 1$
end while

where N/M is the sample number from human-labeled data/search logs, and Δy_u is the prediction error for user behavior samples. The GP based contrastive learning can be invoked every K-iteration to accelerate training process and simultaneously memory bank updates the representations for all the samples.

In sum, GPCL involves and augments external knowledge from human assessment that differs from user click behaviors, which benefits the model to for matching unevenly distributed samples. The pseudo-code of entire training process is clarified as Algorithm 2.

4 Experiments and Deployment

4.1 Implementation

The fundamental model is showed as Fig. 1. We employ a 4-layer BERT with 312-d hidden size and 12-head self-attention as textual encoder and a 4-layer DNN as discrete feature encoder with 128/128/64/64 neural cells in every layer. The outputs of both encoders are 64-d and concatenated together to pass a 4-layer DNN with 128/128/64/64 neural cells. For online service, ANN search engine is deployed to index document embeddings using IVF-PQ algorithm [12,25] with 32-byte quantization.

4.2 Offline Evaluation

We created a new dataset to evaluate EBR-based ranking fairness, and made it publicly available[1]. The dataset consists two parts: *DataClick* and *DataHuman*. *DataClick* provides large samples from query logs of Xiaohongshu's search engine. *DataHuman* is conducted via the query-document relevance assessment from human experts. The documents are classified into three different categories

[1] Please contact any author to access the data.

Algorithm 2. Gaussian Process based Contrastive Learning

Input: an embedding model with learnable parameter θ_t;
a set of human-labeled samples Ω;
a set of random samples from user click behaviors Φ;
the hyper-parameters: t_{max}, α, K, ϵ, m;
Output: optimized θ_t with contrastive learning
set $t = 1$, $\lambda_0 = 0$, $g = 0$; randomize θ_0
initialize memory bank with θ_0
while $t < t_{max}$ && L is not converged **do**
  sample X_u from Φ
  for each $(q_u, d_u) \in X_u$ **do**
    calculate L_u
  end for
  if t % K == 0 **then**
    update memory bank
    sample X from Ω
    sample $\tilde{X} \in \Phi$ based on X by Algo.1
    calculate λ as eq.8
    for each pair $(\tilde{q}, \tilde{d}) \in \tilde{X}$ **do**
      make contrastive pairs: $(\tilde{q}, d), (q, \tilde{d})$
      calculate L_c as eq.7
      if $\Delta\tilde{y} > max(\Delta Y)$ **then**
        update observation: $\tilde{x}, \Delta\tilde{y} \rightarrow \Omega$
      end if
    end for
  end if
  calculate loss: $\mathcal{L} = (1 - \lambda)L_u + \lambda L_c$
  backward propagation
  update θ_t
  $t = t + 1$
end while

clicked documents (user may be interested), exposed but un-clicked documents (user may not be interested) and random documents. Table 1 reports the dataset statistics.

We examine the recall results produced by our model on test sets extracted from *DataClick* and *DataHuman* for 3 aspects: relevance, diversity and utility.

1) *Relevance* to the query is the fundamental requirement for the search quality and the accuracy is presented as AUC_{rel} which is performed on a human-assessed test set.
2) *Diversity* of the retrieved items is also important to EBR system since it reflects the fairness with respect to the exposure of different categories or topics, which is even harder for representative learning. We follow [4] to employ an indicator as the measurement:

$$Div = NDCG(L) \cdot AWRF(L) \tag{9}$$

Table 1. Scale of each dataset.

dataset	clicked	un-clicked	random
DataClick	71M	71M	2272M
DataHuman	88000	22000	NULL

Table 2. Evaluation with precision metrics. Superscript * denotes $p < 0.05$ under t-test.

methods	Relevance	Diversity	Utility		
	AUC_{rel}	Div	AUC_{clk}	$NDCG$	$RECALL@1K$
Base_{clk}	0.739	0.0921	0.764	0.053	0.606
$\text{Base}_{clk+rel}$	0.748	0.0923	0.761	0.052	0.596
+DA	0.763	0.0914	0.762	0.053	0.59
+AL	0.748	0.0915	0.767	0.06	0.619
+ANCE	0.741	0.0920	0.768	0.065	0.637
+FAIR-PG	0.703	0.0924	0.731	0.049	0.563
+GPCL	**0.788***	0.0924	**0.812***	**0.072***	**0.651***

3) *Utility* of the search results can be measured via series of click-based metrics, i.e., $NDCG@10$ for 10 nearest neighbors retrieved from ANN engine. AUC_{clk} and $RECALL@1000$ are also reported in our experiments. AUC_{clk} is performed on the test set which is sampled from $DATA_{clk}$ and $RECALL@1000$ exhibits the proportion of clicked documents in top-1000 of the nearest neighbors as mentioned in [12].

To prove the effectiveness of relevance based contrastive learning with collaborative intelligence, we set a collection of contrast methods referring to 4 aspects: data augmentation, active learning, contrastive learning, and fair exposure. All of the contrast methods are built on the fundamental funnel-shaped neural network as presented in Sect. 3.2 which had been serving online before GPCL was launched. The base model is also tested as a baseline.

1) *Data Augmentation.* Assessing the relevance of q-d pairs by human is expensive and time-consuming, hence, data augmentation is a regular method to expand training set automatically in machine/deep learning. In this test, the manually-labeled data are augmented by the text reformulation and mixed with the log-based samples to feed the neural model.
2) *Active Learning.* As the fundamental model is a representative learning framework, we set a threshold τ to judge the relevance for given query-document pair using their cosine similarity, and consequently non-relevant documents based on this judgment can be excluded when sampling from *DataClick*. In this test τ is set to be 0.7 which performs best.
3) *Contrastive Learning.* ANCE [28] is well known as a contrastive learning method in IR field and its core task is to learn from global negatives which

can be easily transferred to the other neural models. We construct an asynchronously updated ANN index for the global negatives and make the fundamental model learn the contrastive loss as ANCE.

4) *Fair exposure*: Addressing individual unfairness for EBR model is a brand new goal that differs from prior studies. To the best of our knowledge, Fair-PG-Rank [24] is the most similar work to GPCL since it is an in-process method and can optimize both group and individual fairness. Hence Fair-PG-Rank is employed as the baseline for fairness comparison, and we tailored it as an EBR method by utilizing the disparity loss D as an additional loss in the basic EBR model.

Table 2 reports the experimental results along with different metrics and baselines. The proposed GPCL outperforms other baselines on all sides. Note that, as the disparity loss from FAIR-PG-RANK participates in model training, the diversity index Div is obviously improved while other metrics decrease a lot. The rationale is that disparity loss makes a trade-off between utility and diversity. In sum, GPCL achieves the convincing performance of relevance and utility, and performs equally with the FAIR-PG-RANK in aspect of individual fairness (diversity). Therefore we can conclude that GPCL can effectively enhance EBR system in terms of relevance, fairness and utility.

4.3 Online A/B Testing

As GPCL had successfully passed offline evaluations, we launch it on Xiaohongshu's search and conduct online A/B testing between previous EBR model and GPCL with same settings for ANN search engine. Table 3 reports the improvements for CTR (Click-Through-Rate), CES (an aggregative indicator in our business) and CRR (Customer-Retention-Rate) according to our monitor on 7-day online traffic.

Table 3. Improvements in online A/B testing

method	CTR	CES	CRR
Base+GPCL	+1.58%	+2.16%	+0.1%

5 Conclusion

In this paper, we investigate the effectiveness of human-in-the-loop contrastive learning for de-biased EBR representation learning. Comparing with existing data augmentation and contrastive learning methods, the proposed GPCL can effectively estimate the discrepancy of two distributions from model prediction and true relevance with non-parametric Gaussian method. Moreover, this framework can be injected into various EBR models to generate de-biased representations, and enable effective and low-cost human-AI collaboration. Extensive experiments, with new dataset, validate the effectiveness and superiority of the proposed method. To motivate other scholars further investigate this novel but important problem, we make our dataset (search log plus human labeled query-document pairs with relevance assessment) publicly available.

References

1. Aumüller, M., Bernhardsson, E., Faithfull, A.: ANN-benchmarks: a benchmarking tool for approximate nearest neighbor algorithms. In: Beecks, C., Borutta, F., Kröger, P., Seidl, T. (eds.) SISAP 2017. LNCS, vol. 10649, pp. 34–49. Springer, Cham (2017). https://doi.org/10.1007/978-3-319-68474-1_3
2. Beutel, A., et al.: Fairness in recommendation ranking through pairwise comparisons. In: Proceedings of the 25th ACM SIGKDD International Conference on Knowledge Discovery & Data Mining, pp. 2212–2220 (2019)
3. Biega, A.J., Gummadi, K.P., Weikum, G.: Equity of attention: amortizing individual fairness in rankings. In: The 41st International ACM SIGIR Conference on Research & Development in Information Retrieval, pp. 405–414 (2018)
4. Ekstrand, M.D., McDonald, G., Raj, A., Johnson, I., Warncke-Wang, M.: TREC 2021 fair ranking track participant instructions (2021)
5. Fan, M., Guo, J., Zhu, S., Miao, S., Sun, M., Li, P.: MOBIUS: towards the next generation of query-ad matching in Baidu's sponsored search. In: Proceedings of the 25th ACM SIGKDD International Conference on Knowledge Discovery & Data Mining, pp. 2509–2517 (2019)
6. Feldman, M., Friedler, S.A., Moeller, J., Scheidegger, C., Venkatasubramanian, S.: Certifying and removing disparate impact. In: Proceedings of the 21th ACM SIGKDD International Conference on Knowledge Discovery and Data Mining, pp. 259–268 (2015)
7. Grbovic, M., Cheng, H.: Real-time personalization using embeddings for search ranking at Airbnb. In: Proceedings of the 24th ACM SIGKDD International Conference on Knowledge Discovery & Data Mining, pp. 311–320 (2018)
8. Grover, A., Leskovec, J.: node2vec: scalable feature learning for networks. In: Proceedings of the 22nd ACM SIGKDD International Conference on Knowledge Discovery and Data Mining, pp. 855–864 (2016)
9. Hardt, M., Price, E., Srebro, N.: Equality of opportunity in supervised learning. In: Advances in Neural Information Processing Systems, vol. 29 (2016)
10. Huang, J.T., et al.: Embedding-based retrieval in Facebook search. In: Proceedings of the 26th ACM SIGKDD International Conference on Knowledge Discovery & Data Mining, pp. 2553–2561 (2020)
11. Huang, P.S., He, X., Gao, J., Deng, L., Acero, A., Heck, L.: Learning deep structured semantic models for web search using clickthrough data. In: Proceedings of the 22nd ACM International Conference on Information & Knowledge Management, pp. 2333–2338 (2013)
12. Jegou, H., Douze, M., Schmid, C.: Product quantization for nearest neighbor search. IEEE Trans. Pattern Anal. Mach. Intell. **33**(1), 117–128 (2010)
13. Joachims, T., et al.: Evaluating retrieval performance using clickthrough data (2003)

14. Johnson, J., Douze, M., Jégou, H.: Billion-scale similarity search with GPUs. IEEE Trans. Big Data **7**, 535–547 (2019)
15. Kamishima, T., Akaho, S., Asoh, H., Sakuma, J.: Fairness-aware classifier with prejudice remover regularizer. In: Flach, P.A., De Bie, T., Cristianini, N. (eds.) ECML PKDD 2012. LNCS (LNAI), vol. 7524, pp. 35–50. Springer, Heidelberg (2012). https://doi.org/10.1007/978-3-642-33486-3_3
16. Le, Q., Mikolov, T.: Distributed representations of sentences and documents. In: International Conference on Machine Learning, pp. 1188–1196. PMLR (2014)
17. Malkov, Y.A., Yashunin, D.A.: Efficient and robust approximate nearest neighbor search using hierarchical navigable small world graphs. IEEE Trans. Pattern Anal. Mach. Intell. **42**(4), 824–836 (2018)
18. Mikolov, T., Chen, K., Corrado, G., Dean, J.: Efficient estimation of word representations in vector space. arXiv preprint arXiv:1301.3781 (2013)
19. Palangi, H., et al.: Semantic modelling with long-short-term memory for information retrieval. arXiv preprint arXiv:1412.6629 (2014)
20. Pedreshi, D., Ruggieri, S., Turini, F.: Discrimination-aware data mining. In: Proceedings of the 14th ACM SIGKDD International Conference on Knowledge Discovery and Data Mining, pp. 560–568 (2008)
21. Shen, Y., He, X., Gao, J., Deng, L., Mesnil, G.: A latent semantic model with convolutional-pooling structure for information retrieval. In: Proceedings of the 23rd ACM International Conference on Conference on Information and Knowledge Management, pp. 101–110 (2014)
22. ACM SIG-KDD: DeepWalk: online learning of social representations (2014)
23. Singh, A., Joachims, T.: Fairness of exposure in rankings. In: Proceedings of the 24th ACM SIGKDD International Conference on Knowledge Discovery & Data Mining, pp. 2219–2228 (2018)
24. Singh, A., Joachims, T.: Policy learning for fairness in ranking. arXiv preprint arXiv:1902.04056 (2019)
25. Sivic, J., Zisserman, A.: Video google: a text retrieval approach to object matching in videos. In: IEEE International Conference on Computer Vision, vol. 3, pp. 1470–1470. IEEE Computer Society (2003)
26. Sun, F., et al.: BERT4Rec: sequential recommendation with bidirectional encoder representations from transformer, pp. 1441–1450 (2019)
27. Vaswani, A., et al.: Attention is all you need. arXiv preprint arXiv:1706.03762 (2017)
28. Xiong, L., et al.: Approximate nearest neighbor negative contrastive learning for dense text retrieval. arXiv preprint arXiv:2007.00808 (2020)
29. Zafar, M.B., Valera, I., Gomez Rodriguez, M., Gummadi, K.P.: Fairness beyond disparate treatment & disparate impact: learning classification without disparate mistreatment. In: Proceedings of the 26th International Conference on World Wide Web, pp. 1171–1180 (2017)
30. Zehlike, M., Castillo, C.: Reducing disparate exposure in ranking: a learning to rank approach. In: Proceedings of The Web Conference 2020, pp. 2849–2855 (2020)

SemPool: Simple, Robust, and Interpretable KG Pooling for Enhancing Language Models

Costas Mavromatis[1(✉)], Petros Karypis[2], and George Karypis[1]

[1] University of Minnesota, Minneapolis, USA
{mavro016,karypis}@umn.edu

[2] University of California San Diego, La Jolla, USA
pkarypis@ucsd.edu

Abstract. Knowledge Graph (KG) powered question answering (QA) performs complex reasoning over language semantics as well as knowledge facts. Graph Neural Networks (GNNs) learn to aggregate information from the underlying KG, which is combined with Language Models (LMs) for effective reasoning with the given question. However, GNN-based methods for QA rely on the graph information of the candidate answer nodes, which limits their effectiveness in more challenging settings where critical answer information is not included in the KG. We propose a simple graph pooling approach that learns useful semantics of the KG that can aid the LM's reasoning and that its effectiveness is robust under graph perturbations. Our method, termed SemPool, represents KG facts with pre-trained LMs, learns to aggregate their semantic information, and fuses it at different layers of the LM. Our experimental results show that SemPool outperforms state-of-the-art GNN-based methods by 2.27% accuracy points on average when answer information is missing from the KG. In addition, SemPool offers interpretability on what type of graph information is fused at different LM layers.

1 Introduction

Question answering (QA) is a complex reasoning task that requires understanding of a given natural language query, as well as domain-specific knowledge. For instance, answering biomedical questions requires understanding of biomedical terms as well as knowledge about biomedicine. Language models (LMs) [3,22] are pre-trained on large corpora to understand their underlying semantics. Thus, fine-tuning LMs for the given reasoning tasks [9,14] is the dominant paradigm in NLP for QA.

Despite their success, LMs struggle on intensive reasoning tasks that require good in-domain knowledge [15]. As a result, recent methods incorporate Knowledge Graphs (KGs) during the QA task [4,19], which are graphs that capture factual knowledge explicitly as triplets. Each triplet consists of two entities and their corresponding relation. Most successful KG-based methods [34] leverage

D.-N. Yang et al. (Eds.): PAKDD 2024, LNAI 14648, pp. 154–166, 2024.
https://doi.org/10.1007/978-981-97-2238-9_12

Graph Neural Networks (GNNs) [28], which have shown remarkable performance at reasoning tasks with graph information [17,37].

Nevertheless, GNNs operate on graph data while LMs use natural language sequences, which makes information exchange between the two modalities challenging. In fact, our empirical findings (Sect. 4) suggest that GNNs mainly provide graph statistical information for the QA task [29] rather than information that grounds the LM's reasoning and is robust under graph perturbations. In addition, the representation space mismatch between graph (KGs are usually represented with external node embeddings) and language (represented with pre-trained LMs) does not aid the information exchange between the two modalities.

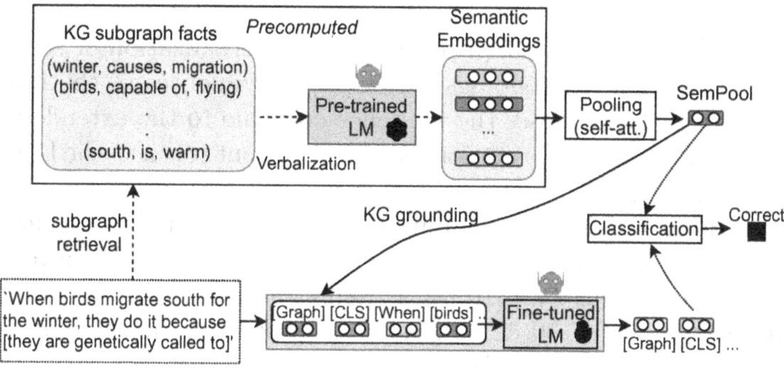

Fig. 1. Our SemPool method performs simple graph pooling to enhance the LM's reasoning. Facts of the KG are represented by their semantic information with pre-trained LMs. SemPool aggregates the graph's semantic information into a single representation that is fed into the LM for QA.

In this work, we present SemPool, a simple graph pooling method that enhances the LM's reasoning with KG textual information. As illustrated in Fig. 1, SemPool represents each fact in the KG with the pre-trained LM, aiming at semantic alignment between graph and language. SemPool then performs a global graph pooling operation in order to aggregate semantic information from the whole graph into a single representation. The aggregated representation is fused as input into the fine-tuned LM for QA, which grounds the LM's reasoning to the information provided. Moreover, we extend SemPool to fuse different type of semantic information into different LM's layers (Sect. 5.3), providing more flexibility during learning.

SemPool demonstrates robust performance under different settings. We experiment with standard QA benchmarks (OpenbookQA, RiddleSense, MedQA-USMLE), (i) when complemented by complete in-domain KGs, and (ii) when complemented by in-domain KGs where critical information about the candidate answers is missing. SemPool outperforms the *best* performing GNN-based approach by 2.27% accuracy points in the challenging case, while it is competitive (second-best) in the easier case. In addition, our experiments show that

SemPool is effective under different LMs (Sect. 7.1), highlight the importance of semantic alignment between language and graph, and illustrate SemPool's interpretability (Sect. 7.2).

2 Related Work

Question Answering with KGs. Many KGs have been employed to improve QA for different domains, such as ConceptNet [24] for commonsense QA. Graph neural networks [23,28] have been widely used to combine KG information with language models [4,10,25,26,34,35] leading to SOTA QA systems. In this work, we give new insights on the sensitivity of GNN-based methods with respect to the provided graph and propose a simple approach that improves robustness for QA. Other methods have explored to provide the verbalization of the retrieved KG facts [1,31] or their embeddings [21] as input sequences to the LM, which, however, considerably increases the inference cost due to the extended context. SemPool integrates graph information as a special input token to the LM, offering low computational cost.

Graph-Augmented LMs. Combining LMs with graphs that include textual features is an emerging research area [13,20]. Recent methods have explored fine-tuning LMs on graph data [16,36] as well as aiding the pre-training of LMs with graph information [7,30,32,33]. SemPool provides a new simple approach to fuse graph information into the LM, easily integrated to existing approaches.

3 Problem Statement and Preliminaries

Multi-choice QA. We study the problem of multiple-choice question answering (QA), where given an optional context c, a question q, and a set of candidate answers $\mathcal{A}$, the goal is to select the correct answer $a^* \in \mathcal{A}$. Multiple-choice QA is transformed into a classification problem by (i) concatenating the question's context with each of the candidate answer $a \in \mathcal{A}$ into a statement q_a, e.g., $q_a = [c, q, a]$ = "*When birds migrate south for the winter, they do it because (A) they are genetically called to*", and (ii) selecting the most probable statement from $\{[c, q, a] : a \in \mathcal{A}\}$. Given each input $[c, q, a]$, a fine-tuned LM is used to determine whether the textual input is plausible. The output token representations $\boldsymbol{q}^{(L)} = \text{LM}([c, q, a])$ are used for classifying the statement as the correct one, usually followed by a pooling operation. We point the readers for more details to the corresponding papers [3,14].

Knowledge Graphs (KGs). Knowledge Graph (KG) powered QA aims at leveraging external factual information from a KG in order to improve the LM's reasoning ability. For instance, the KG might contain the factual information (*winter, causes, bird migration*), which is relevant for the question q = "*When birds migrate south for the winter, they do it because?*". Formally, KG is a multi-relational graph $\mathcal{G} := (\mathcal{V}, \mathcal{E})$ that contains a set nodes (entities) $\mathcal{V}$ and a set of edges (facts) $\mathcal{E}$. Set $\mathcal{E} \subseteq \mathcal{V} \times \mathcal{R} \times \mathcal{V}$ contains facts in the tuple form (h, r, t), where $h, t \in \mathcal{V}$ and $r \in \mathcal{R}$ is the relation between nodes h and t ($\mathcal{R}$ denotes the relation set).

Subgraph Retrieval. For each statement $[c, q, a]$, a subgraph $\mathcal{G}_{q_a} \subseteq \mathcal{G}$ is retrieved based on the input's context, which may include nodes that correspond to question entities or answer entities. For example, [34] performs entity linking between the question's and the KG's entities, and extracts the two-hop nodes between the linked entities, filtering out question-irrelevant edges. The context-specific subgraph $\mathcal{G}_{q_a} = (\mathcal{V}_{q_a}, \mathcal{E}_{q_a})$ contains a set of nodes $\mathcal{V}_{q_a}$, the set of relations $\mathcal{R}$ and a set of facts $\mathcal{E}_{q_a} \subseteq \mathcal{V}_{q_a} \times \mathcal{R} \times \mathcal{V}_{q_a}$. Note that different candidate answers $a \in \mathcal{A}$ for question q lead to different subgraphs $\mathcal{G}_{q_a}$ as the context changes. Similar to previous works [35], a virtual question node is added to each subgraph and is linked to question and answer entities, by adding *(question, entity, birds)* and *(question, a_entity, children)* to the edge set $\mathcal{E}_{q_a}$, for example.

Graph Neural Networks (GNNs). GNNs learn to update the representation of a node v by aggregating representations of its neighbors, set $\mathcal{N}(v)$, in a recursive manner. Following the message passing strategy [5], GNNs update the representation $\boldsymbol{h}_v^{(l)}$ of node v at layer l as

$$\boldsymbol{h}_v^{(l)} = \psi\Big(\boldsymbol{h}_v^{(l-1)}, \phi(\{\boldsymbol{m}_{(v,v')}^{(l)} : v' \in \mathcal{N}(v)\})\Big), \tag{1}$$

where $\boldsymbol{m}_{(v,v')}$ is the message between two entities v and v', linked with a relation r, and depends on their corresponding representations. Function $\phi(\cdot)$ is an aggregation, e.g., sum, of all neighboring messages and function $\psi(\cdot)$ is a neural network. In order to enable language to graph information fusion, many QA GNN-based approaches [34,35] set the question node's embedding to the question representation obtained by the LM.

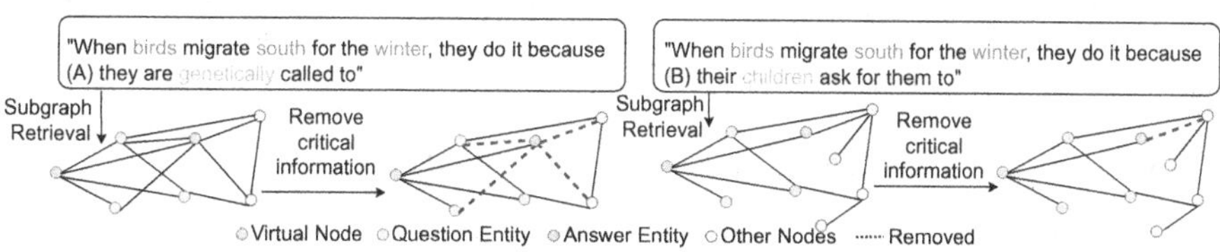

Fig. 2. Setting when critical answer information is removed from the KG. Originally, GNNs propagate information from the answer nodes (light-red color) to other nodes of the graph, and the candidate answer with more links is more likely to be the correct answer [29]. If we remove the answer node's edges, information propagation becomes challenging and GNNs struggle to discriminate between correct and incorrect answers (Sect. 4). (Color figure online)

4 Empirical Findings on Robustness

As discussed in Sect. 3, GNNs leverage the graph information of the retrieved KG to update the node embeddings. However, current QA GNN-based approaches use external node embeddings to represent the nodes' information. The representation space of these embeddings is not necessarily aligned with the representation space of the LM, which limits the effectiveness of fusing semantic information between natural language and graph. For example, [21,29] show that replacing the node embeddings with simple node features, such as node types (node coloring in Fig. 2), leads to better QA performance. These findings indicate that GNNs rely on the underlying graph statistics, e.g., the number of connections between answer nodes and other graph nodes [29], to discriminate between correct and incorrect answers.

To test our hypothesis, we experiment with a setting for multiple-choice QA, where critical information is missing from the KG. For each retrieved KG subgraph, we remove the facts (edges) that include the candidate answer a, similar to the case where answer entities are not linked in the graph. The setting becomes challenging as GNNs cannot easily propagate answer-specific information and need to leverage information about the remaining entities to improve the LM's reasoning. The studied setting is illustrated in Fig. 2.

We present the results for OBQA and RiddleQA datasets in Table 1. When removing answer information from the KG, GNNs show significant performance degradation and cannot effectively discriminate between correct and incorrect answers. The relative performance degradation is up to 4.8% for OBQA and using the external KG improves over the LM (w/o KG) by only 0.2% accuracy points. The experiment suggests that the message passing of GNNs learns to propagate answer-specific information, depends on the connectivity of the answer nodes, and is limited when this information is not present or is removed from the graph.

Table 1. QA performance comparison when complete information about candidate answers are in the graph (w/ ans) and when their edges are removed (w/o ans). Δ_{acc} denotes the relative performance degradation.

	OBQA			RiddleQA		
	w/ ans	w/o ans	Δ_{acc}	w/ ans	w/o ans	Δ_{acc}
LM (w/o KG)	64.8	64.8	0.0	60.7	60.7	0.0
LM + GNN*	68.3	65.0	−4.8%	66.7	64.8	−2.8%

*Results are averaged over three representative QA-GNNs (Sect. 7.1). The seed LM is `RoBERTa-Large`.

5 SemPool: Semantic Graph Pooling

We present a graph-based pooling method, termed SemPool, that aims at robustness during QA with KGs. Unlike message passing methods that depend on *local graph* information around the answer nodes, SemPool leverages *global textual* information from the KG to represent its semantics. Global information is more robust under local graph perturbations, e.g., incomplete edges around nodes, while the textual representation of the KG aids its integration to the LM. SemPool's overall framework is depicted in Fig. 1. Next, we provide SemPool's components in detail.

5.1 KG Initialization

SemPool retrieves a subgraph $\mathcal{G}_{q_a}$ for context $[c, q, a]$ following existing works [34] (Sect. 3). From now on, we denote the retrieved subgraph as $\mathcal{G}_q$ (instead of $\mathcal{G}_{q_a}$) for better readability. SemPool uses the LM to encode the textual information of each fact $(h, r, t) \in \mathcal{E}_q$ in the retrieved subgraph, aiming at semantic alignment between language and graph, as follows. We transform each edge in the subgraph, $(h, r, t) \in \mathcal{E}_q$, into natural language based on predefined templates for each relation $r \in \mathcal{R}$. For instance, (*winter, causes, bird migration*) is transformed to "winter causes bird migration". Then the verbalized fact, e, for edge (h, r, t) is encoded by the *pre-trained* LM. In order to compute a single edge embedding $\boldsymbol{h}_e$ for each edge e, we use mean-pooling or cls-pooling of the computed token embeddings (Fig. 3).

5.2 Pooling

SemPool performs global pooling over the edge embeddings $\{\boldsymbol{h}_e : e \in \mathcal{E}_q\}$ of the retrieved subgraph $\mathcal{G}_q$. However, $\mathcal{G}_q$ is determined based on the linked nodes and their neighbors (Sect. 3) and as a result, it may contain some noisy facts in the set $\mathcal{E}_q$. Thus, we propose a self-attention pooling layer that weights the importance of each fact with respect to the semantics of the subgraph. Specifically, we compute a global graph representation $\boldsymbol{g}_q$ with a weighted aggregation by

$$\boldsymbol{g}_q = \sum_{e \in \mathcal{E}_q} a_e f_v(\boldsymbol{h}_e), \tag{2}$$

where $f_v : \mathbb{R}^d \rightarrow \mathbb{R}^d$ is a linear projection and $a_e \in [0, 1]$ measures the importance of each fact $e \in \mathcal{E}_q$. Weight a_e is computed by a softmax$(\cdot)$ operation as

$$a_e = \text{softmax}_{e \in \mathcal{E}_q}\big(f_k(\boldsymbol{h}_e)\big), \tag{3}$$

where $f_k : \mathbb{R}^d \rightarrow \mathbb{R}^d$ is a neural network. Learning the importance of each fact facilitates the identification of facts that provide new information to the LM for the QA task.

5.3 KG Grounding

We ground the LM to the subgraph's semantic information $\boldsymbol{g}_q$ by inserting a special `[Graph]` token in the beginning of the question q. The embedding of the `[Graph]` token is set to $\boldsymbol{g}_q$. During training, the LM's transformer layers [27] learn to fuse information between the question's tokens and the graph token.

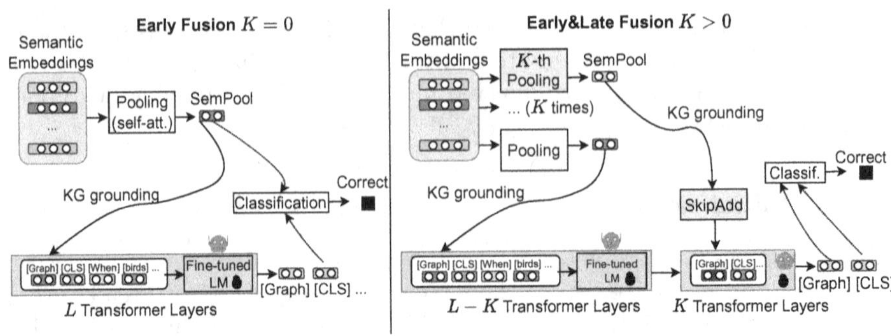

Fig. 3. SemPool architecture with early (left) and late (right) fusion. Number K represents the number of late fusion layers.

Early Fusion. In the early fusion approach, the graph representation $\boldsymbol{g}_q$ is prepended to the query. After L transformer layers, the LM outputs the final hidden states $[\boldsymbol{h}_{\texttt{Graph}}^{(L)}, \boldsymbol{h}_{\texttt{CLS}}^{(L)}, \ldots, \boldsymbol{h}_T^{(L)}]$. We use the [CLS] token as the final question representation $\boldsymbol{q}^{(L)} := \boldsymbol{h}_{\texttt{CLS}}^{(L)}$ for answer classification. Moreover, we use $\boldsymbol{g}_q$ for the answer classification loss so that the pooling module gives more attention to facts useful for the QA task. Given an answer candidate $a \in \mathcal{A}$, its probability $p(a|q)$ of being the correct answer for question q is computed by

$$p(a|q) = \exp\big(f_q(\boldsymbol{q}^{(L)}) + f_g(\boldsymbol{g}_q)\big), \tag{4}$$

where $f_q, f_g : \mathbb{R}^d \to \mathbb{R}^1$ are MLP networks. During training, we optimize the parameters via the cross entropy loss.

Early&Late Fusion. The late fusion approach has skip connections [6] that fuse graph information into deeper layers of the LM. This encourages the LM to mix useful graph semantics with language before predictions. The hyperparameter K denotes the K last transformer layers where graph information is fused. For each transformer layer, we have a dedicated pooling module that computes $\boldsymbol{g}_q^{(k)}$, where $k \in \{0, \ldots, K\}$. Similar to Eq. 2 and Eq. 3, each $\boldsymbol{g}_q^{(k)}$ is obtained via

$$\boldsymbol{g}_q^{(k)} = \sum_{e \in \mathcal{E}_q} a_e^{(k)} f_v^{(k)}(\boldsymbol{h}_e), \quad a_e^{(k)} = \text{softmax}_{e \in \mathcal{E}_q}(f_k^{(k)}(\boldsymbol{h}_e)). \tag{5}$$

In the beginning, we set the embedding of the graph token $\boldsymbol{h}_g^{(0)}$ to $\boldsymbol{g}_q^{(0)}$. At the $(L-k)$-th layer of the LM, $\boldsymbol{h}_{\texttt{GRAPH}}^{(L-k)}$ is updated via a skip connection as

$$\boldsymbol{h}_{\texttt{GRAPH}}^{(L-k)} = \boldsymbol{h}_{\texttt{GRAPH}}^{(L-k)} + \boldsymbol{h}_g^{(k)}. \tag{6}$$

We compute the final answer probabilities as

$$p(a|q) = \exp\big(f_q(\boldsymbol{q}^{(L)}) + f_g(\boldsymbol{h}_{\texttt{GRAPH}}^{(L)})\big), \tag{7}$$

where $f_q, f_g : \mathbb{R}^d \rightarrow \mathbb{R}^1$ are MLP networks. Note that the final representation $g_q^{(L)}$ used for answer classification depends on the previous states of $\{g_q^{(k)}\}_{k=1}^K$, which are optimized altogether during training.

6 Experimental Setting

QA Datasets. We evaluate SemPool on three multiple-choice question-answering datasets across two domains. OpenBookQA (**OBQA**; [18]) dataset is a 4-way multiple-choice QA dataset that requires reasoning with elementary science knowledge. It contains 5,957 questions along with an optional open book of scientific facts. We use the official data split. RiddleSense (**RiddleQA**; [11]) dataset is a 5-way multiple-choice task testing complex riddle-style commonsense reasoning. It has 5,715 questions and we split the dev set in half to make in-house dev/test sets. MedQA-USMLE (**MedQA**; [8]) dataset is 4-way multiple-choice task that originates from the USMLE practice sets, requiring biomedical and clinical knowledge. The dataset has 12,723 questions and we use the original data splits.

Table 2. Test performance comparison on QA datasets. Purple color denotes performance degradation at the adversarial setting, while teal color denotes improvement.

	(w/ ans.)				(w/o ans.)				Avg.
	OBQA	RiddleQA	MedQA	Avg.	OBQA	RiddleQA	MedQA	Avg.	
message passing									
LM + QAGNN	67.8 (±2.8)*	67.0*	38.0*	57.60	67.0 (±0.7)	65.2 (±0.4)	36.8	56.33	56.97
LM + GreaseLM	66.9*	67.2*	38.5*	57.53	64.4 (±3.2)	63.7 (±1.7)	39.0	55.70	56.61
LM + GSC	70.3 (±0.8)*	66.0 (±1.5)	38.0	**58.10**	63.6 (±2.6)	65.6 (±0.7)	37.8	55.66	56.88
no message passing									
LM (w/o KG)	64.8 (±2.4)*	60.7*	37.2*	54.23	64.8 (±2.4)*	60.7*	37.2*	54.23	54.23
LM + SemPool	67.7 (±1.2)	67.3 (±0.4)	38.9	57.96	69.5 (±0.5)	67.2 (±1.2)	39.4	**58.70**	**58.33**

*Published results. We use the `RoBERTa-Large` LM for OBQA and RiddleQA (commonsense). We use the `SapBERT-Base` LM for MedQA (biomedical).

Knowledge Graphs. Following prior works, we use ConceptNet [24], a general-domain knowledge graph, as our external knowledge source $\mathcal{G}$ for OBQA and RiddleQA. ConceptNet has 799,273 nodes and 2,487,810 edges in total. For MedQA-USMLE, we use the KG provided by [8]. This KG contains 9,958 nodes and 44,561 edges. For each question, we retrieve subgraphs following the algorithm of [34]. We set the default subgraph size to 32 nodes, which empirically performs well in all datasets. In addition, we study the setting in Fig. 2.

Language and Graph Encoding. We use (i) RoBERTa-Large [14] and AristoRoBERTa [2] for the experiments on OBQA and RiddleQA, and (ii) SapBERT [12] and BioLinkBERT-Base [33] for MedQA, demonstrating SemPool's effectiveness with respect to different LM initializations. We encode KG facts via the respective pretrained LMs for each case: (i) RoBERTa-Large for OBQA and RiddleQA, (ii) SapBERT and BioLinkBERT-Base for MedQA.

SemPool Implementation. We follow prior work's implementation for the QA task [35]. We use RAdam optimizer with learning rates selected from $\{1, 2, 5\} \times$ e−5 for the LM and set to 1e−3 for our pooling encoder with a batch size of 32, training epochs are selected from $\{20, 30, 70\}$. For SemPool, we tune K from $\{0, 2, 3, 5\}$ and select cls or mean pooling for the KG edge representation based on the dev set. Experiments are conducted on a Nvidia GeForce RTX 3090 24 GB machine.

Compared Methods. We compare SemPool with representative LM+KG GNN methods: QAGNN [34], GreaseLM [35], and GSC [29]. QAGNN uses the question's representation to guide the GNN updates. GreaseLM fuses information from both the language and the graph into the last interaction layers of the LM. QAGNN and GreaseLM use external node embeddings for the KG. GSC treats language and graph separately, and relies on the node/edge types for discriminating between correct and incorrect answers. For a fair comparison, we use the same LM for all compared methods. In addition, we report performance of fine-tuning the LM without using any KG information, 'LM (w/o KG)'.

7 Results

7.1 Main Results

Table 3. Performance comparison on QA datasets with different LMs at the adversarial setting (w/o ans).

	OBQA		RiddleQA		MedQA		Avg.
	RoBERTa	AristoRoBERTa	RoBERTa	AristoRoBERTa	SapBERT	BioLinkBERT	
	(dev/test)	(dev/test)	(dev/test)	(dev/test)	(dev/test)	(dev/test)	(dev/test)
LM + GNN♡	71.4/67.0	71.4/74.0	65.7/66.5	66.1/69.0	38.9/39.0	40.7/40.9	59.03/59.40
LM + SemPool	70.4/69.6	73.0/75.2	66.2/67.7	68.2/69.2	37.9/39.4	42.3/41.6	**59.66/60.45**

♡ We use the best performing GNN from Table 2: QAGNN for OBQA, GSC for RiddleQA, GreaseLM for MedQA.

We present the results when comparing SemPool with existing GNN-based (message passing) approaches. Table 2 shows that SemPool is the most robust method under different configurations and datasets, although it does not involve any complex message passing. SemPool improves over GNNs by 1.45–1.72% accuracy points on average, while GNNs struggle on the setting when answer node facts are removed from the KG (w/o ans.). Moreover, the benefit of grounding the LM's reasoning to the KG becomes clear when comparing SemPool with LM (w/o KG), where SemPool significantly improves performance by 4.1% points.

In Table 3, we present results when using different LMs for the QA task. SemPool outperforms the *best* performing GNN by 0.63% and 1.05% accuracy points for the dev and test set, respectively, at the critical setting, when answer information is missing. We observe that SemPool's improvements increase with more powerful LMs: AristoRoBERTa for OBQA and RiddleQA, and BioLinkBERT for MedQA. In these cases, SemPool outperforms the GNNs by 1.6% (OBQA), 2.1% (RiddleQA), and 1.7% (MedQA) accuracy points at the dev set.

7.2 Ablation Studies and Analysis

Semantic Alignment. Table 4 shows the importance of semantic alignment between language and graph representations. When using RoBERTa for QA and SBERT for computing graph embeddings, language and graphs semantics are not aligned, which leads to poor performance. On the other hand, using RoBERTa as the graph encoder improves performance by up to 17% accuracy points. AristoRoBERTa is pre-tuned for QA tasks and thus, it benefits from both RoBERTa and SBERT graph embeddings.

Graph Fusion. Figure 4 shows the importance of graph to language fusion. In most cases, late fusion ($K > 0$) outperforms early fusion ($K = 0$) as it injects graph information in multiple LM's layers. The optimal number of fusion layers K is model and task specific, but can be tuned based on the dev set. For example, RiddleQA has more complex questions and requires $K = 5$ fusion layers for the RoBERTa LM.

Table 4. Dev set performance of different embedding models, averaged over OBQA and RiddleQA datasets.

	Language Encoder	
	RoBERTa	AristoRoBERTa
Graph Encoder		
RoBERTa	68.3	70.6
SBERT	52.3	70.8

Interpretability. Figure 5 illustrates the working mechanism of SemPool in one examples case from the OBQA dataset. We observe that different layers of SemPool extract different semantics. At $K = 0$, SemPool focuses on question entities (relation: 'entity'), which helps the LM give additional importance to the linked entities during its first layers of reasoning. At $K = 1$, SemPool focuses on both question and answer entities (relation: 'a_entity'). Thus, the LM uses additional semantics for the candidate answers. At the last fusion layer, SemPool learns to aggregate new information for the LM, e.g., *(bird, related_to, chirp)*. This helps the LM to ground its predictions based on the global KG semantics. For the incorrect answer (B), SemPool identifies the irrelevant concepts 'ask_for' and 'get_money' that do not provide any useful information to the LM.

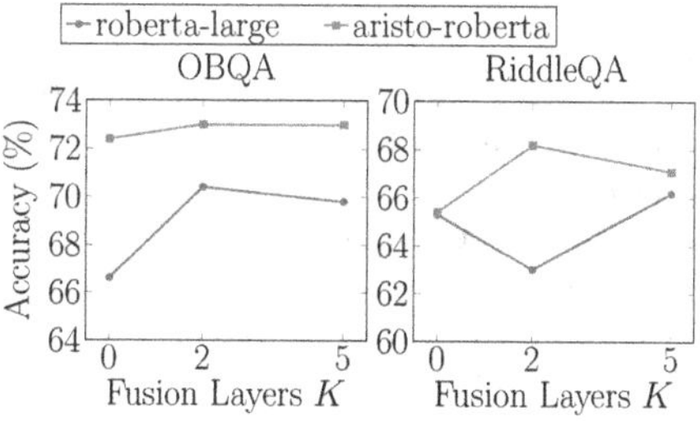

Fig. 4. Dev set performance with respect to the number K of fusion layers, using two different LMs.

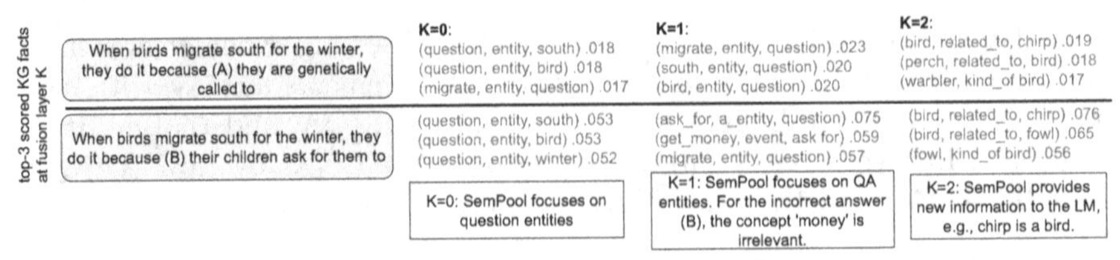

Fig. 5. Working mechanism of SemPool: Top-3 scored facts at each $K \in \{0, 1, 2\}$, along with their attention weights.

8 Conclusions

We study a critical setting for KG-based QA, where information about the candidate answer entities is missing from the KG. Our empirical results showed that graph-based (message passing) approaches struggle on the QA task under answer-based graph perturbations (Sect. 4). We propose SemPool, a graph pooling approach, that is more robust on KG-based QA task as it treats the graph as a set of facts. Experimental results show that SemPool outperforms competing methods by 2.27% accuracy points, while offering interpretability during inference (Sect. 7.2).

Acknowledgements. This work was supported in part by NSF (1447788, 1704074, 1757916, 1834251, 1834332), Army Research Office (W911NF1810344), Intel Corp, and Amazon Web Services. Access to research and computing facilities was provided by the Minnesota Supercomputing Institute.

References

1. Agarwal, O., Ge, H., Shakeri, S., Al-Rfou, R.: Knowledge graph based synthetic corpus generation for knowledge-enhanced language model pre-training. In: NAACL (2021)
2. Clark, P., et al.: From 'F' to 'A' on the NY regents science exams: an overview of the aristo project. AI Mag. **41**(4), 39–53 (2020)
3. Devlin, J., Chang, M.W., Lee, K., Toutanova, K.: BERT: pre-training of deep bidirectional transformers for language understanding. arXiv preprint arXiv:1810.04805 (2018)
4. Feng, Y., Chen, X., Lin, B.Y., Wang, P., Yan, J., Ren, X.: Scalable multi-hop relational reasoning for knowledge-aware question answering. In: EMNLP (2020)
5. Gilmer, J., Schoenholz, S.S., Riley, P.F., Vinyals, O., Dahl, G.E.: Neural message passing for quantum chemistry. In: ICML (2017)
6. He, K., Zhang, X., Ren, S., Sun, J.: Deep residual learning for image recognition. In: IEEE CVPR (2016)
7. Ioannidis, V.N., et al.: Efficient and effective training of language and graph neural network models. arXiv preprint arXiv:2206.10781 (2022)
8. Jin, D., Pan, E., Oufattole, N., Weng, W.H., Fang, H., Szolovits, P.: What disease does this patient have? a large-scale open domain question answering dataset from medical exams. Appl. Sci. **11**, 6421 (2021)

9. Khashabi, D., et al.: UNIFIEDQA: crossing format boundaries with a single QA system. In: EMNLP Findings (2020)
10. Lin, B.Y., Chen, X., Chen, J., Ren, X.: KagNet: knowledge-aware graph networks for commonsense reasoning. In: EMNLP-IJCNLP (2019)
11. Lin, B.Y., Wu, Z., Yang, Y., Lee, D.H., Ren, X.: RiddleSense: reasoning about riddle questions featuring linguistic creativity and commonsense knowledge. In: ACL Findings (2021)
12. Liu, F., Shareghi, E., Meng, Z., Basaldella, M., Collier, N.: Self-alignment pretraining for biomedical entity representations. In: NAACL-HLT (2021)
13. Liu, J., et al.: Towards graph foundation models: a survey and beyond. arXiv preprint arXiv:2310.11829 (2023)
14. Liu, Y., et al.: RoBERTa: a robustly optimized BERT pretraining approach. arXiv preprint arXiv:1907.11692 (2019)
15. Mallen, A., Asai, A., Zhong, V., Das, R., Hajishirzi, H., Khashabi, D.: When not to trust language models: Investigating effectiveness and limitations of parametric and non-parametric memories. In: ACL (2023)
16. Mavromatis, C., et al.: Train your own GNN teacher: graph-aware distillation on textual graphs. In: ECML-PKDD (2023)
17. Mavromatis, C., Karypis, G.: ReaRev: adaptive reasoning for question answering over knowledge graphs. In: EMNLP Findings (2022)
18. Mihaylov, T., Clark, P., Khot, T., Sabharwal, A.: Can a suit of armor conduct electricity? A new dataset for open book question answering. In: EMNLP (2018)
19. Mihaylov, T., Frank, A.: Knowledgeable reader: enhancing cloze-style reading comprehension with external commonsense knowledge. In: ACL (2018)
20. Pan, S., Luo, L., Wang, Y., Chen, C., Wang, J., Wu, X.: Unifying large language models and knowledge graphs: a roadmap. arXiv preprint arXiv:2306.08302 (2023)
21. Park, J., et al.: Relation-aware language-graph transformer for question answering. In: AAAI (2023)
22. Raffel, C., et al.: Exploring the limits of transfer learning with a unified text-to-text transformer. JMLR **21**, 1–67 (2020)
23. Schlichtkrull, M., Kipf, T.N., Bloem, P., Van Den Berg, R., Titov, I., Welling, M.: Modeling relational data with graph convolutional networks. In: ESWC (2018)
24. Speer, R., Chin, J., Havasi, C.: ConceptNet 5.5: an open multilingual graph of general knowledge. In: AAAI (2017)
25. Sun, Y., Shi, Q., Qi, L., Zhang, Y.: JointLK: joint reasoning with language models and knowledge graphs for commonsense question answering. In: NAACL-HLT (2022)
26. Tian, Y., et al.: Graph neural prompting with large language models. arXiv preprint arXiv:2309.15427 (2023)
27. Vaswani, A., et al.: Attention is all you need. In: NeurIPS (2017)
28. Veličković, P., Cucurull, G., Casanova, A., Romero, A., Lio, P., Bengio, Y.: Graph attention networks. In: ICLR (2018)
29. Wang, K., Zhang, Y., Yang, D., Song, L., Qin, T.: GNN is a counter? Revisiting GNN for question answering. In: ICLR (2022)
30. Xie, H., et al.: Graph-aware language model pre-training on a large graph corpus can help multiple graph applications (2023)
31. Xie, T., et al.: UnifiedSKG: unifying and multi-tasking structured knowledge grounding with text-to-text language models. In: EMNLP (2022)
32. Yasunaga, M., et al.: Deep bidirectional language-knowledge graph pretraining. In: NeurIPS (2022)

33. Yasunaga, M., Leskovec, J., Liang, P.: LinkBERT: pretraining language models with document links. In: ACL (2022)
34. Yasunaga, M., Ren, H., Bosselut, A., Liang, P., Leskovec, J.: QA-GNN: reasoning with language models and knowledge graphs for question answering. In: NAACL (2021)
35. Zhang, X., et al.: GreaseLM: graph reasoning enhanced language models for question answering. In: ICLR (2022)
36. Zhao, J., et al.: Learning on large-scale text-attributed graphs via variational inference. In: ICLR (2023)
37. Zhu, Z., Galkin, M., Zhang, Z., Tang, J.: Neural-symbolic models for logical queries on knowledge graphs. In: ICML (2022)

Medical and Biological Data

Spatial Gene Expression Prediction Using Multi-Neighborhood Network with Reconstructing Attention

Panrui Tang, Zuping Zhang(✉), Cui Chen, and Yubin Sheng

Central South University, Changsha 410083, China
{224711119,zpzhang,214701013,ybsheng}@csu.edu.cn

Abstract. Spatial transcriptomics (ST) has made it possible to link local spatial gene expression with the properties of tissue, which is very helpful to the research of histopathology and pathology. To obtain more ST data, we utilize deep learning methods to predict gene expression on tissue slide images. Considering the importance of the dependence of local tissue images on their neighborhoods, we propose the novel Multi-Neighborhood Network (MNN), composed of down-sampling module and vanilla Transformer blocks. Moreover, to satisfy the needs of architecture and address the computational and parameter challenges arising from it, we introduce dual-scale attention block and reconstructing attention block. To demonstrate the effectiveness of this network structure and the superiority of attention mechanisms, we conducted comparative experiments, where MNN achieved optimal PCC@M (1×10^1) of 9.23 and 8.54 for the lung cancer and mouse brain datasets of 10x Genomics website, respectively, outperforming several state-of-the-art (SOTA) methods. This reveals the superiority of our method in terms of spatial gene prediction.

Keywords: Spatial Transcriptomics · Gene Expression Prediction · Tissue Slide Image · Reconstructing Attention · Vision Transformer

1 Introduction

Spatial Transcriptomics (ST), which measures spatially resolved messenger RNA (mRNA) profiles in tissue sections using unique DNA barcodes to map the expression of thousands of genes at each spot on a tissue slide, is a relatively new technique capable of simultaneously obtaining spatial positional information of cells and gene expression data [16,17]. This kind of information is highly beneficial for researchers, helping them understand variations in gene expression across different regions within a given tissue sample.

However, acquiring ST data requires specific hardware means (10X Genomics Visium) and the density of probe clusters is limited. So, how can we reduce the requirements for the number of probes, lower experimental costs, and obtain more ST data? Deep learning has already been applied in gene prediction and

D.-N. Yang et al. (Eds.): PAKDD 2024, LNAI 14648, pp. 169–180, 2024.
https://doi.org/10.1007/978-981-97-2238-9_13

RNA sequence expression [1,14]. While due to dataset limitations (bulk RNA-seq data), these works cannot link tissue images with specific gene expressions [4]. ST data, consisting of pixel positions of spots and labels for each spot, can precisely facilitate supervised learning in models, thereby predicting gene expression at other undetected locations, as shown in Fig. 1. Deep learning methods in image processing are a hot topic, with extensive research on structures like CNNs and Transformers, especially in natural images represented by the ImageNet [3,5,8–10,13,18–21,24]. However, these methods ignore neighborhoods and is ill-suited for larger receptive field. If attempting to expand the model's receptive field would significantly increase training complexity significantly, leading to unacceptable computational complexity and memory cost.

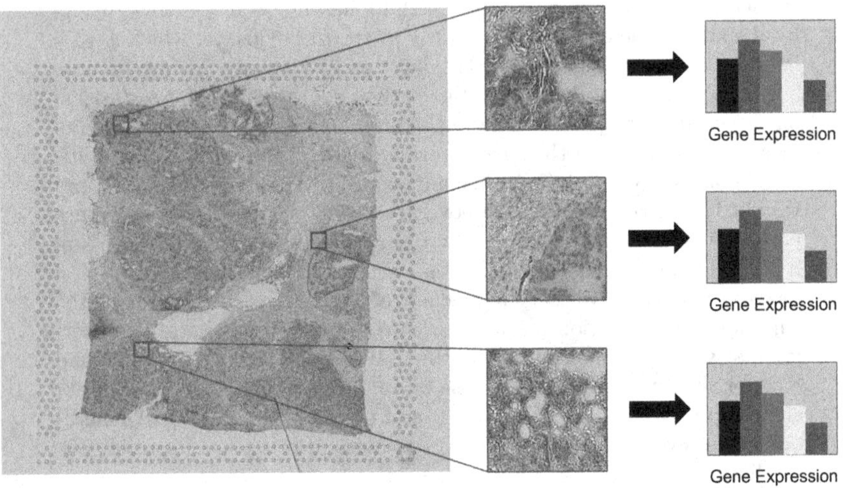

Fig. 1. Overview of a tissue slice. Each window of the tissue slice image corresponds to distinct gene expression one by one. As shown in picture, there are 3 windows and each of them is with 5 gene types. Predicting the gene expression of each window is our task.

To tackle these limitations, seizing the characteristics of biological images, we introduce a novel framework with larger receptive field, called the Multi-Neighborhood Network (MNN), as shown in the Fig. 2. It considers information from neighborhoods beyond the input image to predict the gene expression. MNN consists of two parts, with the first divided into four stages, and the second fusing neighborhoods.

Our contributions are listed as follows:

- To integrate information from a larger receptive field, we propose the Multi-Neighborhood Network, which can predict gene expression in slide image windows more accurately and efficiently.

- To handle the significant computation and memory usage brought by the larger receptive field, we suggest employing parameter reuse, Dual-Scale Attention (DSA) and Reconstructing Attention (RA). RA can automatically assign tokens and reconstruct tokens.
- We conducted comparative experiments, and the performance of our model achieved the best results compared to many SOTA methods, demonstrating the effectiveness of expanding the model's receptive range. We also carried out ablation studies to explore the optimal number of neighborhoods, showcasing the superiority of the newly proposed attention mechanism.

2 Related Work

2.1 Gene Expression Prediction

Measuring gene expression is crucial across various fields, including tissue cell functionality, micro-environment interactions, lineage tracing during developmental processes, and pathology of diseases. In recent years, the application of bulk sequencing or single-cell sequencing has significantly broadened our horizons [6]. Subsequently, there are deep learning methods that focus on bulk RNA-Seq and single-cell RNA [12]. However, such methods inevitably lose the original positional information of cells within tissues, making it difficult to acquire information about the cellular composition and gene expression statuses in different regions of the tissues, as well as the gene expression differences between various functional areas.

Spatial transcriptomics measures spatially resolved messenger RNA (mRNA) profiles in tissue sections using unique DNA barcodes, which are used to map the expression of thousands of genes at each spot on a tissue slide. This advances the research into the authentic gene expression of cells in tissues. Addressing this type of data, He et al. [7] designed ST-Net, which is the first method to combine and analyze slice data and ST. However, only convolutions are used as backbones, and only a single window is considered in this work.

2.2 Vision Transformer (ViT)

Initially, the Transformer was designed for natural language processing. Later, Vaswani et al. [19] applied it to image processing, marking the first application of a pure Transformer architecture in the field of computer vision. It segments images into patches and treats them as tokens. The most significant departure from CNNs is its replacement of convolution with attention for global information modeling.

However, the self-attention mechanism that gives it superior performance is also the aspect most in need of improvement, as the vanilla attention generates heavy computational complexity and substantial storage footprints when dealing with high-resolution images. This constitutes the primary challenge in expanding the receptive field in spatial gene prediction tasks. Addressing this drawback, many researchers have proposed their own solutions, a comprehensive survey of which can be found in [18]. Sparse attention is one such solution, its main idea being the introduction of prior knowledge to eliminate some of the attention [5,8,9,13,21]. Other researchers have attempted to tailor sparse features more closely to specific tasks, as seen in models such as DAT [21], DPT [3], and TCFormer [24]. However, this type of attention typically requires designers to provide prior knowledge, whereas our proposed attention can autonomously select the more important parts to focus on. Similar to [3,21,24], we also reduce complexity by decreasing the number of tokens for keys and values. Uniquely, we draw on the concept of pooling from graph neural networks to restructure tokens, thereby avoiding the information loss that comes from outright discarding. This process adheres to a principle: it aims to ensure tokens contain more valuable information, by allocating more tokens to important areas and fewer tokens to less important ones.

3 Method

The overall architecture is depicted in Fig. 2. Consistent with the down-sampling frameworks of most image models, an image input, $X \in \mathbb{R}^{H \times W \times 3}$, first undergoes a 4x down-sampling, followed by three instances of 2x down-sampling [5,10,13,20]. Between each pair of down-sampling layers, there are S_i pairs of Transformer blocks, each consisting of dual-scale Transformer (DST) block and reconstructing Transformer (RT) block. This process is repeated $N+1$ times, after which the resulting $N+1$ feature maps are fed into the vanilla Transformer blocks, with N representing the number of neighborhoods. The 7×7 feature map corresponding to the central region is extracted and sent to a linear layer for the final prediction. The attention of DST and RT is designed to generate masks and reconstruct tokens, respectively. The details of attention are elaborated upon in Sect. 3.1 and Sect. 3.2.

3.1 Dual-Scale Attention

Dual-Scale Attention (DSA) operates on two scales, comprising two components, namely global attention and local attention. Their structure is depicted in Fig. 3(a). Global attention adopts a different way from that of vanilla attention to get *key* and *values* (Eq. (1)). Tokens is merged through convolution, the kernel size of which is 7.

$$
\begin{aligned}
Q_{DS-G} &= XW_i^{Q_{DS-G}}, \\
K_{DS-G} &= Conv(X)W_i^{K_{DS-G}}, \\
V_{DS-G} &= Conv(X)W_i^{V_{DS-G}}, \\
A_{DS-G} &= Softmax(\frac{Q_{DS-G}K_{DS-G}}{\sqrt{d_i}}), \\
A'_{DS-G} &= UpSample(A_{DS-G})
\end{aligned}
\tag{1}
$$

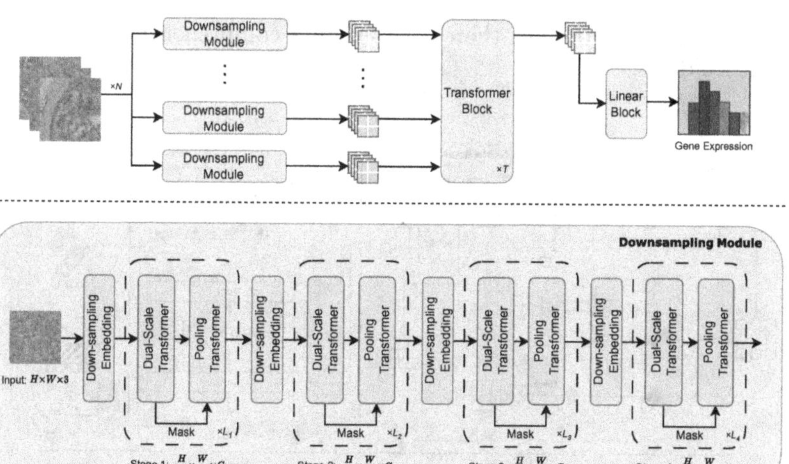

Fig. 2. An overview of Multi-Neighborhood Network. (a) illustrates the network architecture of the model. (b) depicts the structure of the down-sampling module, which consists of the down-sampling embedding, the dual-scale Transformer block, and the reconstructing Transformer block. The dual-scale Transformer block provides a mask for the reconstructing attention.

$DS-G$ refers to the global attention within the dual-scale attention, and $W_l^{Q_{DS-G}}$ represents the trainable parameters used to calculate the *query* in global attention of l-th layer. Consequently, the attention matrix $A_{DS-G} \in \mathbb{R}^{(H_i \times W_i) \times (\frac{H_i}{7} \times \frac{W_i}{7})}$ is reduced by 7×7 compared to its original size. Then, it is up-sampled to restore to its original size for mask computation.

The computation of *query*, *key* and *value* in the local attention mechanism is the same as that of vanilla attention. However, the computation of attention is confined to a small region of the feature map. To facilitate computation, we set the side length of the region to be 7, consistent with the kernel size of the convolution, as illustrated by Eq. (2). The attention matrix $A_{DS-L} \in \mathbb{R}^{(49 \times 49) \times (\frac{H_i}{7} \times \frac{W_i}{7}) \times (\frac{H_i}{7} \times \frac{W_i}{7})}$ is reshaped to restore its original size, which are then used in the mask computation.

$$
\begin{aligned}
Q_{DS-L} &= X^{7\times7} W_i^{Q_{DS-L}}, \\
K_{DS-L} &= X^{7\times7} W_i^{K_{DS-L}}, \\
V_{DS-L} &= X^{7\times7} W_i^{V_{DS-L}}, \\
A_{DS-L} &= Softmax(\frac{Q_{DS-L} K_{DS-L}}{\sqrt{d_i}}), \\
A'_{DS-L} &= Reshape(A_{DS-L})
\end{aligned}
\tag{2}
$$

The ultimate output of dual-scale attention is constituted by the concatenation of the yields from two scales. This resultant output incorporates both global information and local details, thereby ensuring a comprehensive representation of the all windows.

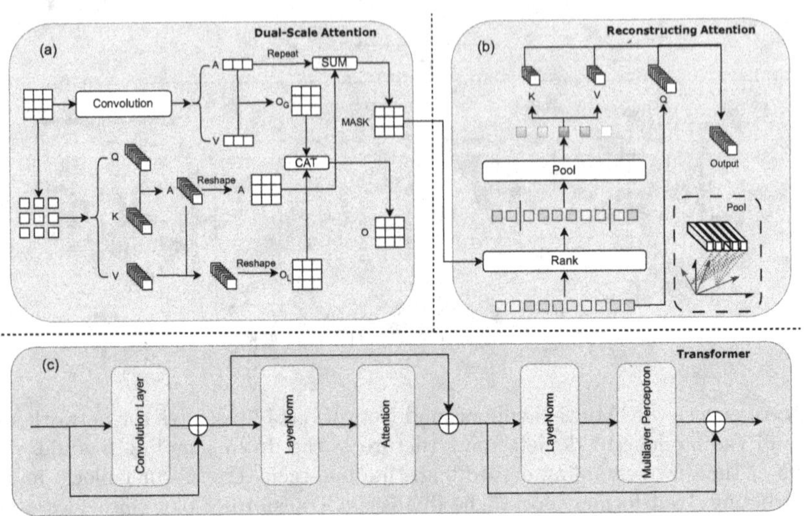

Fig. 3. Design of down-sampling module. (a) The mechanism of the dual-scale attention, wherein O_G and O_L denote the outputs of global and local attention, respectively, with O representing the output of the dual-scale attention. (b) The mechanism of reconstructing attention, with the lower right corner portraying the projection process of the pooling operation. (c) Details of 3 kinds of Transformer block.

Table 1. 3 patterns of pooling.

Pattern	Pooling Ratio		
	Group 1	Group 2	Group 3
1	28	56	196
2	7	14	49
3	1	2	–

3.2 Reconstructing Attention

Vanilla attention is often inefficient in image processing tasks due to its equal treatment of every token, consuming identical computational resources even for relatively insignificant segments. Previous works [5,8,9,21] contemplated reducing computational overhead by sparse attention. Alternatively, we propose that attention should be adaptively allocated to areas warranting focus. Specifically, a greater number of tokens should be assigned to portions deserving attention, and the number of tokens allotted to each region should not be uniform. This method is depicted in the Fig. 3(b), namely Reconstructing Attention (RA). Based on the summation of the two types of attention matrices generated in Sect. 3.1 (A'_{DS-G} and A'_{DS-L}), we rank the significance of each token. For tokens of high importance, a lower pooling ratio is applied during projection, while those deemed less crucial are subjected with a higher pooling ratio. In this experiment, we establish three distinct pooling schemes for tokens, as shown in the Table 1. The importance of tokens gradually decreases from Group 1 to Group 3, hence Group 1 has the lowest pooling ratio. We randomly generate K orthogonal unit vectors $u_k(k = 1, 2, ..., K)$, project the tokens onto these vectors through dot multiplication and weighted summation to acquire K reconstructed tokens u'_k:

$$u_k' = \sum_{j=1}^{n} (x_j \cdot u_k) u_k, \tag{3}$$

The *key* and *value* are computed against reconstructed tokens U', significantly reducing the size of the attention matrix.

3.3 Transformer Block

In this study, based on the distinctions in attention, Transformers can be categorized into three types: multi-scale Transformer, reconstructing Transformer, and vanilla Transformer. Their shared structures are depicted in Fig. 3(c). The calculation process employed are also identical:

$$\begin{aligned} \widehat{X^i} &= Attention\left(LayerNorm\left(X^{i-1}\right)\right) + X^{i-1}, \\ X^i &= MLP\left(LayerNorm\left(\widehat{X^i}\right)\right) + \widehat{X^i} \end{aligned} \tag{4}$$

As illustrated in the Fig. 1(a), DST and RT are utilized for feature extraction in the down-sampling module, while the vanilla Transformer is employed for the extraction of information of neighboring spots. Directly inputting a large-scale image into the down-sampling module would not only entail substantial computational load (given the quadratic relationship between the attention computation complexity and image size) but also introduce considerable noise. Within the down-sampling module, the number of layers at each of the four stages is denoted as S_1, S_2, S_3, and S_4, is set to $[1, 1, 1, 1]$. The hidden dimensions for the 4 stages are $[64, 128, 128, 256]$. The vanilla Transformer comprises a total of V blocks, with the number set to 3. Based on the experiments conducted in Sect. 5.2, we set the number of neighborhoods at 4.

4 Experiment

4.1 Datasets

We conducted experiments using the publicly available 10x Genomics dataset[1]. 10x Genomics is a dataset website that encompasses a diverse array of spatial gene expression datasets, including those of mouse brains, mouse kidneys, breast cancer, lung cancer, and more. We selected datasets from lung cancer and mouse brains for our analysis. The former contains 2 slide images, composed of 10,053 windows, while the latter includes 3 slide images, consisting of 20,569 windows. Drawing on previous tasks [7,23] involving gene transcriptomes, we predicted the categories of 250 gene types with the highest mean values in the datasets. For the integrity and fairness of the experiments, all model testing was based on a five-fold cross-validation.

4.2 Experiment Setup

Evaluation Metrics: Following [23], we employed three criteria for comparing: PCC (Pearson Correlation Coefficient), MSE (Mean Squared Error), and MAE (Mean Absolute Error). Specifically, for comprehensive comparison of performance of PCC, it can be further refined into 3 metrics: PCC@F, PCC@S, and PCC@M, representing the quartile, median, and mean values of PCC, respectively. For ease of comparison, all the experimental data are presented 1×10^1. Given the predicted values and ground truth, higher values of PCC@F, PCC@S, and PCC@M indicate superior model capabilities. Furthermore, MSE and MAE describe the disparity in distribution between the model's predicted values and the ground truth. Lower levels of MSE and MAE denote better model performance.

Table 2. Comparison to prior methods.

Methods	Params(m)	Lung Cancer					Mice Brain				
		MAE	MSE	PCC@F	PCC@S	PCC@M	MAE	MSE	PCC@F	PCC@S	PCC@M
Biformer	13.10M	0.76	0.11	8.76	9.08	9.18	0.92	0.16	7.72	8.07	7.93
CycleMLP	12.82M	0.80	0.12	8.66	9.19	8.99	0.79	0.11	7.78	8.10	7.98
MPViT	5.93M	0.72	0.11	8.81	9.29	9.12	0.72	0.09	8.17	8.36	8.32
ViTAE	10.59M	0.75	0.11	8.75	9.25	9.08	0.72	0.09	8.14	8.34	8.29
ViT-small	21.66M	0.83	0.13	8.63	9.17	8.97	0.81	0.12	7.73	8.01	7.93
Shunted-T	11.17M	**0.71**	**0.10**	8.81	9.29	9.13	0.72	0.09	8.14	8.33	8.29
MNN	9.1M	0.83	0.12	**8.96**	**9.37**	**9.23**	**0.71**	**0.08**	**8.42**	**8.59**	**8.54**

Implementation Details: The size of windows is 224×224. All windows are standardized to the mean and standard deviation of ImageNet. Before training,

[1] https://www.10xgenomics.com/resources/datasets.

each of them undergoes random geometric transformations such as horizontal and vertical flips, and 90-degree rotations. For all experiments, the lung cancer dataset is trained for 40 epochs, and the mouse brain dataset for 80 epochs. The learning rate is set at 5×10^{-5}, with a cosine annealing scheduler used during training and weight decay set at 1×10^{-5}. The batch size is 16. All experiments were conducted using two NVIDIA A10 GPUs.

5 Results and Discussion

5.1 Baseline Experiments for Multi-Neighborhood Network

We selected models that have been state-of-the-art (SOTA) in ImageNet classification in recent years, including ViT [19], CycleMLP [2], ViTAE [22], Biformer [25], MPViT [11], and ShuntedTrans [15]. Table 2 reports the average values of five-fold cross-validation for each method. With a comparable number of parameters, our model surpassed these SOTA methods on PCC-related metrics (the most crucial indicators in this task). The reason why we did not obtain the best results of MAE and MSE of Lung Cancer is that this model is fed with a larger area of tissue to better compare different regions to capture the overall characteristics of gene expression, but it is not conducive to the fitting of the number of each gene. To more directly showcase the superiority of our approach, we ranked the prediction of 250 genes and issued ten deciles. We plot a 3D bar chart of PCC@M (Fig. 4). It is evident that our method has better performance at each decile. This indicates that our approach does not amplify the predictive disparity among the 250 gene types; instead, it enhances the overall predictive

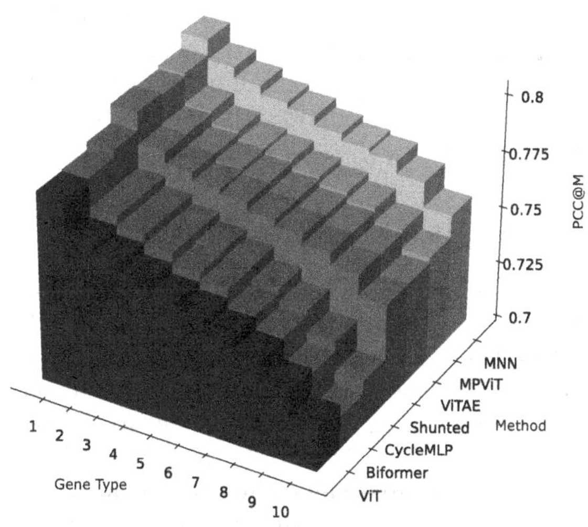

Fig. 4. PCC@M of 7 methods

Table 3. Ablation study on numbers of neighborhoods and attention.

Metrics	MAE	MSE	PCC@F	PCC@S	PCC@M
Number of Neighbours					
0	0.817	0.124	8.60	9.16	8.94
4	0.832	0.119	8.96	9.37	9.23
8	0.813	0.119	8.85	9.31	9.14
Methods					
Vanilla Attention	1.029	0.215	8.46	9.06	8.78
Bi-Level Routing Attention	**0.796**	0.128	8.64	9.15	8.96
Reduction Cell	0.827	0.120	8.80	9.28	9.09
MNN	0.832	**0.119**	**8.96**	**9.37**	**9.23**

performance of the model, regardless of the difficulty of predicting different types of genes.

We believe the improved performance can be attributed to the down-sampling module's ability to perceive an $N+1$ windows due to parameter sharing, making it less prone to over-fitting compared to other models.

5.2 Influence of Number of Neighborhoods

Given that the number of neighborhoods is a hyperparameter critical to the performance of the model, it's worth exploring to assess the influence of different numbers of neighborhoods. Considering spatial symmetry, we selected the number of neighborhoods N from the set $\{0, 4, 8\}$. According to Table 3, it can be concluded that the absence of neighborhoods results in the loss of vital information, leading to a decrease of 0.29 PCC@M. However, an excessive number of neighborhoods does not substantially enhance performance and may potentially impede the training process. More importantly, it leads to a significant increase in computational demands.

5.3 Influence of Attention Mechanisms

To demonstrate the superiority of DSA and RA, we selected vanilla attention and bi-level routing attention as comparative attention. Additionally, we juxtaposed the combination of DST and RT against the reduction cell [22]. As illustrated in the Table 3, our methods achieved superior results. Comparison with the reduction cell indicates that through enhancements in the attention mechanism, it is feasible to surpass the effects brought forth by the Parallel Convolution Module (PCM). Given that current patch embedding predominantly utilizes convolutional operations, the convolution modules running in parallel with attention may not significantly contribute when the expressive capacity of the attention mechanism is sufficiently robust.

Vanilla attention fails to show improved performance due to its uniform treatment of tokens. Vanilla and bi-level routing attention's performance declined compared to the contrast experiments, likely due to a reduction in the number of layers. This underscores the necessity for differential treatment in refining attention mechanisms. Bi-Level routing attention is designed to reduce computation complexity through the sorting and discarding of tokens. Discarding might lead to the loss of crucial information. Instead, projection and addition emerge as more appropriate pooling methods for this task.

6 Conclusion

This paper introduces a novel method tailored for spatial gene prediction tasks, termed the multi-neighborhood network. Taking into account the characteristics of tissue images, the model's receptive field is expanded. To reduce computing consumption, features from various windows are extracted independently, and a more efficient, powerful form of dual-scale attention and reconstructing attention is proposed to fit this framework. Our experiments validate the necessity of enlarging the receptive area for gene expression prediction. Through comparative experiments with numerous SOTA methods, we have demonstrated the superiority of our model. Subsequent ablation studies also confirm the advantages of DSA and RA. Because of its capacity of effective and reliable gene expression prediction, MNN has the potential to facilitate the acquisition of ST data and promote spatial gene expression research.

References

1. Chen, M., Zhang, B., Topatana, W., Cao, J., Zhu, H., Juengpanich, S., Mao, Q., Yu, H., Cai, X.: Classification and mutation prediction based on histopathology h&e images in liver cancer using deep learning. NPJ Precis. Oncol. **4**(1), 14 (2020)
2. Chen, S., Xie, E., Ge, C., Chen, R., Liang, D., Luo, P.: CycleMLP: a MLP-like architecture for dense prediction. arxiv 2021. arXiv preprint arXiv:2107.10224
3. Chen, Z., et al.: DPT: deformable patch-based transformer for visual recognition. In: Proceedings of the 29th ACM International Conference on Multimedia, pp. 2899–2907 (2021)
4. Dawood, M., Branson, K., Rajpoot, N.M., Minhas, F.U.A.A.: All you need is color: image based spatial gene expression prediction using neural stain learning. In: Kamp, M., et al. Machine Learning and Principles and Practice of Knowledge Discovery in Databases. ECML PKDD 2021. Communications in Computer and Information Science, vol. 1525, pp. 437–450. Springer, Cham (2021). https://doi.org/10.1007/978-3-030-93733-1_32
5. Dong, X., et al.: CSWIN transformer: a general vision transformer backbone with cross-shaped windows. In: Proceedings of the IEEE/CVF Conference on Computer Vision and Pattern Recognition, pp. 12124–12134 (2022)
6. Gerlinger, M., et al.: Intratumor heterogeneity and branched evolution revealed by multiregion sequencing. N. Engl. J. Med. **366**(10), 883–892 (2012)
7. He, B., et al.: Integrating spatial gene expression and breast tumour morphology via deep learning. Nat. Biomed. Eng. **4**(8), 827–834 (2020)

8. Ho, J., Kalchbrenner, N., Weissenborn, D., Salimans, T.: Axial attention in multidimensional transformers. arXiv preprint arXiv:1912.12180 (2019)
9. Huang, Z., Wang, X., Huang, L., Huang, C., Wei, Y., Liu, W.: CCNet: CRISS-cross attention for semantic segmentation. In: Proceedings of the IEEE/CVF International Conference on Computer Vision, pp. 603–612 (2019)
10. Jian, S., Kaiming, H., Shaoqing, R., Xiangyu, Z.: Deep residual learning for image recognition. In: IEEE Conference on Computer Vision & Pattern Recognition, pp. 770–778 (2016)
11. Lee, Y., Kim, J., Willette, J., Hwang, S.J.: Mpvit: Multi-path vision transformer for dense prediction. In: Proceedings of the IEEE/CVF Conference on Computer Vision and Pattern Recognition, pp. 7287–7296 (2022)
12. Li, X., Wang, C.Y.: From bulk, single-cell to spatial RNA sequencing. Int. J. Oral Sci. **13**(1), 36 (2021)
13. Liu, Z., et al.: SWIN transformer: hierarchical vision transformer using shifted windows. In: Proceedings of the IEEE/CVF International Conference on Computer Vision, pp. 10012–10022 (2021)
14. Lu, W., Graham, S., Bilal, M., Rajpoot, N., Minhas, F.: Capturing cellular topology in multi-gigapixel pathology images. In: Proceedings of the IEEE/CVF Conference on Computer Vision and Pattern Recognition Workshops, pp. 260–261 (2020)
15. Ren, S., Zhou, D., He, S., Feng, J., Wang, X.: Shunted self-attention via multi-scale token aggregation. In: Proceedings of the IEEE/CVF Conference on Computer Vision and Pattern Recognition, pp. 10853–10862 (2022)
16. Rodriques, S.G., et al.: Slide-SEQ: a scalable technology for measuring genome-wide expression at high spatial resolution. Science **363**(6434), 1463–1467 (2019)
17. Ståhl, P.L., et al.: Visualization and analysis of gene expression in tissue sections by spatial transcriptomics. Science **353**(6294), 78–82 (2016)
18. Tay, Y., Dehghani, M., Bahri, D., Metzler, D.: Efficient transformers: a survey (2022)
19. Vaswani, A., et al.: Attention is all you need (2017). arXiv preprint arXiv:1706.03762 (2019)
20. Wang, W., et al.: Pyramid vision transformer: a versatile backbone for dense prediction without convolutions. In: Proceedings of the IEEE/CVF International Conference on Computer Vision, pp. 568–578 (2021)
21. Xia, Z., Pan, X., Song, S., Li, L.E., Huang, G.: Vision transformer with deformable attention (2022)
22. Xu, Y., Zhang, Q., Zhang, J., Tao, D.: Vitae: vision transformer advanced by exploring intrinsic inductive bias. Adv. Neural. Inf. Process. Syst. **34**, 28522–28535 (2021)
23. Yang, Y., Hossain, M.Z., Stone, E.A., Rahman, S.: Exemplar guided deep neural network for spatial transcriptomics analysis of gene expression prediction. In: Proceedings of the IEEE/CVF Winter Conference on Applications of Computer Vision, pp. 5039–5048 (2023)
24. Zeng, W., et al.: Not all tokens are equal: human-centric visual analysis via token clustering transformer (2022)
25. Zhu, L., Wang, X., Ke, Z., Zhang, W., Lau, R.W.: Biformer: vision transformer with bi-level routing attention. In: Proceedings of the IEEE/CVF Conference on Computer Vision and Pattern Recognition, pp. 10323–10333 (2023)

APFL: Active-Passive Forgery Localization for Medical Images

Nan Wang, Jiaqi Shi, Liping Yi, Gang Wang(✉), Ming Su, and Xiaoguang Liu

College of C.S., ICIC, Nankai-Huapai Joint Lab, TMCC, SysNet, Nankai University, Tianjin, China
{wangn,shijq,yiliping,wgzwp,suming,liuxg}@nbjl.nankai.edu.cn

Abstract. Medical image forgery has become an urgent issue in academia and medicine. Unlike natural images, images in the medical field are so sensitive that even minor manipulation can produce severe consequences. Existing forgery localization methods often rely on a single image attribute and suffer from poor generalizability and low accuracy. To this end, we propose a novel active-passive forgery localization (APFL) algorithm to locate the forgery region of medical images attacked by three common forgeries: splicing, copy-move and removal. It involves two modules: a) active forgery localization, we utilize reversible watermarking to locate the fuzzy forgery region, and b) passive forgery localization, we train a lightweight model named KDU-Net through knowledge distillation to precisely locate the forgery region in the fuzzy localization result extracted by active forgery localization. The lightweight KDU-Net as a student model can achieve similar performance to RRU-Net as a teacher model, while its model capacity is only 24.6% of RRU-Net, which facilitates fast inference for medical diagnostic devices with limited computational power. Since there are no publicly available medical tampered datasets, we manually produce tampered medical images from the real-world Ophthalmic Image Analysis (OIA) fundus image dataset. The experimental results present that APFL achieves satisfied forgery localization accuracy under the three common forgeries and shows robustness to rotation and scaling post-processing attacks.

Keywords: Medical image · Forgery localization · Digital watermarking · Neural network · Knowledge distillation

Supported in part by the National Science Foundation of China under Grant 62272252 and Grant 62272253, the Science and Technology Development Plan of Tianjin under Grant 20JCZDJC00610 and Grant 19YFZCSF00900, the Key Research and Development Program of Guangdong under Grant 2021B0101310002, Foundation of State Key Laboratory of Public Big Data (No. PBD2022-12) and the Fundamental Research Funds for the Central Universities.

First Author and Second Author contribute equally to this work.

D.-N. Yang et al. (Eds.): PAKDD 2024, LNAI 14648, pp. 181–193, 2024.
https://doi.org/10.1007/978-981-97-2238-9_14

1 Introduction

Medical images are essential tools for disease diagnosis, such as X-CT, MRI and etc [32]. Forged medical images may cause misdiagnose and hinder medical progress [5]. Image forgery aims to intentionally alter the semantics of images, mainly including splicing, copy-move, removal and etc [6].

Existing forgery localization methods for natural images are generally divided into two categories: active forgery localization mainly based on digital signature and watermarking, and passive forgery localization mainly includes traditional feature extraction methods and deep learning-based methods.

Compared with natural images, medical images have some special characteristics such as fixed human body structure, low contrast, delicate texture and the need for high-resolution detail information to diagnose diseases, which render forgery localization for medical images more challenging than for natural images. Existing forgery localization methods for natural images only perform well on the grayscale smooth regions in medical images. And only a few studies focus on forgery localization for medical images, using either traditional methods which based on the difference of local texture features between regions [10,22,27] or machine learning-based methods [3,24]. However, their pixel-level localization results are imprecise, can only locate approximate regions, and the applicable forgery type is mostly copy-move only.

To improve the performance and generality of forgery localization for medical images, we propose a novel active-passive forgery localization (APFL) algorithm, which involves two detection steps: 1) *active forgery localization* utilizes a reversible watermarking technology to extract fuzzy forgery regions; 2) *passive forgery localization* further refines forgery regions through a designed lightweight KDU-Net model trained as a knowledge distillation form. The reversible watermarking in the active phase guarantees image authenticity and enables fuzzy localization, upon which the passive phase uses knowledge distillation to simplify the network structure for faster localization refinement and resist three common forgeries. Besides, the lightweight KDU-Net is helpful to speed up inference on medical devices with limited computing power.

Our main contributions are summarized as follows:

- We propose a novel medical image forgery localization algorithm with two-step detections: active forgery localization based on reversible watermarking and passive forgery localization with a designed lightweight KDU-Net model.
- We provide a medical image forgery dataset by splicing, copy-moving, or removing lesion regions on a real-world Ophthalmic Image Analysis (OIA) fundus image dataset.
- Experimental results present that APFL achieves compared forgery localization accuracy to the state-of-the-art baseline with only 24.6% model size and shows robustness to forgery post-processing attacks of rotation and scaling.

2 Related Work

2.1 Active Forgery Localization

Digital watermarking methods [1,13] pre-embedded a digital watermarking into a generated image, then the image is judged as tampered if the integrity of the embedded digital watermarking has been disrupted, and the tampered regions can be fuzzily located by comparing the pre-embedded and disrupted digital watermarking. Since then, reversible digital watermarking techniques [7,8,14,15, 18,25,26,33] have been proposed to detect forgery for medical images with high resolution. Although these methods improve reversibility and imperceptibility, they sacrifice several embedded information to reduce image distortion, thus they can only locate the forgery regions fuzzily.

2.2 Passive Forgery Localization

Traditional passive forgery localization methods [2,9,19,28,37] mostly depend on extracting manual features or statistical features. Recently, deep learning techniques such as convolutional neural networks (CNN) have been applied for image forgery detection [4,16,17,20,21,29–31,34,38], and they perform significantly surpass traditional methods. The above two categories of passive forgery localization methods both can precisely locate the forgery regions. But above mentioned methods all target natural images, which perform normally when detecting grayscale smooth regions in medical image.

In particular, the RRU-Net structure enhances the U-Net model by incorporating residual propagation and feedback modules to address the issue of gradient degradation problem and strengthen the representation of features. However, experimental evaluations were solely conducted on natural images, neglecting to leverage the excellent performance of U-Net in the field of medical image segmentation. However, experiments exclusively on natural images overlooked the exceptional performance of U-Net in medical image segmentation. This prompts our study to consider adopting it as a teacher model.

3 Methodology

Overview. As the workflow of APFL shown in Fig. 1, it consists of two steps: (1) *active forgery localization* based on reversible watermarking to locate fuzzy tampered regions. (2) *passive forgery localization* with a lightweight model to locate the refined tampered regions.

3.1 Active Fuzzy Localization with Reversible Watermarking

Watermarking Embedding. After medical image collection, it is necessary to embed watermarks to protect the image integrity. Firstly, we divide region of interest (ROI) and region of non interest (RONI) according to the diagnostic

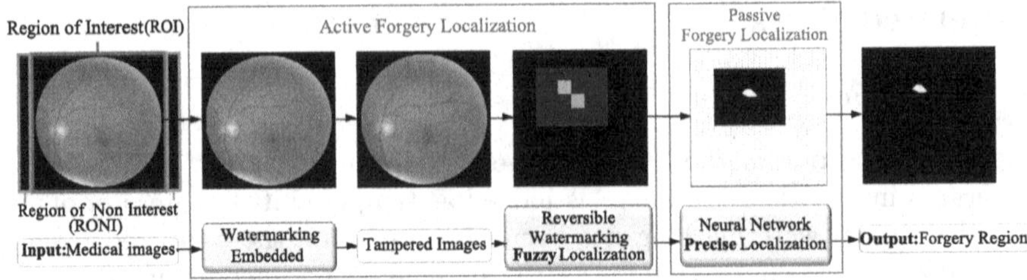

Fig. 1. Workflow of APFL.

importance of different locations. The visual quality of ROI cannot undergo any influence. As it contains important lesion information in medical images, it becomes indispensable for diagnosis. RONI is generally a meaningless black area which is hardly subject to tampering. So in APFL algorithm active forgery localization step the medical images are partitioned. The ROI is divided into regular and non-overlapping patches of size $m \times m$, and in each image patch, the pixels are sorted according to Zigzag rule starting from the top left corner to obtain the pixel sequence, and each pixel is recorded as $x_i, i \in [1, 32^2]$ in turn.

Four pixels are selected in sequence as the discriminant pixel patch in the pixel sequence, and C_T is the texture complexity. According to the two preset thresholds T_1 and T_2 ($T_1 < T_2$), the discriminant pixel patches are classified as flat, rough or extended. If $C_T < T_1$, the discriminant pixel patch is a flat, and feature information can be embedded. If $C_T > T_2$, it's a rough patch that needs to be discarded. If $T_1 \leq C_T \leq T_2$, it's an extended patch. We should add pixel one by one and recalculate C_T and classify it again until the patch is recognized as flat or rough. Each flat patch can embed up to 2 bits of feature information, respectively embedded in the maximum pixel value and the minimum pixel value.

When embedding feature in a flat patch, it is necessary to sort the pixels $\{x_1, x_2, ..., x_n\}$ in the patch in ascending order to obtain a new pixel sequence $\{x_{\varphi(1)}, x_{\varphi(2)}, ..., x_{\varphi(n)}\}$, where $x_{\varphi(1)} \leq x_{\varphi(2)} \leq ... \leq x_{\varphi(n)}$. $(\varphi(1), \varphi(2), ..., \varphi(n))$ and $(1, 2, ..., n)$ are mapping relations. When $x_{\varphi(i)} = x_{\varphi(j)}$ and $i < j$, $\varphi(i) < \varphi(j)$. The feature information is embedded in the maximum and minimum values of the flat block pixel sequence respectively.

The maximum pixel value translation expansion method is as follows, and the prediction error of the flat patch is $d_{max} = x_u - x_v$,where

$$\begin{cases} u = min(\varphi(n), \varphi(n-1)), \\ v = max(\varphi(n), \varphi(n-1)). \end{cases} \tag{1}$$

By shifting d_{max}, a new prediction error d'_{max} can be obtained. The new maximum pixel value $x'_{\varphi(n)}$ which embedded with feature information is:

$$x'_{\varphi(n)} = x_{\varphi(n)} + |d'_{max}| = \begin{cases} x_{\varphi(n)} + b, & d_{max} = 1, \\ x_{\varphi(n)} + 1, & d_{max} > 1, \\ x_{\varphi(n)} + b, & d_{max} = 0, \\ x_{\varphi(n)} + 1, & d_{max} < 0. \end{cases} \quad (2)$$

where $b \in \{0, 1\}$ represents the feature information to be embedded. Embedding information into the minimum pixel value is done in a similar way.

Auxiliary information includes the boundary coordinates of the medical image partition, the compressed location map are embedded in the RONI to replace the least significant bit of each pixel value in turn.

Watermarking Extraction and Fuzzy Localization. APFL algorithm embeds a reversible watermarking in the ROI to recover original image completely and losslessly when an image is without tampering. While the medical image is subject to tampering attack, the watermarking can be used to approximately identify the forgery region. Our watermarking scheme ensures the reversibility of the ROI region and also enables the fuzzy localization when forgery occurs.

Extracting features is similar to embedding features. Only flat patches need to further extract feature information. The pixels $\{x'_1, x'_2, ..., x'_n\}$ of the flat patch are sorted ascending, and obtain a new sequence $\{x'_{\varphi(1)}, x'_{\varphi(2)}, ..., x'_{\varphi(n)}\}$. When $x'_{\varphi(i)} = x'_{\varphi(j)}$and $i < j$,$\varphi(i) < \varphi(j)$. The feature information is extracted from the maximum and minimum values of the flat patch pixels respectively. The method of restoring the original maximum pixel value is as follows, and the prediction error of the flat patch is calculated as: $d'_{max} = x'_u - x'_v$.

When $d'_{max} > 0$, $x'_u > x'_v$, so $\varphi(n-1) > \varphi(n)$. When $d'_{max} \in \{1, 2\}$, the original maximum pixel value $x_{\varphi(n)} = x'_u - d'_{max} + 1$. When $d'_{max} > 2$, no feature information is embedded, so $x_{\varphi(n)} = x'_u - 1$. When $d'_{max} \leq 0$, $x'_u \leq x'_v$, so $\varphi(n-1) < \varphi(n)$. When $d'_{max} \in \{0, -1\}$, the original maximum pixel value $x_{\varphi(n)} = x'_v + d'_{max}$. When $d'_{max} < -1$, $x_{\varphi(n)} = x'_v - 1$.

The method of restoring the minimum pixel value in the flat patch is similar to the maximum pixel value.

The features of the ROI after extraction is recalculated, comparing the size of the extracted feature with the recalculated feature. If same, it means there is no forgery and the ROI has been restored to the original image. If different, it means the image patch may suffer from forgery attack. To generate a fuzzy localization result of the forgery region, it is necessary to set all pixel values to 255 in the image patches with different feature information and all pixel values to 0 in the remaining patches accordingly.

3.2 Passive Precise Localization with Lightweight KDU-Net

Forgery Focus. Since active fuzzy localization with reversible watermarking has located the fuzzy forgery region, we only focus on the fuzzy forgery region in active fuzzy localization to refine the forgery region.

Specifically, as shown in Fig. 1, we crop the rectangular which wraps the fuzzy forgery region and input it after patching into a neural network for locating forgery precisely. The focus on the fuzzy forgery region reduces the redundant information input into the network, which is beneficial to a precise location. Cropping the region into $k \times k$ patches with a stride s not only directs the model's attention to detailed information but also alleviates computational burden.

Lightweight KDU-Net via Knowledge Distillation. Since medical diagnostic devices in hospitals often have limited computing power and neither the doctors nor the patients can tolerate the long waiting time for images, it's necessary to train a lightweight model for inference. To this end, we utilize knowledge distillation to train a lightweight KDU-Net as the student model guided by an RRU-Net [4] as the teacher model. The training process and the structures of the two models are shown in Fig. 2. We construct the distillation loss (soft loss) proposed in [11]:

$$L_{kd} = T^2 KL(\delta(\frac{y^t}{T}), \gamma(\frac{y^s}{T})), \tag{3}$$

where KL represents the KL divergence between the prediction y^t of the teacher model after softmax δ and the prediction y^s of the student model after log softmax γ. T is a temperature factor for smoothing output.

The hard loss of the teacher model RRU-Net is the sum of a Cross-Entropy loss [36] and a Dice loss [35], i.e.,

$$L_{teacher} = L^t_{CE} + L^t_{Dice}. \tag{4}$$

The loss of the student model KDU-Net consists of hard loss (also the sum of Cross-Entropy loss and Dice loss) and distillation (soft) loss, i.e.,

$$L_{student} = (1 - \alpha)(L^s_{CE} + L^s_{Dice}) + \alpha L_{kd}. \tag{5}$$

The specific knowledge distillation process consists of three steps: (i) Pre-training teacher model: using the training set to pre-train a teacher model with good performance on forgery localization. (ii) Training the student model: during the training process, the same medical image is input into both two models. The output of teacher model is used as the soft label and the ground truth is used as the hard label to calculate the distillation loss and hard loss, respectively, and the weighted sum is used to generate the student model loss. (iii) Prediction: locate the forgery of the test set using only the student model KDU-Net.

After training through knowledge distillation, the lightweight KDU-Net can perform similar to RRU-Net while KDU-Net only has 24.6% size of RRU-Net, which can be deployed on medical devices with constrained computing power and storage for precise forgery localization.

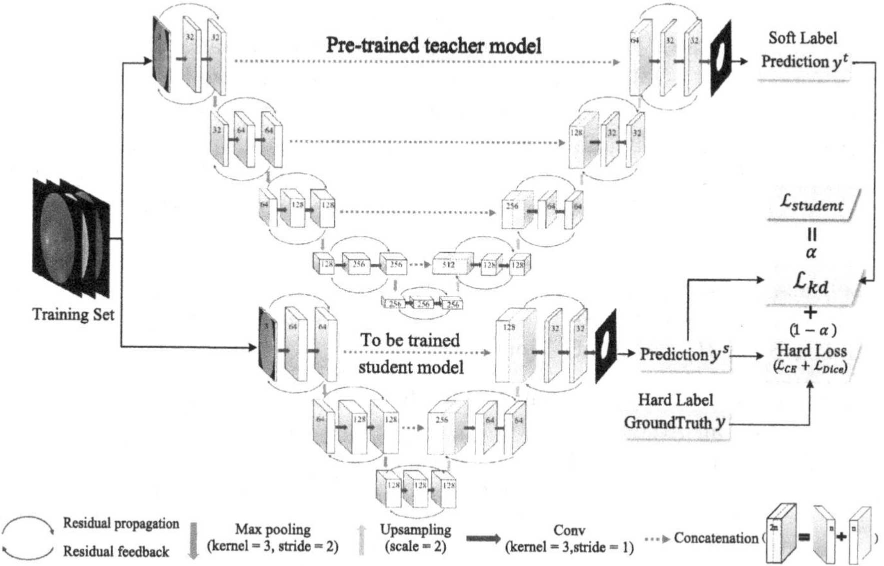

Fig. 2. Knowledge distillation with a lightweight KDU-Net as the student model (used for inference) and an RRU-Net as the teacher model.

4 Experiments

We implement APFL and baselines with PyTorch on NVIDIA GeForce RTX 3090 GPUs with 24G memory and evaluate them on our tampered TOIA datasets.

4.1 Settings

Datasets. Since there is no publicly available dataset of medical image forgery, We manually tamper the medical images from a real-world Ophthalmic Image Analysis (OIA) fundus datasett[1] by randomly selecting positions within lesion areas to enlarge or conceal abnormalities. The tampered dataset is named Tampered Ophthalmic Image Analysis (TOIA). In practice, for 10000 real fundus images with 768×512 pixels in the OIA dataset, digital watermarking is embedded into all the images. And only 80% of the images are forged by splicing, copy-move and removal forgery. The processed TOIA dataset is then divided into the training set, validation set and test set with a ratio of 8:1:1.

Evaluation Metrics. We choose pixel-level metrics including F_1 *score, True Positive Rate (TPR) and False Positive Rate (FPR)* [23]. Higher F_1, higher TPR and lower FPR mean better forgery localization performance. F_1 takes both precision and recall into account, so is a better measure of localization

[1] https://github.com/nkicsl/OIA.

performance compared to accuracy in the case of sample imbalance. TPR is the ratio between the number of pixels detected as forged and the number of actually tampered pixels, reflecting the missed detection rate. FPR is the ratio between the number of normal pixels wrongly detected as forged and the number of actually normal pixels, reflecting the false detection rate. Besides, we evaluate the watermarking perceptibility and image distortion with *Peak Signal Noise Ratio (PSNR)* [12]. A higher PSNR indicates less image distortion and better watermarking imperceptibility.

Baselines. We compare APFL with four baselines: 1) classical fully convolutional neural networks **FCN** [17] and **U-Net** [21]. 2) forgery localization models: **ManTra-Net** [31] and **RRU-Net** [4].

Hyperparameters. Hyperparameters of all algorithms are tuned to be optimal. All models use an Adam optimizer, a training epoch of 100, a batch size of 16, a learning rate of 0.0001, and a group normalization for stable performances. In our APFL, the distillation loss's coefficient $\alpha = 0.7$ and temperature $T = 2$.

Table 1. Results on TOIA. Bold and italic mean best and second best.

Method	F_1(%)	TPR(%)	FPR(%)
FCN	76.80	82.95	5.03
U-Net	81.08	89.11	6.38
ManTra-Net	86.69	76.98	*1.61*
RRU-Net	**96.13**	**95.23**	2.07
KDU-Net(undistilled)	87.60	90.29	3.17
KDU-Net(distilled)	*92.67*	*92.73*	**1.51**

4.2 Comparison Results

Table 1 shows that our APFL achieves the lowest FPR and comparable F_1 score and TPR with the best-performed RRU-Net, and the visualized results in Fig. 3 present that our APFL can precisely locate the forgery regions. Although APFL performs lower F_1 score and TPR than RRU-Net, the lightweight KDU-Net of APFL only has 3.86 MB parameters due to knowledge distillation while RRU-Net has 15.67 MB parameters, i.e., the model size of our lightweight KDU-Net is only 24.6% of RRU-Net, which speeds up inference on medical devices with limited storage and computing power.

4.3 Robustness Evaluation

In this section, we test the robustness of APFL to various numbers of forgery regions and forgery post-processing attacks.

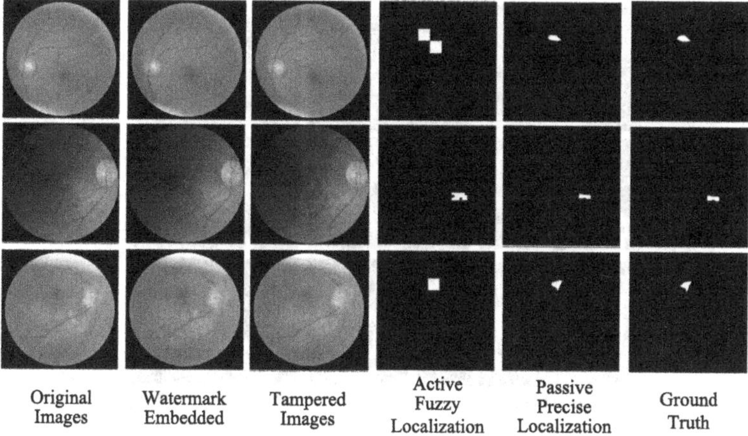

Fig. 3. Visualized forgery localization process of APFL.

Robustness to Numbers of Forgery Regions. The visualized results in Fig. 4 show that APFL can precisely locate the forgery regions with varied shapes whether one or multiple regions are forged, verifying its robustness to various numbers of forgery regions.

Robustness to Post-processing Attacks. To test the robustness of APFL to rotation and scaling post-processing attacks, we construct two specified test sets: **Rotation test set**, we randomly choose 50 untampered images from the above-divided test set, and for each image, we apply one of three forgery means (splicing, copy-move and removal) to the tampered regions, and then we add tampered region with varied rotation degrees (from 30° to 180°, step by 30°) on images, i.e., there are $50 \times 7 = 350$ tampered images generated. **Scaling test set**, we also randomly choose 50 untampered images from the above-divided test set, and for each image, we apply one of three forgery means to the tampered regions, and then we add tampered region with varied scaling factors (from 91% to 109%, step by 2%) on images, i.e., there are $50 \times 10 = 500$ tampered images.

Figure 5 (a) and (b) present that our APFL performs stable with varied rotation degrees and scaling factors, showing robustness to rotation and scaling post-processing attacks.

4.4 Hyperparameter Evaluation

The active tamper detection part embeds feature information and auxiliary information respectively to the ROI and RONI of the medical image. The features are embedded in a reversible watermark form, and the images need to be divided into patches before embedding. The number of patches leads to the difference in the amount of information to be embedded. The smaller the patch, the more information is embedded. The auxiliary information is embedded in an irre-

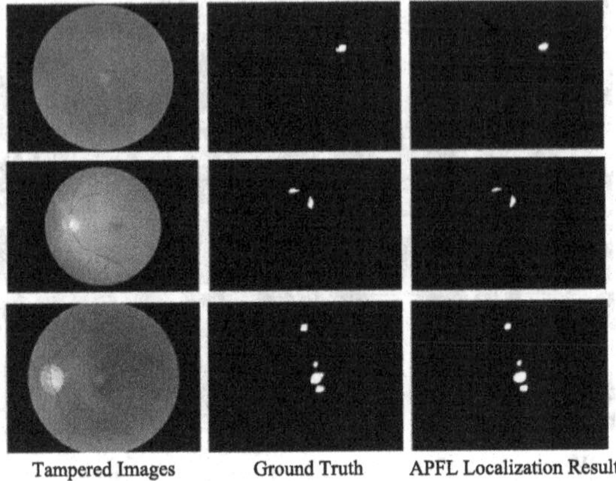

Fig. 4. Visualized results of APFL under single or multiple forgery regions.

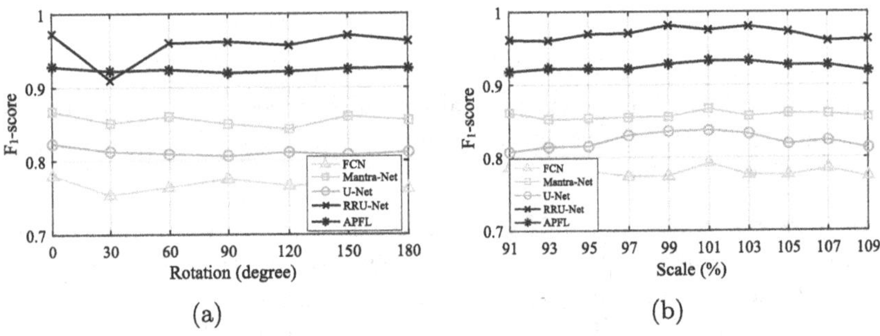

Fig. 5. F_1 score varies with different rotation degrees and scaling factors.

versible form, so after the watermarking is extracted, only the ROI can restore the original image losslessly, and the RONI still has distortion.

To evaluate the watermarking embedding capacity and perceptibility of the APFL algorithm and baselines for different patch sizes $m \times m$. As shown in table 2, we conducted experiments with different medical image patch sizes.

It can be seen that the watermarking of the APFL algorithm performs well in both imperceptibility and embedding capacity. Due to the large size of medical images in the TOIA dataset, in order to increase the capacity of watermarking embedding as much as possible while reducing the distortion of medical images, we chose 32×32 as the final image patch size.

4.5 Ablation Study

To verify the effects of active and passive steps in APFL, we test three cases: (A) only active forgery localization; (B) only passive forgery localization by inputting uncropped images into KDU-Net; (C) APFL with both two steps.

Table 2. Watermarking performance of APFL under different image patch sizes.

image patch size(pixels)	PSNR after embedding the watermarking(dB)	Embedded information(bits)	PSNR after extracting the watermarking(dB)
16 × 16	66.31	**96632**	82.39
32 × 32	70.01	23758	88.27
64 × 64	78.46	6013	90.51
96 × 96	**81.99**	2601	**92.04**

Table 3. Results of ablation studies. A higher TPR (lower missed detection rate) and a lower FPR (lower false detection rate) mean better performances.

Case	Active	Passive	F_1(%)	TPR(%)	FPR(%)
A	✓	✗	*78.93*	*89.30*	7.84
B	✗	✓	76.96	76.17	**0.56**
C	✓	✓	**92.67**	**92.73**	*1.51*

Table 3 shows that case-(A) performs higher TPR and FPR than case-(B). Since active forgery localization can only locate fuzzy forgery regions, most tampered pixels can be located but several normal pixels may also be detected as tampered, which verifies the effectiveness of active forgery localization in reducing missed detection rate. Because passive forgery localization fed images with uncropped complete size into neural networks and the tampered regions only occupy a small part of one image, the redundant information from un-tampered regions disturb forgery detection, thus leading to a larger missed detection rate. But the false detection rate is reduced for case-(B) since neural networks can learn more detailed information. Case-(C) involving the two steps performs the highest TPR and low FPR due to the necessity of the missed detection rate reduced by active forgery localization and the false detection rate dropped by passive forgery localization.

5 Conclusion

In this paper, we proposed APFL, a novel medical image forgery localization algorithm combining active and passive tamper detection to locate tampered regions by three common forgery means: splicing, copy-move and removal. In the active fuzzy localization step, we use reversible watermarking to locate fuzzy forgery regions. In the passive precise localization step, we train a lightweight KDU-Net through knowledge distillation to locate precise forgery regions further. The experimental results verify the effectiveness of APFL in locating regions tampered with the common three forgery means and the robustness to varied numbers of forgery regions, rotation and scaling post-processing attacks.

References

1. Ansari, I.A., et al.: SVD based fragile watermarking scheme for tamper localization and self-recovery. Int. J. Mach. Learn. Cybern. **7**, 1225–1239 (2016)
2. Ardizzone, E., et al.: Copy-move forgery detection by matching triangles of keypoints. IEEE Trans. Inf. Forensics Secur. **10**(10), 2084–2094 (2015)
3. Arun Anoop, M., Poonkuntran, S.: LPG: a novel approach for medical forgery detection in image transmission. J. Ambient. Intell. Humaniz. Comput. **12**(5), 4925–4941 (2021)
4. Bi, X., et al.: Rru-net: The ringed residual u-net for image splicing forgery detection. In: Proceedings of the IEEE/CVF Conference on Computer Vision and Pattern Recognition Workshops. pp. 0–0 (2019)
5. Bik, E.M., et al.: The prevalence of inappropriate image duplication in biomedical research publications. MBio **7**(3), 16 (2016)
6. Castillo Camacho, I., et al.: A comprehensive review of deep-learning-based methods for image forensics. J. Imaging **7**(4), 69 (2021)
7. Coatrieux, G., et al.: Reversible watermarking based on invariant image classification and dynamic histogram shifting. IEEE Trans. Inf. Forensics Secur. **8**(1), 111–120 (2012)
8. Feng, B., et al.: A reversible watermark with a new overflow solution. IEEE Access **7**, 28031–28043 (2018)
9. Fridrich, J.: Detection of copy-move forgery in digital images. In: Proceedings of Digital Forensic Research Workshop, vol. 2003 (2003)
10. Ghoneim, A., et al.: Medical image forgery detection for smart healthcare. IEEE Commun. Mag. **56**(4), 33–37 (2018)
11. Hinton, G., et al.: Distilling the knowledge in a neural network. Comput. Sci. **14**(7), 38–39 (2015)
12. Hore, A., Ziou, D.: Image quality metrics: PSNR vs. SSIM. In: 2010 20th International Conference on Pattern Recognition, pp. 2366–2369. IEEE (2010)
13. Hsu, C.S., Tu, S.F.: Digital watermarking scheme for copyright protection and tampering detection. Int. J. Inf. Technol. Secur. **11**(1) (2019)
14. Ishtiaq, M., et al.: Hybrid predictor based four-phase adaptive reversible watermarking. IEEE Access **6**, 13213–13230 (2018)
15. Lei, B., et al.: Reversible watermarking scheme for medical image based on differential evolution. Expert Syst. Appl. **41**(7), 3178–3188 (2014)
16. Liu, B., Pun, C.M.: Deep fusion network for splicing forgery localization. In: Proceedings of the European Conference on Computer Vision (ECCV) Workshops (2018)
17. Long, J., et al.: Fully convolutional networks for semantic segmentation. In: Proceedings of the IEEE Conference on Computer Vision and Pattern Recognition, pp. 3431–3440 (2015)
18. Luo, L., et al.: Reversible image watermarking using interpolation technique. IEEE Trans. Inf. Forensics Secur. **5**(1), 187–193 (2009)
19. Luo, W., et al.: Robust detection of region-duplication forgery in digital image. In: 18th International Conference on Pattern Recognition (ICPR 2006), vol. 4, pp. 746–749. IEEE (2006)
20. Rao, Y., Ni, J.: A deep learning approach to detection of splicing and copy-move forgeries in images. In: 2016 IEEE International Workshop on Information Forensics and Security (WIFS), pp. 1–6. IEEE (2016)

21. Ronneberger, O., Fischer, P., Brox, T.: U-net: convolutional networks for biomedical image segmentation. In: Navab, N., Hornegger, J., Wells, W., Frangi, A. (eds.) Medical Image Computing and Computer-Assisted Intervention - MICCAI 2015, MICCAI 2015. LNCS, vol. 9351, pp 234–241. Springer, Cham (2015). https://doi.org/10.1007/978-3-319-24574-4_28
22. Sharma, S., Ghanekar, U.: A rotationally invariant texture descriptor to detect copy move forgery in medical images. In: 2015 IEEE International Conference on Computational Intelligence & Communication Technology, pp. 795–798. IEEE (2015)
23. Shi, J., Wang, G., Su, M., Liu, X.: Effective medical image copy-move forgery localization based on texture descriptor. In: Goel, S., Gladyshev, P., Johnson, D., Pourzandi, M., Majumdar, S. (eds.) Digital Forensics and Cyber Crime. ICDF2C 2020. LNICS, Social Informatics and Telecommunications Engineering, vol. 351, pp. 62–77. Springer, Cham (2021). https://doi.org/10.1007/978-3-030-68734-2_4
24. Suganya, D., et al.: Copy-move forgery detection of medical images using most valuable player based optimization. Sens. Imaging **22**(1), 1–18 (2021)
25. Thabit, R., Khoo, B.E.: A new robust lossless data hiding scheme and its application to color medical images. Digit. Sig. Process. **38**, 77–94 (2015)
26. Thodi, D.M., Rodríguez, J.J.: Expansion embedding techniques for reversible watermarking. IEEE Trans. Image Process. **16**(3), 721–730 (2007)
27. Ulutas, G., et al.: Medical image tamper detection based on passive image authentication. J. Digit. Imaging **30**(6), 695–709 (2017)
28. Wang, W., et al.: Effective image splicing detection based on image chroma. In: ICIP, pp. 1257–1260. IEEE (2009)
29. Wei, Y., et al.: C2r net: the coarse to refined network for image forgery detection. In: TrustCom/BigDataSE, pp. 1656–1659. IEEE (2018)
30. Wu, Y., et al.: BusterNet: detecting copy-move image forgery with source/target localization. In: Proceedings of the European Conference on Computer Vision (ECCV), pp. 168–184 (2018)
31. Wu, Y., et al.: Mantra-net: Manipulation tracing network for detection and localization of image forgeries with anomalous features. In: Proceedings of CVPR, pp. 9543–9552 (2019)
32. Yi, L., Zhang, J., Zhang, R., Shi, J., Wang, G., Liu, X.: SU-Net: an efficient encoder-decoder model of federated learning for brain tumor segmentation. In: Farkaš, I., Masulli, P., Wermter, S. (eds.) Artificial Neural Networks and Machine Learning - ICANN 2020, ICANN 2020, LNCS, vol. 12396, pp 761–773. Springer, Cham (2020). https://doi.org/10.1007/978-3-030-61609-0_60
33. Zhang, X., et al.: Reversible fragile watermarking for locating tampered blocks in jpeg images. Signal Process. **90**(12), 3026–3036 (2010)
34. Zhang, Y., et al.: Image region forgery detection: a deep learning approach. SG-CRC **2016**, 1–11 (2016)
35. Zhang, Y., et al.: Rethinking the dice loss for deep learning lesion segmentation in medical images. J. Shanghai Jiaotong Univ. (Science) **26**, 93–102 (2021)
36. Zhang, Z., Sabuncu, M.: Generalized cross entropy loss for training deep neural networks with noisy labels. In: Advances in Neural Information Processing Systems, vol. 31 (2018)
37. Zhu, Y.: Sothers: covert copy-move forgery detection based on color LBP. Acta Automatica Sinica **43**(3), 390–397 (2017)
38. Zhu, Y., et al.: Hrda-net: image multiple manipulation detection and location algorithm in real scene. J. Commun. **43**(1), 217–226 (2022)

A Universal Non-parametric Approach for Improved Molecular Sequence Analysis

Sarwan Ali[1](✉), Tamkanat E Ali[2], Prakash Chourasia[1], and Murray Patterson[1]

[1] Georgia State University, Atlanta, GA, USA
{sali85,pchourasia1}@student.gsu.edu, mpatterson30@gsu.edu
[2] Lahore University of Management Science, Lahore, Pakistan
20100159@lums.edu.pk

Abstract. In the field of biological research, it is essential to comprehend the characteristics and functions of molecular sequences. The classification of molecular sequences has seen widespread use of neural network-based techniques. Despite their astounding accuracy, these models often require a substantial number of parameters and more data collection. In this work, we present a novel approach based on the compression-based Model, motivated from [1], which combines the simplicity of basic compression algorithms like Gzip and Bz2, with Normalized Compression Distance (NCD) algorithm to achieve better performance on classification tasks without relying on handcrafted features or pre-trained models. Firstly, we compress the molecular sequence using well-known compression algorithms, such as Gzip and Bz2. By leveraging the latent structure encoded in compressed files, we compute the Normalized Compression Distance between each pair of molecular sequences, which is derived from the Kolmogorov complexity. This gives us a distance matrix, which is the input for generating a kernel matrix using a Gaussian kernel. Next, we employ kernel Principal Component Analysis (PCA) to get the vector representations for the corresponding molecular sequence, capturing important structural and functional information. The resulting vector representations provide an efficient yet effective solution for molecular sequence analysis and can be used in ML-based downstream tasks. The proposed approach eliminates the need for computationally intensive Deep Neural Networks (DNNs), with their large parameter counts and data requirements. Instead, it leverages a lightweight and universally accessible compression-based model. Also, it performs exceptionally well in low-resource scenarios, where limited labeled data hinder the effectiveness of DNNs. Using our method on the benchmark DNA dataset, we demonstrate superior predictive accuracy compared to SOTA methods.

Keywords: Classification · Sequence Analysis · Compression · Gzip

1 Introduction

Molecular sequence analysis stands as a pivotal pursuit in contemporary research, holding the key to unraveling the intricate language encoded in molecular sequences, such as DNA and proteins. The accurate comprehension and classification of these

S. Ali and T.E. Ali—Joint First Authors.

D.-N. Yang et al. (Eds.): PAKDD 2024, LNAI 14648, pp. 194–206, 2024.
https://doi.org/10.1007/978-981-97-2238-9_15

molecular sequences offer profound insights into their structural, functional, and evolutionary characteristics. As the foundation of numerous biological studies, including functional gene annotation, drug discovery, and evolutionary biology, molecular sequence analysis plays an indispensable role in advancing our understanding of the fundamental processes governing life. The pursuit of innovative methodologies in this realm is driven by the quest for more accurate, efficient, and resource-effective approaches to decipher the rich information concealed within the sequences, ultimately contributing to transformative breakthroughs in the broader landscape of molecular biology.

Several methods have been used for the classification of molecular sequences involving Neural Networks, language models, Feature Embedding, and Kernel functions. All these methods face certain challenges when it comes to achieving good accuracy in cases when the available data is less. Neural Network (NN) based methods are one of the most widely employed in molecular sequence classification and have demonstrated impressive accuracy in many cases [2]. However, these methods come with significant limitations including the requirement of a substantial number of parameters and long training times, making them computationally expensive and resource-demanding. Additionally, neural networks and NN-based language models heavily rely on large-scale training data, which may not be readily available for certain biological datasets, particularly in low-resource or rare species scenarios.

Designing low-dimensional embedding for molecular sequences is a challenging task. One of the most feasible solutions to this challenge is sequence data compression. Some of the well-known compression methods include Gzip, zlib, Bz2 [3] etc. Gzip is used on a large scale for lossless data compression [4], it became popular because of certain characteristics which include being free, open source, robust, compact, portable, has low memory overhead, and has reasonable speed [5]. Due to its inertia and its integration with so many sequence analysis tools, even today most of the sequence databases rely on Gzip [5]. The zlib and Bz2 compression algorithms efficiently detect non-randomness and low information content [6]. Their performance gets better as the string length increases. Bz2 compression is used to compress the strings and is not affected by the mass ratios, it does not include character order information due to the process of permuting characters during compression, this negatively affects the accuracy [1]. Our compression-based model offers several notable advantages over traditional neural network-based approaches. Firstly, it eliminates the computationally intensive nature of deep neural networks, reducing the parameter requirements and making them more lightweight and accessible. Secondly, by leveraging Gzip compression, our approach can efficiently handle low-resource biological datasets where labeled data is scarce or limited. This enables us to analyze molecular sequences even in resource-constrained scenarios. Our contributions can be summarized as follows:

- We propose a novel approach for analyzing and classifying molecular sequences, using compression-based models including Gzip and Bz2.
- We develop an algorithm for Distance Matrix computation, in which we take a set of sequences as input and output a non-symmetric Distance matrix using Normalized Compression Distance (NCD) and different compressors.

- We convert the distance matrix into a kernel matrix and extract the low dimensional numerical representation in the end, which can be used as input to any linear and nonlinear machine learning model for supervised and unsupervised analysis. In this way, we also addressed the limitation in [1] where they were only able to apply the k-nearest neighbor classifier for the classification purpose. Hence we show that our proposed method can generalize better for sequence classification.
- We also discuss the theoretical justifications for the proposed pipeline including the symmetry of the distance matrix and reproducing Kernel Hilbert Space for the kernel matrix along positive semi-definite, smoothness, continuity, and sensitivity.

The rest of the paper is organized as follows: In Sect. 2 we will give details of the literature review followed by the proposed method along with the experimental setup in Sect. 3. Our results and experimental evaluation are reported in Sect. 4. Finally, Sect. 5 concludes the paper, emphasizing the effectiveness and potential of our compression-based model in advancing molecular sequence analysis and classification.

2 Related Work

Molecular sequence analysis is based on two types of methods, alignment-based and alignment-free. Alignment-based is suitable for small sequences (due to higher dimensionality). The alignment-free method works well for both short and long sequences [7]. Several methods have been used for the analysis of molecular sequences, including neural networks (NN) [8], language models [9], feature embedding [10], and kernel functions [11]. In recent work, it was seen that the performance of protein prediction tasks can be improved by training language models on protein sequences [12]. Pretrained language models and word embedding methods have proved to be successful in embedding molecular sequences forming easy-to-process representations that are context-sensitive [13]. In the analysis of big data in proteomics, SeqVec provides a scalable approach for the analysis of the protein data [12], which has improved the study of the structure and composition of proteins. ProtBert is a transformer-based model, that uses a masked language model [9] it requires positional encoding and has a high memory requirement. In the case of NN, several methods have been proposed for sequence analysis [14]. Variational AutoEncoder-based methods are also used in the literature for molecular sequence analyses [8]. These NN-based methods prove to be computationally expensive, face increased risk of overfitting, and are resource-demanding. In recent works, deep learning-based feature representation methods have been proposed [15,16]. Several authors proposed feature engineering-based methods to design embeddings for the molecular sequences [17]. Although such methods are efficient in terms of predictive performance, they usually face the problem of the "curse of dimensionality" due to the higher dimensions of the generated vectors. Another method used for the sequence analysis is to project the data into high dimensional feature space using kernel matrices [11,18]. However, these methods could cause an overfitting problem along with scalability issues (memory intensive) [19]. Some prominent sequence comparison methods include Normalized Compression Distance (NCD) [20], Normalized Information Distance (NID) [21], Euclidean Distance, and Manhattan Distance. The NCD, derived from the concept of Kolmogorov complexity [22], provides a measure of similarity between sequences by considering their compressed file sizes. However, such methods are not used in the literature for representation learning, specifically for molecular sequence analysis.

3 Proposed Approach

We propose an Embedding generation method based on lossless compressors and Normalized Compression Distance (NCD) metric. We start this section by discussing the lossless compression methods below.

3.1 Compression Methods

Bz2 Compressor: Bz2 compressor is a general-purpose lossless compressor, based on the Burrows-Wheeler transform (BWT) and Huffman coding [23]. The BWT is a permutation of the letters of the text (in our case characters/nucleotides of the sequence). After applying BWT to the input, an easily compressible form is generated as it groups symbols into runs of similar units, more precisely the input is divided into blocks of at most 900 kB, which is compressed separately, keeping in regard to the local similarities in the data. The compression of the transform includes an initial move-to-front encoding with run length encoding and then Huffman encoding.

Gzip Compressor: Gzip uses very few bits to represent information, which is based on the lossless compression algorithm LZ77 (Lempel-Ziv 77 compression algorithm) [24] and dynamic Huffman algorithm [4]. LZ77-Lempel-Ziv compression algorithm encodes a string based on sequential processing. If the present substring was encountered earlier as well then it is encoded with reference to the previous one. A sliding window is used for every new sequence encountered. Huffman coding is statistical-based compression, where the symbols are encoded using statistical information such as frequency distribution. There are two types of Huffman coding, dynamic and static. As the data we use is not real-time we use the Dynamic Huffman algorithm, which is a two-pass algorithm. In the first pass, the frequency distribution of symbols is calculated and in the second pass, symbols are encoded. In this technique, depending on the occurrence of the symbols variable length codes are assigned to symbols such that symbols with less occurrence are encoded with more significant bits and symbols with high frequency are encoded with fewer bits, as a result, a good compression ratio is obtained.

Remark 1. Note that our proposed method uses both the compression methods described above separately.

3.2 Problem Formulation

Given a pair of sequences s_1 and s_2, where $s_1, s_2 \in S$ (S is a set of all sequences), we first encode s_1 and s_2 using UTF-8 encoding [25], which will give us E_{s_1} and E_{s_2}. After encoding, the E_{s_1} and E_{s_2} are compressed using Gzip or Bz2. We will then get the compressed form, denoted by C_{s_1} and C_{s_2}. In the next step, we compute the length L_{s_1} and L_{s_2} of the compressed sequences. In a similar way, we compute $L_{s_1 s_2}$, which denotes the length of compressed encoded form for the concatenated sequence $s_1 s_2$. We then use L_{s_1}, L_{s_2}, and $L_{s_1 s_2}$ as input to Normalized Compression Distance (NCD) approach to get the final distance value, which is calculated using the following expression:

$$NCD(s_1, s_2) = \frac{Ls_1 s_2 - min\{Ls_1, Ls_2\}}{max\{Ls_1, Ls_2\}} \tag{1}$$

In the condition where $s_1 \neq s_2$, the bytes B needed to encode s_2 based on s_1 information, i.e. B_{12} can be computed using the following expression:

$$B_{12} = Ls_1 s_2 - Ls_1 \tag{2}$$

Similarly, for s_1 and s_3 sequences, we have

$$B_{13} = Ls_1 s_3 - Ls_1 \tag{3}$$

where B_{13} represents the number of bytes needed to encode s_3 based on s_1 information. Given a scenario where s_1 and s_2 belong to the same category but s_3 belongs to a different category than s_1 and s_2, we have the following expression:

$$B_{12} < B_{13} \tag{4}$$

The formulation of the concept mentioned in Equation (4) can be linked to Kolmogorov complexity Z_s (where $s \in S$) and its derived distance metric [22]. Z_s is the lower bound for measuring information as it represents the length of the shortest binary program that outputs s, but there is a limitation that it cannot be used to measure the information content shared between two objects due to the incomputable nature of Z_s [1]. To overcome this limitation, Normalized Compression Distance (NCD) is proposed [26], which is computable and uses compressed length L_s to approximate Kolmogorov complexity Z_s.

The underlying concept of using compressed length in Eq. 1 is that the compressed length is close to Z_s. The general rule says, that the higher the compression ratio, the closer L_s is to Z_s. Using the NCD-based distance (from Eq. 1), we compute pairwise distances for a set of sequences to generate the required distance matrix.

3.3 Our Algorithm

In our algorithmic approach (i.e. in Algorithm 1), we take in a set of sequences (S) as input and output a Distance Matrix (D). We iterate through the Set S, for every sequence referred to as s in our data S and carry out the following steps:

1. Encoded form is generated and stored in a variable Es_1 (line number 2 of Algorithm 1 and step c(i) of Fig. 1)
2. Encoded Es_1 is further compressed using Gzip compressor and fed into Cs_1 (line number 3 of Algorithm 1 and step d(i) of Fig. 1)
3. Calculate the length of the compressed Cs_1 and store in a variable referred to as Ls_1.(line number 4 of Algorithm 1 and step e(i) of Fig. 1)

To save the Normalized Compression Distance (NCD) between every s and the rest of the sequences in the Set S, we initialize an array termed as D_ local as shown in the line number 5 of Algorithm 1.

In another sub-iterative loop, we repeat the steps from 1 to 3 mentioned above for every other sequence in set S (line number 6 of Algorithm 1). To calculate NCD we first require concatenation of s_1 and s_2 (line number 10 of Algorithm 1 and step (b) of Fig. 1), followed by encoding, compression, and calculating its length which is stored in a variable Ls_1s_2 (line numbers 11–13 of Algorithm 1 and steps (c)-(e) of Fig. 1).

Now using the length of the compressed encoded sequences Ls_1, Ls_2 and Ls_1s_2, we calculate NCD (line number 14 of Algorithm 1 and step (f) of Fig. 1) and store in the list referred to as D_ local(line number 15 of Algorithm 1). At the end of the inner iterative loop, this list is appended in a distance matrix D which would contain the NCD values between every sequence in the set of sequences(S)(line number 17 of Algorithm 1).

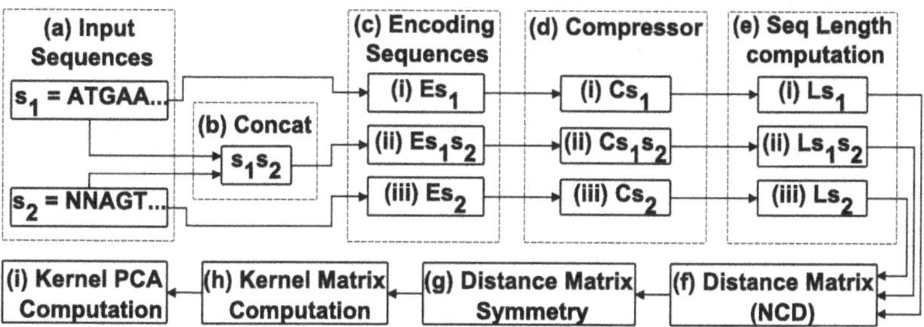

Fig. 1. Overview of the proposed approach.

The Fig. 1 shows the overview of the proposed approach.

Remark 2. Our method is better than Deep Neural Networks as there is no need for preprocessing or training, making it simpler. Secondly, fewer parameters with no GPU resources are needed for distance matrix computation, making it lighter, and the absence of underlying assumptions (e.g., assumptions about the data) makes it universal.

Remark 3. To further understand the idea of NCD-based pairwise distance computation between text/molecular sequences, readers are referred to [1],

3.4 Distance Matrix Symmetry

The Distance matrix (D) obtained using Normalized Compression Distance (NCD) is of size $n \times n$ where n represents the cardinality of the input set S. This Distance Matrix D is non-symmetric, so to convert it to symmetric matrix D' we take the average of upper and lower triangle values and replace the original values of the matrix with the average values.

Algorithm 1. Distance matrix computation with Gzip

```
Input:Set of sequences(S)
Output:Distance Matrix(D)
for s1 in S do
    Es1 ← encoded s1
    Cs1 ← Gzip compressed Es1
    Ls1 ← length of Cs1
    D_local ← [ ]
    for s2 in S do
        Es2 ← encoded s2
        Cs2 ← Gzip compressed Es2
        Ls2 ← length of Cs2
        s1s2 ← Concatenate(s1, s2)
        Es1s2 ← encoded s1s2
        Cs1s2 ← Gzip compressed Es1s2
        Ls1s2 ← length of Cs1s2
        NCD ← (Ls1s2 − Min(Ls1, Ls2)) / Max(Ls1, Ls2)
        D_local.append(NCD)
    end for
    D.append(D_local)
end for
return D
```

3.5 Kernel Matrix Computation

We generate a Kernel matrix from the symmetric distance matrix (D') of size $n \times n$ using a Gaussian Kernel, where.

$$d'_{ij}, d'_{ik} \in D' \tag{5}$$

Euclidean Distance (E) between two pairs of distances d'_{ij} and d'_{ik} (i.e. distance values computed from pairs of sequences) is calculated using the following equation:

$$E_{d'_{ij},d'_{ik}} = ||d'_{ij} - d'_{ik}|| \tag{6}$$

The Gaussian Kernel (K) is defined as a measure of similarity between d'_{ij} and d'_{ik}. It is represented by the equation below:

$$K(d'_{ij}, d'_{ik}) = exp(\frac{-||d'_{ij} - d'_{ik}||^2}{\sigma^2}) \tag{7}$$

where σ^2 represents the bandwidth of the kernel. The kernel value is computed as follows:

$$\begin{cases} K = 1 & \text{if } d'_{ij} \text{ and } d'_{ik} \text{ are identical} \quad (8) \\ K \rightarrow 0 & \text{if } d'_{ij} \text{ and } d'_{ik} \text{ move further apart} \quad (9) \end{cases}$$

The kernel value is computed for each pair of distances in D' to get the $n \times n$ dimensional kernel matrix. Once the kernel matrix is computed, we can leverage kernel Principal Component Analysis (PCA) to derive a lower-dimensional representation of the data. The resulting embeddings, known as kernel principal components, effectively preserve the essential information while retaining the relationships among the molecular sequences including non-linear relations. This representation proves valuable for various downstream tasks, including classification.

3.6 Experimental Setup

Here we describe dataset statistics and evaluation metrics in detail. The experiments are performed on a computer running 64-bit Windows 10 with an Intel(R) Core i5 processor running at 2.10 GHz and 32 GB of RAM. For experiments, we randomly split data into 60–10-30% for training-validation-testing purposes. The experiments are repeated 5 times, and we report average results. Our code is available online for reproducibility[1]. We use real-world molecular sequence data comprised of nucleotide sequences. The summary of the dataset used for experimentation is given in Table 1.

Table 1. Dataset Statistics.

Name	| Seq. |	| Classes |	Sequence Statistics			Reference	Description
			Max	Min	Mean		
Human DNA	4380	7	18921	5	1263.59	[27]	Unaligned nucleotide sequences to classify gene family to which humans belong

We used a variety of ML models for classification, including Support Vector Machine (SVM), Naive Bayes (NB), Multi-Layer Perceptron (MLP), K-Nearest Neighbours (KNN), Random Forest (RF), Logistic Regression (LR), and Decision Tree (DT). For performance evaluation, we used average accuracy, precision, recall, F1 (weighted), F1 (macro), Receiver Operator Characteristic Curve (ROC), Area Under the Curve (AUC), and training runtime. The baseline models we used for results comparisons include PWM2Vec [10] which gives each amino acid in the k-mers a weight based on where it is located in a k-mer position weight matrix (PWM), String Kernel [11] determines the similarity between the two sequences based on the total number of k-mers that are correctly and incorrectly aligned between two sequences, WDGRL [15] a neural network based unsupervised domain adoption technique that uses Wasserstein distance (WD) for feature extraction from input data, Autoencoder [28] that uses a deep neural network to encode data as features which involves iterative optimization of the objective through non-linear mapping from data space X to a smaller-dimensional feature space Z, SeqVec [12] an ELMO (Embeddings from Language Models) based method for representing biological sequences as continuous vectors, and Protein Bert [29] which is an end-to-end model that does not design explicit embeddings but directly performs classification using a previously learned language model.

[1] https://github.com/sarwanpasha/Non-Parametric-Approach.

3.7 Justification of Employing the Kernel Matrix

The generation of a kernel matrix from the NCD-based distance matrix offers several technical justifications and significant benefits:

1. **Nonlinearity:** Constructing a kernel matrix based on NCD distances implicitly maps the data into a higher-dimensional feature space, enabling the capture of intricate nonlinear relationships. This is particularly advantageous when dealing with the complex non-linear interactions often present in molecular sequences.
2. **Capturing Complex Relationships:** Utilizing the Gaussian kernel in generating the kernel matrix allows for the capture of intricate relationships between sequences. It assigns higher similarity values to similar sequences and lower values to dissimilar ones. This capability enables the representation of complex patterns and structures in the data, surpassing the limitations of linear methods.
3. **Theoretical Properties of Gaussian Kernel:** The use of the widely adopted Gaussian kernel leverages several underlying theoretical properties. These include the Reproducing Kernel Hilbert Space (RKHS) [30], enabling the application of efficient kernel methods like support vector machines. The Universal Approximation property [31] allows the kernel to approximate any continuous function arbitrarily well, making it a powerful tool for modeling complex relationships and capturing nonlinear patterns. Mercer's Theorem [32] guarantees that the kernel matrix is positive semi-definite, while the smoothness, continuity, and sensitivity to variations further enhance its ability to capture local relationships and adapt to variations in the data distribution.
4. **Flexibility and Generalization:** The kernel matrix derived from the NCD-based distance matrix can be effectively employed with various machine learning algorithms that operate on kernel matrices. This flexibility allows for the application of a wide range of techniques, including kernel PCA and kernel SVM. Leveraging these algorithms enables the exploitation of the expressive power of the kernel matrix to address diverse tasks, such as dimensionality reduction and classification.
5. **Preserving Nonlinear Information:** Applying kernel PCA to the kernel matrix captures crucial nonlinear information embedded in the data. This process facilitates the extraction of low-dimensional embeddings that preserve the underlying structure and patterns. Projecting the data onto the principal components retains discriminative information while reducing dimensionality. This proves particularly valuable for handling high-dimensional datasets.
6. **Enhanced Performance:** The utilization of the NCD-based distance matrix and kernel matrix can lead to improved performance in various tasks. The incorporation of NCD distance and the capturing of complex relationships in the data enhance the discriminative power of the embeddings. This results in a more accurate classification of the sequences.

4 Results and Discussion

The classification results that are averaged over 5 runs are reported in Table 2 for the Human DNA dataset. For the evaluation metrics including average accuracy, precision,

Table 2. Classification results (averaged over 5 runs) on **Human DNA** dataset for different evaluation metrics. The best classifier performance for every embedding is shown with the underline. Overall best values are shown in bold.

Embeddings	Algo.	Acc. ↑	Prec. ↑	Recall ↑	F1 (Weig.) ↑	F1 (Macro) ↑	ROC AUC ↑	Train Time (sec.) ↓
PWM2Vec	SVM	0.302	0.241	0.302	0.165	0.091	0.505	10011.3
	NB	0.084	0.442	0.084	0.063	0.066	0.511	4.565
	MLP	0.310	0.350	0.310	0.175	0.107	0.510	320.555
	KNN	0.121	0.337	0.121	0.093	0.077	0.509	2.193
	RF	0.309	0.332	0.309	0.181	0.110	0.510	65.250
	LR	0.304	0.257	0.304	0.167	0.094	0.506	23.651
	DT	0.306	0.284	0.306	0.181	0.111	0.509	1.861
String Kernel	SVM	0.618	0.617	0.618	0.613	0.588	0.753	39.791
	NB	0.338	0.452	0.338	0.347	0.333	0.617	0.276
	MLP	0.597	0.595	0.597	0.593	0.549	0.737	331.068
	KNN	0.645	0.657	0.645	0.646	0.612	0.774	1.274
	RF	0.731	0.776	0.731	0.729	0.723	0.808	12.673
	LR	0.571	0.570	0.571	0.558	0.532	0.716	2.995
	DT	0.630	0.631	0.630	0.630	0.598	0.767	2.682
WDGRL	SVM	0.318	0.101	0.318	0.154	0.069	0.500	0.751
	NB	0.232	0.214	0.232	0.196	0.138	0.517	**0.004**
	MLP	0.326	0.286	0.326	0.263	0.186	0.535	8.613
	KNN	0.317	0.317	0.317	0.315	0.266	0.574	0.092
	RF	0.453	0.501	0.453	0.430	0.389	0.625	1.124
	LR	0.323	0.279	0.323	0.177	0.095	0.507	0.041
	DT	0.368	0.372	0.368	0.369	0.328	0.610	0.047
Autoencoder	SVM	0.621	0.638	0.621	0.624	0.593	0.769	22.230
	NB	0.260	0.426	0.260	0.247	0.268	0.583	0.287
	MLP	0.621	0.624	0.621	0.620	0.578	0.756	111.809
	KNN	0.565	0.577	0.565	0.568	0.547	0.732	1.208
	RF	0.689	0.738	0.689	0.683	0.668	0.774	20.131
	LR	0.692	0.700	0.692	0.693	0.672	0.799	58.369
	DT	0.543	0.546	0.543	0.543	0.515	0.718	10.616
SeqVec	SVM	0.656	0.661	0.656	0.652	0.611	0.791	0.891
	NB	0.324	0.445	0.312	0.295	0.282	0.624	0.036
	MLP	0.657	0.633	0.653	0.646	0.616	0.783	12.432
	KNN	0.592	0.606	0.592	0.591	0.552	0.717	0.571
	RF	0.713	0.724	0.701	0.702	0.693	0.752	2.164
	LR	0.725	0.715	0.726	0.725	0.685	0.784	1.209
	DT	0.586	0.553	0.585	0.577	0.557	0.736	0.24
Protein Bert	-	0.542	0.580	0.542	0.514	0.447	0.675	58681.57
Gzip (ours)	SVM	0.692	0.844	0.692	0.699	0.692	0.771	2.492
	NB	0.464	0.582	0.464	0.478	0.472	0.704	0.038
	MLP	**0.831**	0.833	**0.831**	**0.830**	**0.813**	**0.890**	7.546
	KNN	0.773	0.792	0.773	0.776	0.768	0.856	0.193
	RF	0.810	**0.858**	0.810	0.812	0.811	0.858	6.539
	LR	0.621	0.822	0.621	0.616	0.581	0.712	0.912
	DT	0.648	0.651	0.648	0.648	0.624	0.780	2.590
Bz2 (ours)	SVM	0.545	0.769	0.545	0.524	0.501	0.669	2.856
	NB	0.403	0.577	0.403	0.411	0.410	0.653	0.034
	MLP	0.696	0.702	0.696	0.698	0.670	0.809	7.601
	KNN	0.697	0.715	0.697	0.699	0.677	0.813	0.215
	RF	0.720	0.804	0.720	0.722	0.721	0.798	6.000
	LR	0.488	0.721	0.488	0.449	0.401	0.626	0.899
	DT	0.574	0.577	0.574	0.574	0.547	0.735	2.290

recall, weighted and macro F1, and ROC-AUC, our proposed Gzip-based representation outperformed all baselines. For classification training runtime, WDGRL with Naive Bayes performs the best due to the minimum size of embedding compared to other embedding methods.

To test the statistical significance of results, we used the student t-test and observed the p-values using averages and SD results of 5 runs. We observed that the SD values for all datasets and metrics are very small i.e. mostly < 0.002, we also noted that p-values were < 0.05 in the majority of the cases (because SD values are very low). This confirmed the statistical significance of the results.

From the overall average classification results, SD results, and statistical significance results, we can conclude that the proposed NCD compression-based method can outperform the SOTA for predictive performance on real-world molecular sequence dataset. Moreover, even after fine-tuning the Large language model (LLM) such as SeqVec, the proposed parameter-free method significantly outperforms the LLM for all evaluation metrics. With the theoretical justifications and statistical significance of the results, we can conclude that using the proposed method in a real-world scenario for molecular sequence analysis can help biologists understand different viruses and deal with future pandemics efficiently.

5 Conclusion

In conclusion, we propose lightweight and efficient compression-based models for classifying molecular sequences. By combining the simplicity of the compression methods (e.g., Gzip and Bz2) with a powerful nearest neighbor algorithm, our method achieves state-of-the-art performance without the need for extensive parameter tuning or pre-trained models. The compression-based Model successfully overcomes the limitations of neural network-based methods, offering a more accessible and computationally efficient solution, especially in low-resource scenarios. In the future, we will be exploring the applications of our model in other bioinformatics domains and investigating ways to further optimize and tailor the approach for specific biological datasets.

References

1. Jiang, Z., et al.: Low-resource text classification: a parameter-free classification method with compressors. In: Findings of the Association for Computational Linguistics: ACL 2023, pp. 6810–6828 (2023)
2. Budach, S., Marsico, A.: Pysster: classification of biological sequences by learning sequence and structure motifs with convolutional neural networks. Bioinformatics **34**(17), 3035–3037 (2018)
3. Burrows, M.: A block-sorting lossless data compression algorithm. In: SRS Research Report, vol. 124 (1994)
4. Shah, A.S., Sethi, M.A.J.: The improvised Gzip, a technique for real time lossless data compression. EAI Endorsed Trans. CASA **6**(17), 6 (2019)
5. Kryukov, K., Jin, L., Nakagawa, S.: Efficient compression of SARS-COV-2 genome data using nucleotide archival format. In: Patterns, vol. 3, no. 9 (2022)

6. Vedak, M., Ackland, G.J.: Compression and information entropy of binary strings from the collision history of three hard balls. J. Phys. Commun. **7**(5), 055002 (2023)
7. Ferragina, P., Giancarlo, R., Greco, V., Manzini, G., Valiente, G.: Compression-based classification of biological sequences and structures via the universal similarity metric: experimental assessment. BMC Bioinf. **8**(1), 1–20 (2007)
8. Wei, R., Mahmood, A.: Recent advances in variational autoencoders with representation learning for biomedical informatics: a survey. IEEE Access **9**, 4939–4956 (2021)
9. Devlin, J., Chang, M.-W., Lee, K., Toutanova, K.: Bert: pre-training of deep bidirectional transformers for language understanding (2019)
10. Ali, S., et al.: PWM2Vec: an efficient embedding approach for viral host specification from coronavirus spike sequences. Biology **11**(3), 418 (2022)
11. Ali, S., Sahoo, B., Khan, M.A., Zelikovsky, A., Khan, I.U., Patterson, M.: Efficient approximate kernel based spike sequence classification. In: IEEE/ACM Transactions on Computational Biology and Bioinformatics (2022)
12. Heinzinger, M., et al.: Modeling aspects of the language of life through transfer-learning protein sequences. BMC Bioinf. **20**(1), 1–17 (2019)
13. Yang, K.K., Fusi, N., Lu, A.X.: Convolutions are competitive with transformers for protein sequence pretraining, bioRxiv, pp. 2022–05 (2022)
14. Le, N.Q.K., et al.: ienhancer-5step: identifying enhancers using hidden information of DNA sequences via chou's 5-step rule and word embedding. Anal. Biochem. **571**, 53–61 (2019)
15. Shen, J., Qu, Y., Zhang, W., Yu, Y.: Wasserstein distance guided representation learning for domain adaptation. In: AAAI Conference on Artificial Intelligence (2018)
16. Dong, B., et al.: Antimicrobial peptides prediction method based on sequence multidimensional feature embedding. Front. Genet. **13**, 1069558 (2022)
17. Gemci, F., Ibrikci, T., Cevik, U.: Deep learning algorithm for detection of protein remote homology. Comput. Syst. Sci. Eng. **46**(3) (2023)
18. Ghandi, M., et al.: Enhanced regulatory sequence prediction using gapped k-mer features. In: PLoS Computational Biology, vol. 10, no. 7 (2014)
19. Ali, S., Sardar, U., Patterson, M., Khan, I.U.: Biosequence2vec: efficient embedding generation for biological sequences. In: Pacific-Asia Conference on Knowledge Discovery and Data Mining, pp. 173–185 (2023)
20. Azevedo, D., Rodrigues, A.M., Canhão, H., Carvalho, A.M., Souto, A.: Zgli: a pipeline for clustering by compression with application to patient stratification in spondyloarthritis. Sensors **23**(3), 1219 (2023)
21. Mantaci, S., Restivo, A., Sciortino, M.: Distance measures for biological sequences: some recent approaches. J. Approximate Reasoning **47**, 109–124 (2008)
22. Kolmogorov, A.N.: On tables of random numbers. Sankhyā: Indian J. Stat. Ser. A **25**, 369–376 (1963)
23. Carpentieri, B.: Compression of next-generation sequencing data and of DNA digital files. Algorithms **13**(6), 151 (2020)
24. Ziv, J., Lempel, A.: A universal algorithm for sequential data compression. IEEE Trans. Inf. Theory **23**(3), 337–343 (1977)
25. Yergeau, F.: UTF-8, a transformation format of unicode and ISO 10646, Technical Report (1996)
26. Li, M., Chen, X., Li, X., Ma, B., Vitanyi, P.: The similarity metric. IEEE Trans. Inf. Theory **50**(12), 3250–3264 (2004)
27. Human DNA. https://www.kaggle.com/code/nageshsingh/demystify-dna-sequencing-with-machine-learning/data Accessed 10 October 2022
28. Xie, J., Girshick, R., Farhadi, A.: Unsupervised deep embedding for clustering analysis. In: International Conference on Machine Learning, pp. 478–487 (2016)

29. Brandes, N., Ofer, D., Peleg, Y., Rappoport, N., Linial, M.: ProteinBERT: a universal deep-learning model of protein sequence and function. Bioinformatics **38**(8), 2102–2110 (2022)
30. Xu, J.-W., et al.: An explicit construction of a reproducing gaussian kernel hilbert space. In: Conference on Acoustics Speech and Signal Processing Proceedings, vol. 5, pp. V–V (2006)
31. Hammer, B., Gersmann, K.: A note on the universal approximation capability of support vector machines. Neural Process. Lett. **17**, 43–53 (2003)
32. Minh, H.Q., Niyogi, P., Yao, Y.: Mercer's theorem, feature maps, and smoothing. In: Conference on Learning Theory, pp. 154–168 (2006)

Dynamic GNNs for Precise Seizure Detection and Classification from EEG Data

Arash Hajisafi[1(✉)], Haowen Lin[1], Yao-Yi Chiang[2], and Cyrus Shahabi[1]

[1] University of Southern California, Los Angeles, CA, USA
{hajisafi,haowenli,shahabi}@usc.edu

[2] University of Minnesota, Minneapolis, MN, USA
yaoyi@umn.edu

Abstract. Diagnosing epilepsy requires accurate seizure detection and classification, but traditional manual EEG signal analysis is resource-intensive. Meanwhile, automated algorithms often overlook EEG's geometric and semantic properties critical for interpreting brain activity. This paper introduces NeuroGNN, a dynamic Graph Neural Network (GNN) framework that captures the dynamic interplay between the EEG electrode locations and the semantics of their corresponding brain regions. The specific brain region where an electrode is placed critically shapes the nature of captured EEG signals. Each brain region governs distinct cognitive functions, emotions, and sensory processing, influencing both the semantic and spatial relationships within the EEG data. Understanding and modeling these intricate brain relationships are essential for accurate and meaningful insights into brain activity. This is precisely where the proposed NeuroGNN framework excels by dynamically constructing a graph that encapsulates these evolving spatial, temporal, semantic, and taxonomic correlations to improve precision in seizure detection and classification. Our extensive experiments with real-world data demonstrate that NeuroGNN significantly outperforms existing state-of-the-art models.

Keywords: Dynamic Graph Neural Network (GNN) · Automated Seizure Detection & Classification · EEG Data Analysis

1 Introduction

Seizure detection and classification from EEG data is crucial for diagnosing epilepsy, a neurological disorder affecting around 1% of the global population, with classification being particularly important for precise treatment [6]. The conventional approach to seizure detection and classification is largely manual, requiring skilled personnel to analyze extensive EEG recordings. This manual process is time-consuming, expensive, and prone to human errors [27].

Automated approaches using Convolutional Neural Networks (CNNs) [24] and Recurrent Neural Networks (RNNs) [19] have been explored to tackle these

D.-N. Yang et al. (Eds.): PAKDD 2024, LNAI 14648, pp. 207–220, 2024.
https://doi.org/10.1007/978-981-97-2238-9_16

challenges. However, these architectures do not capture the relationships between EEG observations in various brain regions, a crucial aspect of understanding the underlying neurological dynamics during seizures. Graph Neural Networks (GNNs) can capture relationships among EEG data points using a graph representation, yet existing GNN-based approaches either rely on predefined relations to generate a static graph or only use temporal correlations for dynamic graph generation, overlooking spatial and semantic relationships [8,27]. Such approaches miss out on leveraging multiple informative contexts and often result in poor performance, especially in scenarios marked by data scarcity, such as in classifying rare seizure types. Toward this end, a comprehensive solution is needed to automatically capture both static and dynamic evolving relationships and incorporate temporal, spatial, and semantic contexts simultaneously. This would allow for a detailed representation of the underlying correlations within the brain, enabling an end-to-end methodology for seizure detection and classification while overcoming the limitations of training under data scarcity.

Drawing inspiration from BysGNN [8], a GNN framework that creates a dynamic multi-context graph for forecasting visits to points of interest (POIs) based on spatial, temporal, semantic, and taxonomic contexts, we aim to generate a similar multi-context graph for our EEG data. However, adapting BysGNN as is to our domain presents several challenges. First, the shift from time-series forecasting in BysGNN to a multivariate time-series classification task in our work requires a change in architectural design. While BysGNN focuses on patterns within individual graph entities for node regression, our goal requires learning broader patterns across entities to perform graph classification. Second, EEG signals represent neurological events, which causes the translation of multi-context correlations from the POI domain to EEG data to be complex due to the underlying neurological mechanisms in brain activity and functional dependencies. Specifically, this translation is nontrivial because (1) the semantic contexts and relationships in brain areas are based on the functional roles of the corresponding part of the brain or broader brain region, which differs from the semantics of POI visits; (2) unlike POI's geographical coordinates, spatial contexts in EEG data are based on the brain's anatomy and EEG electrode placements; (3) the higher-level patterns in the brain should be defined based on the field of neuroscience, which is not as straightforward as defining high-level visit patterns to POIs. Finally, there are far fewer training samples for seizure classification, especially for rare seizure types, which necessitates designing a pretraining strategy aligned with the multi-context correlation notion to achieve optimal seizure detection and classification performance.

To address these challenges, we propose NeuroGNN[1], a GNN-based framework designed to dynamically construct a graph encapsulating multi-context correlations in EEG data. The correlations are characterized by spatial proximity of electrode placements on the scalp, temporal dependencies within and between time series from different electrodes, semantic similarities derived from neurological brain functions of respective electrode placements, and taxonomic correlations across broader brain regions. The broader regions are defined through

[1] Our source code is available at https://github.com/USC-InfoLab/NeuroGNN.

neuroscientific meta-nodes generated from EEG electrodes, extending the multi-context definitions to these meta-nodes. Our new design incorporates hierarchical pooling to handle node and meta-node representations and employs Bidirectional Gated Recurrent Units (BiGRUs) to enhance the intra-series capture and characterization of time-series data. To address the challenge of small training samples and achieve optimal performance across both detection and classification, we develop a new pretraining strategy with a learning objective that aligns with our graph representation, encompassing both nodes and meta-nodes. Learning to generate this multi-context graph representation allows NeuroGNN to outperform state-of-the-art methods for seizure detection and classification in real-world EEG diagnosis scenarios. Through comprehensive ablation studies, we demonstrate that each multi-context correlation modality significantly contributes to the overall performance enhancement. This is particularly noticeable in rare classes with training under data scarcity.

2 Related Work

EEG signals are commonly used to detect epileptic seizures (e.g., [24]). Since manual EEG analysis by trained physicians is extremely resource and time-intensive, much recent research has focused on using machine learning approaches to automate the diagnostic process. Diagnosing epileptic seizures can be divided into seizure detection and seizure classification. Seizure detection aims to detect abnormal patterns in EEG data. Initially, for seizure detection, researchers started with conventional machine learning techniques, where they first extracted hand-engineered frequency features from the recordings and applied traditional methods such as K-Nearest Neighbors and random forests to detect abnormal EEG signals [15]. Shortly afterward, Roy et al. suggested employing time-distributed neural networks, such as recurrent neural networks, to learn directly from the data without any explicit pre-processing step [19]. Sharathappriyaa et al. propose to compress EEG signals into a latent space representation using Autoencoders and then detect epileptic seizures [23].

Seizure classification differentiates between seizure types like focal or generalized seizures. Most of the works in seizure classification adopt CNNs for EEG-based seizure prediction [24]. For example, Bhattacharyya et al. first transform the 1D signal of each EEG channel into 2D using wavelet transform and then apply a CNN to detect and classify seizure subtypes [4]. However, CNN-based methods often overlook the complex spatial structures of EEG sensors and brain geometry [27]. To address this, recent efforts have explored graph-based neural networks, which better capture these spatial relationships, achieving enhanced detection and classification results [27]. Yet, existing GNN studies on EEG data present a significant limitation: they either assume a static graph structure based solely on EEG sensor distances or generate dynamic graphs considering only the temporal relationships between EEG. Both approaches are constrained, representing only a single type of relationship in the graph structure, spatial or temporal, and thus fail to fully capture the complex interactions among EEG sensors and their signals. Our model addresses this gap by effectively fusing complex relationships from multiple dimensions to create dynamic graphs that improve automatic seizure detection and classification.

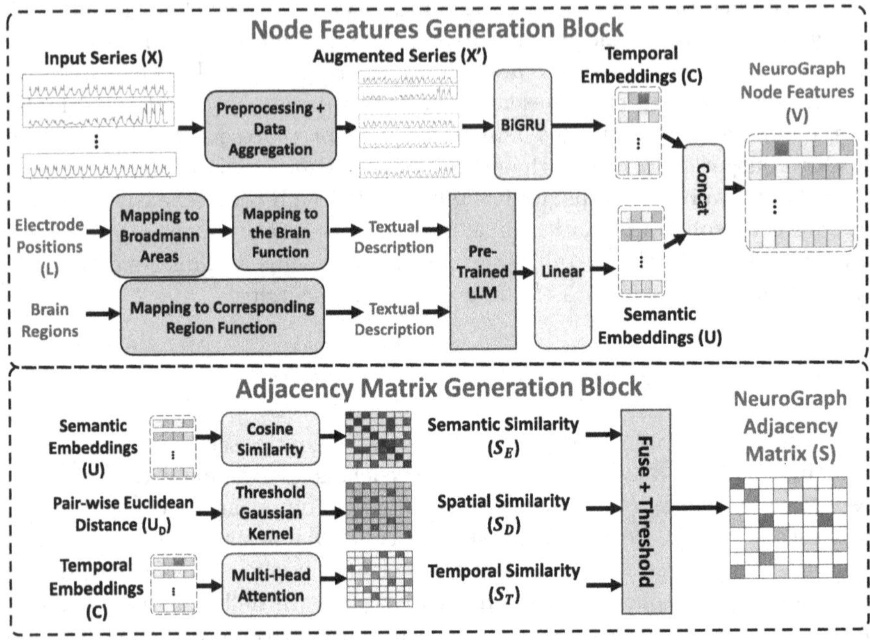

Fig. 1. NeuroGNN Graph Construction

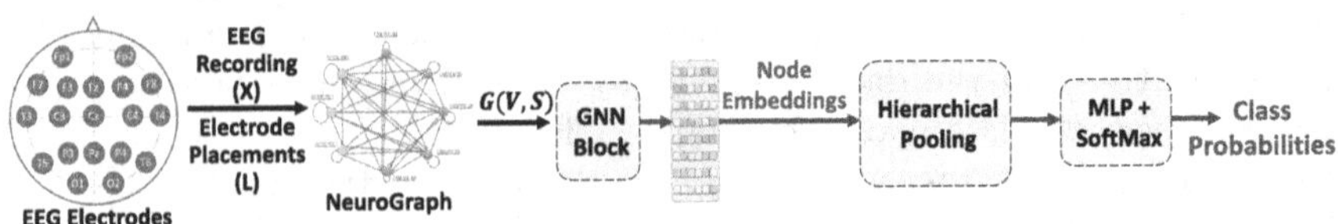

Fig. 2. Seizure Detection and Classification using NeuroGraph

3 Methodology: NeuroGNN

Given EEG recordings within a time-series window X and the 3D coordinates L of electrode placements on the scalp, we first construct the dynamic NeuroGraph G, integrating multi-context relations, as illustrated in Fig. 1. This graph's nodes encapsulate current temporal EEG data and corresponding brain semantics. Edges fuse spatial, temporal, and semantic information to model complex inter-node relationships. Subsequently, as depicted in Fig. 2, the GNN block updates the node embeddings, leveraging these multi-contextual edges. Hierarchical pooling aggregates these embeddings into a single graph-level vector, which is then classified via an MLP, enabling both binary classification for seizure detection and multi-label classification for seizure type prediction. Detailed procedures are in subsequent sections.

3.1 Node Features Generation

The process begins by preprocessing the time-series data and generating new aggregated series to form six meta-nodes, each representing a distinct brain region as identified in [1]. These meta-nodes, representing regional neural activities, complement EEG nodes that capture only localized activities, facilitating a multi-scale representation for learning taxonomic correlations. Meta-nodes aggregate EEG data within their regions through an averaging function on the input $\mathbf{X} \in \mathbb{R}^{N \times T}$, producing $X' \in \mathbb{R}^{N' \times T}$ where $N' = N + 6$ (representing the addition of six brain regions).

Next, to capture intra-series temporal correlations, X' is fed into a Bidirectional Gated Recurrent Unit (BiGRU) [11]. BiGRU's bidirectional processing provides a comprehensive representation of the given time-series window, which is crucial for seizure detection and classification. The BiGRU generates hidden states for each node, denoted as $h_{f,i} \in \mathbb{R}^M$, representing forward temporal dynamics and $h_{b,i} \in \mathbb{R}^M$, representing backward temporal dynamics from the forward and backward passes, respectively, where M shows the hidden dimension. The final node embedding is obtained by concatenating the two hidden states, $c_i = (h_{f,i} \parallel h_{b,i})$, forming an embedding matrix $C \in \mathbb{R}^{N' \times 2M}$.

In parallel, NeuroGNN captures semantic correlations among EEG nodes and meta-nodes using a pre-trained Large Language Model (LLM). The semantic foundations are based on Brodmann areas, which segment the cerebral cortex into distinct functional regions [30]. A mapping between EEG electrode placements L and Brodmann areas is established as per [21]. Textual descriptions encapsulating the functional roles of the corresponding Broadmann areas and brain regions are generated for each EEG node and meta-node based on reputable scientific resources [30]. These descriptions are tokenized and fed into a pre-trained MPNet language model [26], which is fine-tuned via a linear layer during training to optimize semantic embeddings for the downstream task. This process results in a semantic embeddings matrix $U \in \mathbb{R}^{N' \times K}$, where K denotes the semantic embedding dimensionality.

The final feature vector for each node and meta-node is obtained by concatenating the temporal and semantic embeddings, yielding the NeuroGraph node features matrix $V \in \mathbb{R}^{N' \times (2M+K)}$.

3.2 Adjacency Matrix Generation

Next, we construct NeuroGraph's adjacency matrix to dynamically capture multi-context correlations among EEG and meta-nodes.

First, we compute the semantic similarity matrix S_E using cosine distance on semantic embeddings from the node features generation layer. In parallel, we compute spatial and temporal similarities. Spatial similarities are derived by first calculating Euclidean distances among EEG electrodes for EEG node pairs, and average distance for meta-node pairs and meta-node to EEG node pairs. The spatial similarity matrix S_D is then formed by applying a Gaussian kernel [25] to convert distance metrics into similarity scores, controlled by a threshold τ. Temporal similarities are captured by passing temporal embeddings C to a Multi-Head Attention layer [28], resulting in a temporal similarity matrix S_T.

To fuse these similarities, we employ a gating mechanism, $S_{Gate} = (1-\alpha)S_E + \alpha S_D$ to control influence of temporal similarities, forming an un-thresholded adjacency matrix $S' = S_{Gate} \odot S_T$, with α as a learnable parameter between 0 and 1 to adjust the balance between the importance of spatial and semantic similarities. A thresholding step [8] refines S' to obtain the final adjacency matrix S, retaining reliable similarity scores and discarding noisy ones.

3.3 Prediction Using the Generated NeuroGraph

With the generated node features V and adjacency matrix S, the NeuroGraph $G = (V, S)$ for a given window is formed, encapsulating multi-context relationships among EEG nodes and meta-nodes. NeuroGraph is then passed through a GNN block employing a modified Graph Convolutional Networks (GCN) variant [13], omitting the normalization term and adding residual connections between the message passing layers, to preserve directed relationships in NeuroGraph and mitigate oversmoothing [5], respectively. This yields the node embeddings matrix $V' \in \mathbb{R}^{N' \times Z}$, where Z represents the embedding dimension.

Next, a hierarchical pooling mechanism aggregates refined node representations into a single graph embedding vector. Initially, within each brain region, a max-pooling operation aggregates EEG node embeddings. Similarly, a max-pooling operation across all meta-nodes constructs a single embedding vector for meta-nodes. These pooled embeddings are concatenated, forming a collective representation of region-wise and meta-wise node information. A feature-wise mean-pooling operation on this concatenated representation derives the graph embedding vector $\mathbf{g} \in \mathbb{R}^Z$, encapsulating critical graph information.

Finally, vector $\mathbf{g}$ is processed through a Multi-Layer Perceptron (MLP) with a Softmax activation to obtain the final class probabilities. For training detection and classification models, binary and multi-class cross-entropy are utilized as loss functions, respectively.

3.4 Pretraining

Pretraining can mitigate issues related to the scarcity of labeled data and out-of-distribution predictions in many scientific ML applications [9,29]. Therefore, we assess NeuroGNN's performance with weights initialized after a pretraining phase on a self-supervised task. The chosen pretraining task is a node-level regression task, which forecasts future values for the upcoming 12 s from a 60 s preprocessed EEG clip for each EEG node and meta-node. We design two training objectives; **Primary Objective:** Minimizing Forecasting Mean

Squared Error (MSE) Loss across all EEG nodes and meta-nodes, and **Secondary Objective:** Ensuring Consistency Loss to maintain forecast consistency between individual nodes and their corresponding meta-nodes. The overall training objective combines both objectives with a weighted sum.

4 Experiments

4.1 Experimental Setup

Dataset and Data Splits. We conducted our experiments using the Temple University Hospital EEG Seizure Corpus (TUSZ) v1.5.2 [22]. The dataset comprises 3,050 annotated seizure events from over 300 patients across eight seizure types, with data recorded using 19 electrodes from the standard 10–20 EEG system. We follow [27] to preprocess and group the seizures into four classes. Similarly, we excluded five patients from the official TUSZ test set due to their presence in both the official train and test sets [27]. We evaluate the performance of the approaches on the official TUSZ test sets. Table 1 details the statistics of the data used in our experiments. The training set was randomly split into training and validation sets with a 9:1 ratio.

Table 1. Summary of TUSZ v1.5.2 used in our study

Set	EEG Files		Total Seizure Duration		Patients		CF Seizures		GN Seizures		AB Seizures		CT Seizures	
	Non-Seizure	Seizure	Non-Seizure	Seizure	Non-Seizure	Seizure	Seizures	Patients	Seizures	Patients	Seizures	Patients	Seizures	Patients
Train Set	3730	869	705 h 29 m	47 h 26 m	390	202	1,868	148	409	68	50	7	48	11
Test Set	670	230	135 h 46 m	14 h 45 m	10	35	297	24	114	11	49	5	61	4

Preprocessing. Following prior methodologies [2,3,27], EEG recordings were resampled to 200 Hz and divided into 60-s non-overlapping windows (clips). For seizure detection, all EEG clips are utilized; a clip is labeled as a seizure if it contains at least one type of seizure. For seizure classification, only clips with a single seizure type are used. If a seizure ends and another begins within a clip, it is truncated and zero-padded to maintain a 60-second length. Each 60-second clip is further divided into 1-second segments, with the Fast Fourier Transform (FFT) applied to each segment to obtain log amplitudes of non-negative frequency components, as per [27]. Each 60-second clip, now a sequence of 60 log-amplitude representations, is used for classification.

Training. Training was conducted on a RTX 3090 GPU with 24GB of memory under an Ubuntu 20.04 setup with CUDA version 11.4, and PyTorch v1.13.0. Without pretraining, model weights were initialized using the Xavier initialization method [7]. The Adam optimizer [12] was utilized for training, paired with a cosine annealing learning rate scheduler [16]. The training was continued until the evaluation loss increased in 5 consecutive epochs or reached a maximum of 100 epochs. $L2$ regularization was applied to model weights to prevent overfitting. In seizure detection and classification, the model checkpoint achieving the best Area Under the Receiver Operating Characteristic (AUROC) and weighted F1 score on the validation set across different epochs were preserved, respectively. The pretraining task also ran for 300 epochs. The obtained weights for Node Features Generation, Adjacency Matrix Generation, and GNN Block parameters then initialized the model for the following downstream task.

Baselines. We benchmarked NeuroGNN against: (1) **LSTM** [10], for encoding sequential data; (2) **Dense-CNN** [20] that used a densely connected architecture for seizure detection; (3) **CNN-LSTM** [2] integrated 2D convolutions with LSTM for seizure classification; (4) **Corr-DCRNN** [27], a GNN-based model that utilized the DCRNN [14] framework on graphs formed from EEG data cross-correlation at each timestep; and (5) **Dist-DCRNN** [27], another GNN variant leveraging DCRNN on a graph constructed from Euclidean distances of EEG electrode placements. The latter two represent current state-of-the-art (SOTA) in this domain.

Hyperparameter Setting. Cross-validation was used to select hyperparameters for optimal performance. Key settings include an initial learning rate of 0.0002, batch size of 40, hidden BiGRU dimension M and semantics embedding dimension K both at 512. Multi-Head Attention utilized 8 heads, while graph convolution node embedding dimension Z was set to 256. Gaussian kernel threshold τ was set to twice the standard deviation of EEG electrode distances. For pretraining, the MSE loss objective weight was set to 0.9, and the Consistency loss objective weight was adjusted to 0.1. These settings were consistent across both detection and classification tasks.

Table 2. Experiment results. The highest and second highest scores are denoted in bold and italics with underline, respectively. "Improvement" shows the percentage improvement of NeuroGNN compared to the best-performing baseline. The mean and standard deviations are from three random runs.

Method	Seizure Detection (AUROC)		Seizure Classification (Weighted F1)	
	Without Pretraining	With Pretraining	Without Pretraining	With Pretraining
LSTM	0.715 ± 0.016	–	0.686 ± 0.020	–
Dense-CNN	0.796 ± 0.014	–	0.626 ± 0.073	–
CNN-LSTM	0.682 ± 0.003	–	0.641 ± 0.019	–
Corr-DCRNN	$\underline{\mathit{0.804 \pm 0.015}}$	0.850 ± 0.014	$\underline{\mathit{0.701 \pm 0.030}}$	$\underline{\mathit{0.749 \pm 0.017}}$
Dist-DCRNN	0.793 ± 0.022	$\underline{\mathit{0.875 \pm 0.016}}$	0.690 ± 0.035	$\underline{\mathit{0.749 \pm 0.028}}$
NeuroGNN	$\mathbf{0.847 \pm 0.007}$	$\mathbf{0.876 \pm 0.002}$	$\mathbf{0.790 \pm 0.006}$	$\mathbf{0.792 \pm 0.004}$
Improvement	+5.35%	+0.11%	+12.7%	+5.74%

4.2 Results

NeuroGNN Performance. Table 2 presents the results using the evaluation metrics following convention [2,3,17,27]. NeuroGNN outperforms baseline models in seizure detection and classification in both pretraining scenarios. In seizure detection, non-pretrained Corr-DCRNN shows a slight 1.01% improvement over Dense-CNN, while NeuroGNN boosts the AUROC score by over 5% compared to SOTA Corr-DCRNN, underscoring the benefits of capturing multi-context correlations. Both Dist-DCRNN and Corr-DCRNN, tackling different correlation aspects, spatial and temporal, individually, yield similar performance on both tasks. Conversely, NeuroGNN, integrating multi-context correlations within a unified framework, shows substantial performance improvement against DCRNN-based baselines. Particularly in classification, NeuroGNN significantly improves the weighted F1 score of DCRNN-based baselines by 12.7% without pretraining and by 5.74% after pretraining, respectively.

The improvement in seizure detection achieved by NeuroGNN over baselines, while noteworthy, is less significant compared to the enhancement in seizure classification performance. This difference can be attributed to the binary nature of the detection task and the abundance of training data, rendering it a less complex task. Even a simpler model like Dense-CNN holds up well against DCRNN-based models without pretraining. Conversely, the classification task, a 4-class problem with imbalanced class distributions (Table 1), poses more challenges. The limited training samples for certain seizure types highlight the importance of multi-context correlations, illustrated by over 12% improvement in weighted F1 score by NeuroGNN without pretraining.

Impact of Pretraining. Consistent with findings in [27], our results validate the positive effect of pretraining on performance. Our specific pretraining task, aligned with NeuroGNN's design involving EEG nodes and meta-nodes representing brain regions, enhances performance, leading to a 3.42% boost in the AUROC metric for seizure detection. However, the increase in the weighted F1 score for classification is minor. This indicates that while pretraining aids in mitigating data scarcity and out-of-distribution challenges, the multi-context correlations captured by NeuroGNN intrinsically address these issues, making the benefits of pretraining less significant for NeuroGNN compared to DCRNN-

Table 3. Ablation study results, showing median performance over three trials and percentage changes relative to NeuroGNN. Bold and underline percentages highlight the largest and second-largest changes per metric.

	NeuroGNN	w/o. TemporalCorr	w/o. Semantics	w/o. Space	w/o. Meta-Nodes
Detection	**0.847**	0.831	0.828	0.818	0.829
(AUROC)	–	-2.12%	<u>*-2.47%*</u>	**-3.65%**	-2.36%
Classification	**0.790**	0.673	0.719	0.727	0.714
(Weighted F1)	–	**-14.81%**	-8.99%	-7.97%	<u>*-9.62%*</u>

based models. This reduction in pretraining necessity also implies a saving in computational resources and time, positioning NeuroGNN as a more resource-efficient and practical option for real-world applications.

4.3 Ablation Study

In this section, we explore the impact of NeuroGNN's individual components on performance by creating four distinct variants, each removing one aspect of the multi-context correlations. The variants are: (1) **W/O TemporalCorr:** NeuroGNN without temporal correlations for graph edge formation. (2) **W/O Semantics:** NeuroGNN without semantics in node features and graph edge weights. (3) **W/O Space:** NeuroGNN without spatial proximity information between nodes for graph edge formation. (4) **W/O Meta-Nodes:** NeuroGNN without aggregated time-series nodes representing brain regions.

We evaluated each variant on both seizure detection and classification tasks, with results presented in Table 3. As observed from the table, every component improves performance on both tasks. The ablation study shows a noticeable decrease in performance for the classification task in all configurations compared to the detection task. On average, the classification task experiences a 7.7% larger drop in performance in all configurations. This observation supports our claim that multi-context correlations significantly ease the challenges posed by data scarcity in rare classes and the complexity of multi-class classification. For example, the classification task has significantly fewer samples than the detection task (3,730 non-seizure and 869 seizure events), particularly for the AB and CT classes, which have only 50 and 48 training events, respectively.

The analysis indicates that spatial correlations have the most significant impact on the detection task, while temporal correlations affect classification performance the most, leading to a 3.65% and a substantial 14.81% drop in performance, respectively. This can be explained by the nature of the tasks. In seizure detection, the goal is to pinpoint the occurrence of seizures, which often relate to unusual frequency patterns within a close neighborhood of EEG electrodes. On the other hand, classification requires a more detailed examination of frequency anomalies to distinguish the type of seizure, highlighting the importance of temporal relationships among different EEG electrodes and brain regions in associating an EEG clip with a specific seizure class.

4.4 Analysis of Graph Embeddings

This section explores the effectiveness of graph embeddings generated by pre-trained Dist-DCRNN and NeuroGNN on test samples for the classification task. The aim is to assess the effectiveness of the learned graph embeddings by NeuroGNN in comparison with the previous SOTA model in distinguishing different seizure types. After training on seizure classification, both models are used in inference mode to obtain graph embeddings for test samples, with initial weights set based on pretraining for a fair comparison.

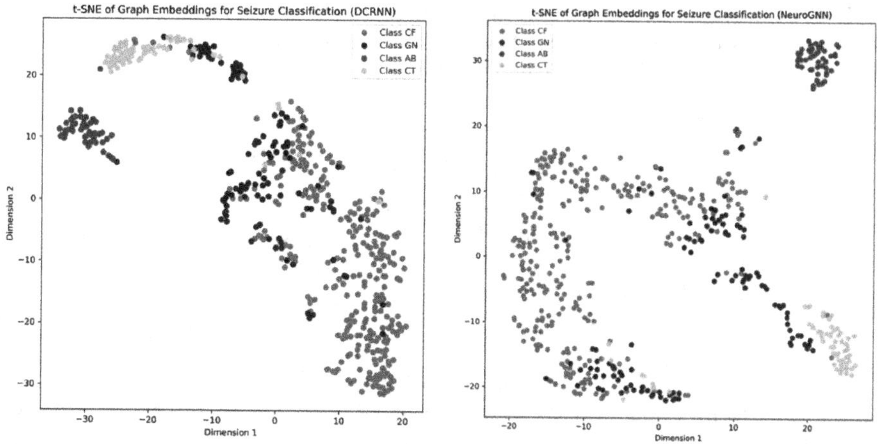

(a) Graph Embeddings from Dist-DCRNN (b) Graph Embeddings from NeuroGNN

Fig. 3. Visualization of T-SNE projections of graph embeddings for seizure classification test samples using trained Dist-DCRNN and NeuroGNN models. Colors represent ground truth seizure labels.

The graph embeddings for the seizure classification test samples are visualized in a 2D space using the t-SNE dimensionality reduction technique [18] as depicted in Fig. 3. These figures reveal a notable pattern of interleaving among samples of the GN seizure class and samples from other classes, which corresponds to a lower True Positive Rate for the GN seizure class, indicating higher classification difficulty for this particular seizure type.

The quality of embeddings is assessed quantitatively by clustering the original high-dimensional graph embeddings into four clusters using the KMeans algorithm, reflecting the number of classes in the dataset. The average clustering purity metric is computed to measure the homogeneity of clusters regarding class membership, defined as the mean of the maximum class proportion across all clusters. The analysis shows that NeuroGNN achieves an average clustering purity value of 0.747, compared to 0.717 for Dist-DCRNN, representing an 11% reduction in clustering impurity by NeuroGNN. This highlights NeuroGNN's superior ability to distinguish samples based on multi-context correlations.

4.5 Handling Scarce Training Data

Experiments were conducted to evaluate NeuroGNN and Dist-DCRNN's performance under different data availability scenarios by using subsampled versions of training data, while keeping the original class distribution through stratified sampling. The results, depicted in Table 4, show the median weighted F1 score in three independent runs for each sampling ratio in the training data, with evaluations performed on the entire test dataset to ensure consistency across all experiments.

Table 4. F1 Score Comparison between NeuroGNN and Dist-DCRNN for Seizure Classification Across Different Training Data Ratios. "Improvement" shows the percentage improvement of NeuroGNN over Dist-DCRNN.

	Training Data Ratio				
	100%	80%	60%	40%	20%
Dist-DCRNN	0.701	0.704	0.679	0.689	0.606
NeuroGNN	0.790	0.795	0.772	0.787	0.781
Improvement	12.7%	12.9%	13.7%	14.2%	28.9%

The table illustrates that as the sampling ratio decreases, the performance gap between NeuroGNN and Dist-DCRNN increases, showcasing the robustness of NeuroGNN to smaller training sizes compared to Dist-DCRNN. Moreover, NeuroGNN's performance exhibits only slight degradation even with reduced training data, emphasizing the model's ability to effectively leverage multi-context correlations to address challenges posed by data scarcity, thereby maintaining a superior performance under data-constrained conditions.

5 Conclusion

We introduced NeuroGNN, a novel GNN framework that significantly enhances seizure detection and classification from EEG data by integrating spatial, temporal, and semantic contexts. NeuroGNN excels particularly in scenarios with limited training data, proving invaluable in distinguishing between rare classes. Given the complexity and cost of collecting health research data from human subjects (e.g., EEG data), techniques that efficiently use sparse training samples are highly desirable, showcasing NeuroGNN's superiority. Even with sufficient training data (e.g., the case of seizure detection), NeuroGNN consistently outperforms prior state-of-the-art methods, albeit with a smaller margin.

Acknowledgments. Research supported by the National Science Foundation (NSF) under CNS-2125530 and the National Institute of Health (NIH) under grant 5R01LM014026.

Disclaimer: The views and conclusions contained herein are those of the authors and should not be interpreted as necessarily representing the official policies or endorsements, either expressed or implied, of NSF or NIH.

References

1. Ackerman, S., et al.: Discovering the brain (1992)
2. Ahmedt-Aristizabal, D., et al.: Neural memory networks for seizure type classification. In: IEEE EMBC (2020)
3. Asif, U., et al.: Seizurenet: multi-spectral deep feature learning for seizure type classification. In: MLCN & RNO-AI Workshops at MICCAI (2020)

4. Bhattacharyya, A., et al.: Tunable-q wavelet transform based multiscale entropy measure for automated classification of epileptic EEG signals. Appl. Sci. **7**(4), 385 (2017)
5. Chen, D., et al.: Measuring and relieving the over-smoothing problem for graph neural networks from the topological view. In: AAAI, vol. 34, pp. 3438–3445 (2020)
6. Falco-Walter, J.: Epilepsy-definition, classification, pathophysiology, and epidemiology. In: Seminars in neurology, vol. 40, pp. 617–623 (2020)
7. Glorot, X., Bengio, Y.: Understanding the difficulty of training deep feedforward neural networks. In: AISTATS, pp. 249–256 (2010)
8. Hajisafi, A., et al.: Learning dynamic graphs from all contextual information for accurate point-of-interest visit forecasting. arXiv preprint arXiv:2306.15927 (2023)
9. Hendrycks, D., et al.: Using pre-training can improve model robustness and uncertainty. In: ICML, pp. 2712–2721 (2019)
10. Hochreiter, S., Schmidhuber, J.: Long short-term memory. Neural Comput. **9**(8), 1735–1780 (1997)
11. K. Cho, K., et al.: Learning phrase representations using RNN encoder-decoder for statistical machine translation. arXiv preprint arXiv:1406.1078 (2014)
12. Kingma, D.P., Ba, J.: Adam: a method for stochastic optimization. arXiv preprint arXiv:1412.6980 (2014)
13. Kipf, T.N., Welling, M.: Semi-supervised classification with graph convolutional networks. In: ICLR (2017)
14. Li, Y., et al.: Diffusion convolutional recurrent neural network: Data-driven traffic forecasting. In: ICLR (2018)
15. Lopez, S., et al.: Automated identification of abnormal adult EEGs. In: Proceedings of IEEE SPMB, pp. 1–5 (2015)
16. Loshchilov, I., Hutter, F.: SGDR: stochastic gradient descent with warm restarts. arXiv preprint arXiv:1608.03983 (2016)
17. Ma, Y., et al.: TSD: transformers for seizure detection. bioRxiv, pp. 2023–01 (2023)
18. van der Maaten, L., Hinton, G.: Visualizing data using t-SNE. J. Mach. Learn. Res. **9**(11), 2579–2605 (2008)
19. Roy, S., et al.: Deep learning enabled automatic abnormal EEG identification. In: Proceedings of IEEE EMBC, pp. 2756–2759 (2018)
20. Saab, K., et al.: Weak supervision as an efficient approach for automated seizure detection in electroencephalography. NPJ Digit. Med. **3**(1), 59 (2020)
21. Scrivener, C.L., Reader, A.T.: Variability of EEG electrode positions and their underlying brain regions: visualizing gel artifacts from a simultaneous EEG-FMRI dataset. Brain Behav. **12**(2), e2476 (2022)
22. Shah, V., et al.: The temple university hospital seizure detection corpus. Front. Neuroinform. **12**, 83 (2018)
23. Sharathappriyaa, V., et al.: Auto-encoder based automated epilepsy diagnosis. In: Proceedings of IEEE ICACCI, pp. 976–982 (2018)
24. Shoeibi, A., et al.: Epileptic seizures detection using deep learning techniques: a review. In: International Journal of Environmental Research and Public Health (2021)
25. Shuman, D., et al.: The emerging field of signal processing on graphs: extending high-dimensional data analysis to networks and other irregular domains. IEEE Signal Process. Mag. **30**(3), 83–98 (2013)
26. Song, K., et al.: MPNet: masked and permuted pre-training for language understanding. NeurIPS **33**, 16857–16867 (2020)
27. Tang, S., et al.: Self-supervised graph neural networks for improved electroencephalographic seizure analysis. In: Proceedings of ICLR (2022)

28. Vaswani, A., et al.: Attention is all you need. In: NeurIPS (2017)
29. Zeiler, M.D., Fergus, R.: Visualizing and understanding convolutional networks. In: ECCV, pp. 818–833 (2014)
30. Zilles, K.: Brodmann: a pioneer of human brain mapping-his impact on concepts of cortical organization. Brain **141**(11), 3262–3278 (2018)

A Novel Population Graph Neural Network Based on Functional Connectivity for Mental Disorders Detection

Yuheng Gu, Shoubo Peng, Yaqin Li, Linlin Gao, and Yihong Dong(✉)

Ningbo University, Ningbo, China
dongyihong@nbu.edu.cn

Abstract. Accurate and rapid clinical confirmation of psychiatric disorders based on imaging, symptom and scale data has long been difficult. Graph neural networks have received increasing attention in recent years due to their advantages in processing unstructured relational data, especially functional magnetic resonance imaging data. However, all existing methods have certain drawbacks. Individual graph methods are able to provide important biomarkers based on functional connectivity modelling, but their accuracy is low. Population graph methods, which improve the prediction performance by considering the similarity between patients, lack clinical interpretability. In this study, we propose a functional connectivity-based population graph (FCP-GNN) approach that possesses excellent classification capabilities while also providing significant biomarkers for clinical reference. The proposed method is divided into two phases. In the first phase, brain region features are learned hemispherically and used to identify biomarkers through a local-global dual-channel pooling layer. In the second phase, a heterogeneous population map is constructed based on gender. The feature information of same-sex and opposite-sex neighbours is learned separately using a hierarchical feature aggregation module to obtain the final embedding representation. The experiment results show that FCP-GNN achieves state-of-the-art performance in classification prediction work on two public datasets.

Keywords: Rs-fMRI data · Mental disorder · GNN · Classification

1 Introduction

Given the intricate etiology of psychiatric disorders [24] and the absence of substantial pathological biomarkers [26], clinicians face considerable challenges in swiftly and accurately diagnosing patients [20]. The advent of computer-assisted diagnostic methodologies has led to the development of numerous models designed to aid physicians in identifying potential risks with enhanced speed and precision, thus facilitating the formulation of timely treatment plans. In the

This work was supported by Ningbo Municipal Public Welfare Technology Research Project(2023S023) and Natural Science Foundation of Ningbo (2023J114).

D.-N. Yang et al. (Eds.): PAKDD 2024, LNAI 14648, pp. 221–233, 2024.
https://doi.org/10.1007/978-981-97-2238-9_17

present study, we propose a novel computer-assisted diagnostic technique aimed at improving patient diagnosis and augmenting clinical interpretability.

The brain is commonly perceived as an intricate network, and functional magnetic resonance imaging (fMRI) technology has recently emerged as an important tool for investigating functional brain networks. Several works have employed the functional connectome (FC), which is obtained by processing resting-state functional MRI images (rs-fMRI), to analyse brain network disorders [21].

In recent years, the graph neural network has emerged as a powerful method for studying the brain functional connectome and advancing computer-aided diagnosis. Functional connectome networks (FCNs) are usually represented as graphs. Nodes in the graph represent functional brain regions, and edges connect these regions. The above modelling approach is known as the individual graph approach [15,25]. However, this approach has not yet been successful in accurately predicting diseases. Population graphs are commonly used in various studies incorporating subjects with similar features. These graphs are created by models that learn the genetic, phenotypic, and imaging data of the subjects and the connections between them to make diagnoses [3,8,19,21,22,30]. However, population graph methods cannot identify reliable biomarkers since their nodes represent only multidimensional features, and the functional brain networks are not modelled, as done by individual graph methods. It reduces the clinical interpretability of these methods and makes them challenging to be trusted by clinicians. Recent studies have integrated functional connectomes as features in population graph models [10,29]. However, these studies did not examine how information flows within the brain, nor did they account for the natural sex differences among the subjects, which significantly impacts the prediction accuracy.

This study proposes a population graph neural network based on the brain functional connectome (FCP-GNN) to address the limitations of existing population graph models. The proposed network utilizes the functional connectivity of each subject to construct graph data and achieves high predictive performance in an end-to-end manner. Moreover, the proposed method identifies significant biomarkers for disease classification. The proposed methodology includes two primary phases. The initial steps include obtaining individual-level features, and the functional connectivity of each subject is modelled as graph. To obtain graph-embedded representations, the model must learn distinguishable features while discarding redundant information. For the first time, we present a graph convolutional neural network that integrates intrahemispheric and interhemispheric information (IIH-GCN). This network fuses information among brain regions in the same hemisphere, which is consistent with the neural information exchange processes. To eliminate regions that are irrelevant to disease prediction while retaining as many essential global features as possible, a local-global dual-channel pooling (LGP) approach is proposed. This approach implements the pooling operation at local and global levels and achieves information complementarity through a cross-channel convolutional layer. The final high-importance nodes retained after the LGP process include biomarkers that are associated with disease prediction. During the second stage, a heterogeneous GNN is created to learn group-level features based on sex. We utilize the phenotypic data of each subject to encode the constructed edges. To aggregate the information of opposite-sex and same-site neighbours in each layer during the feature learning process, a graph transformer is used to learn the feature embeddings of the nodes. Finally, the multilevel features of each node are fused to obtain classification predictions.

Overall, the main contributions of our work can be summarized as follows:

1. In this paper, we propose a population graph neural network (FCP-GNN) based on the functional connectivity of the brain that explicitly extracts features by modelling the connectomes of each sample and maintains excellent predictive performance while identifying significant biomarkers.
2. We propose a novel graph convolutional layer IIH-GCN to obtain graph embeddings for high quality brain networks and a local-global dual-channel pooling (LGP) approach to enhance interpretability in identifying essential biomarkers. A graph transformer module with a multihead attention mechanism is employed to extract features of same-sex and opposite-sex relationships.
3. Compared with various types of methods, FCP-GNN achieves the best performance based on two datasets. The LGP scheme identifies essential biomarkers that distinguish patients from healthy controls. Notably, these regions of interest (ROIs) are consistent with the literature on the aetiology and symptoms of psychiatric disorders.

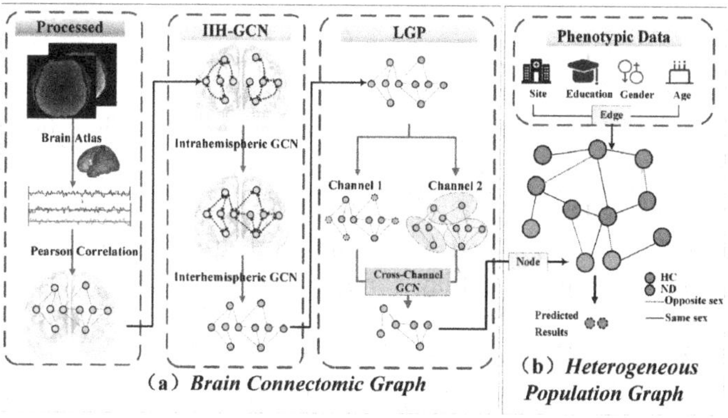

Fig. 1. Overall framework diagram of the proposed method

2 Methods

An overview of the model is illustrated in Fig. 1, which consists of two main parts: the Brain Connectomic Graph and the Heterogeneous Population Graph.

2.1 Brain Connectomic Graph

Graph Construction. Resting-state functional magnetic resonance imaging (rs-fMRI) data were preprocessed using a brain mapping template. The average time series in m ROIs were determined after the data were processed.

The brain networks carry topological information, and GNNs are highly advantageous in dealing with this type of data, which resembles node and edge connections. Therefore, each sample is first transformed into an undirected graph $G_{ind}(V, E)$, where V is the set of vertices, and each vertex represents an ROI. E is the set of edges, which includes the connection information between each vertex. We computed the Pearson correlation coefficient (PCC) between each ROI time series and other ROI time series to obtain a functional connectivity matrix $\mathbf{X}_{G_{ind}}$ that is used as the feature matrix for G_{ind}. The edge weights between nodes are their PCC values. An excessive number of edge connections in a dense graph results in redundant information and increases computational costs. Our model only retains edges with edge weights larger than α composing the adjacency matrix $\mathbf{A}_{G_{ind}} \in \mathbb{R}^{m \times m}$.

IIH-GCN. Numerous studies have shown that information transfer within the same cerebral hemisphere is faster than information transfer across hemispheres and that information transfer within hemispheres is more complex and detailed than that across hemisphere [7]. We propose a graph convolutional layer IIH-GCN. It can learn and combine intra- and interhemispheric information.

Graph convolutional networks (GCNs) [12] are tools for efficiently extracting features from graph that is defined as:

$$\mathbf{X}_G^{(l+1)} = \sigma(\mathrm{GCN}(\mathbf{X}_G, \mathbf{A}_G)) = \sigma\left(\hat{\mathbf{D}_G}^{-\frac{1}{2}} \hat{\mathbf{A}_G} \hat{\mathbf{D}_G}^{\frac{1}{2}} \mathbf{X}_G{}^{(l)} \mathbf{W}_G{}^{(l)}\right) \tag{1}$$

where G is the input graph, $\mathbf{W}_G{}^{(l)}$ denotes the learnable weight matrix, $\hat{\mathbf{A}_G}=\mathbf{A}_G+\mathbf{I}$ is the adjacency matrix with self-loops added to each node, $\mathbf{D}_G$ is the degree matrix, $\hat{\mathbf{D}_G}^{-\frac{1}{2}} \hat{\mathbf{A}_G} \hat{\mathbf{D}_G}^{\frac{1}{2}}$ is the normalized Laplace matrix, σ is the activation function, and $\mathbf{X}_G^{(l)}$ and $\mathbf{X}_G^{(l+1)}$ are the feature representations for the lth layer and the $l+1$th learned layer, respectively.

First, fusing information from brain area nodes within the same hemisphere is notated as $\mathbf{X}_{G_l}^{(l)}$ and $\mathbf{X}_{G_r}^{(l)}$ through Eq. (1), where G_l and G_r denotes the graphs of the left and right hemispheres of the brain from the G_{ind}.

Next, in the cross-hemisphere graph convolution layer, the feature representations of the whole brain are learned as follows:

$$\mathbf{X}_{G_{ind}}^{(l+1)} = \sigma(\mathrm{GCN}(\mathbf{X}_{G_l}^{(l)} \oplus \mathbf{X}_{G_r}^{(l)}, \mathbf{A}_{G_{ind}})) \tag{2}$$

The whole-brain features under this partitioning according to the standard brain atlas are then determined with a slicing operation. Based on the global neighbour-joining matrix, GCN focus on learning information across the hemispheres and the whole brain.

LGP. We propose a local-global dual-channel pooling (LGP) approach to eliminate noisy links while retaining vital vertices and edge connections, LGP help us to retain as much of the essential local and global details in the graph as possible and determine the importance of each ROI for neuropsychiatric disorders (NDs) and healthy controls (HCs) classification.

Channel 1: In the local feature-based pooling layer, we use a method to discard nodes based on local features to create a detailed pooling graph. This is performed by ranking the importance of each node based on a composite score that considers its features and topology. First, we calculate the score S for each node among the n nodes. Then, we use the top-k algorithm to select the k_1 nodes with the highest scores to retain after pooling. We obtain the new neighbour matrix and feature matrix through an indexing operation. The GCN model is used to calculate the node score because it can effectively learn a node's features and topology, including its neighbours.

Channel 2: In this channel of global topology-based pooling, we generate the rough pooled graph G_{coarse} using a node clustering technique. First, a GNN is used to learn the clustering assignment matrix to generate the rough graph, which is defined with the following equation:

$$\mathbf{P} = softmax\left(\text{GCN}\left(\mathbf{X}_{G_{ind}}, \mathbf{A}_{G_{ind}}\right)\right) \in \mathbb{R}^{m \times k_2} \tag{3}$$

where m is the number of nodes in the G_{ind}, k_2 is the number of predefined clusters. GCN is used to cluster the nodes, and the $softmax$ function is used to determine the assignment probability for each node in the G_{ind} for each cluster. Then, the feature matrix $\mathbf{X}_{coarse} \in \mathbb{R}^{k_2 \times D}$ and the adjacency matrix $\mathbf{A}_{coarse} \in \mathbb{R}^{k_2 \times k_2}$ of the rough pooling map: $\mathbf{X}_{coarse} = \mathbf{P}^T \mathbf{X}_{G_{ind}}$, $\mathbf{A}_{coarse} = \mathbf{P}^T \mathbf{A}_{G_{ind}} \mathbf{P}$, where D is the feature dimension of the node in channel 1.

At this point, pooled graphs for two different aspects of the initial graph have been constructed. Then, to efficiently combine the information in the two channels, a cross-channel convolution operation is used, which is defined as follows:

$$\mathbf{X}_{G_{ind}} = \sigma\left(\left[\mathbf{X}_{G_{ind}}^{ch1} + \mathbf{A}_{cross} \cdot \mathbf{X}_{G_{ind}}^{ch2}\right] \cdot \mathbf{V}\right) \tag{4}$$

where $\mathbf{V}$ is the trainable weight matrix, $\mathbf{X}_{G_{ind}}^{ch1} \in \mathbb{R}^{k_1 \times D}$ and $\mathbf{X}_{Gind}^{ch2} \in \mathbb{R}^{k_2 \times D}$ denote the two channel node embedding matrices, respectively, and $\mathbf{A}_{cross} \in \mathbb{R}^{k_1 \times k_2}$ denotes the connectivity relationship between the coarse-grained and fine-grained graphs. Specifically, we obtain the cross-channel connectivity matrix $\mathbf{A}_{cross}$ by using the assignment probabilities of the reserved nodes in channel 1 in the allocation matrix $\mathbf{P}$.

2.2 Heterogeneous Population Graph

Constructing the Graph. Recent research has shown unique differences in brain connectivity between males and females [9]. To account for these differences, we created a sex-based heterogeneous population graph (HPG).

The HPG is represented by $G_{pop}(V', E')$, where the set of nodes V' denotes the subjects, E' represents the set of edges between the vertices, including the same-sex edges $E'_{(s)}$ and the opposite-sex edges $E'_{(o)}$. The features of n subjects are used as the node features $\mathbf{Z} \in \mathbb{R}^{n \times h}$ in the population graph, where $h = k_1 \times D$ is the feature dimension of each subject's fMRI data after Brain Connectomic Graph. Let $\mathbf{A}'$ be the adjacency matrix of V', which shows the connection between the subjects. $\mathbf{A}' = Thr(\mathbf{S}^p)$ is defined as the similarity in the nonimaging information among n subjects, where $\mathbf{S}^p \in \mathbb{R}^{n \times n}$. S^p denotes the similarity score between the two subjects based on their phenotypic data. $Thr(\cdot)$ is a threshold function that controls the sparsity of G_{pop} by retaining only edges with edge weights greater than β.

Information Aggregation and the Classification Layer. The graph transformer [23] model aggregates information about a person's same-sex and opposite-sex neighbours separately. We believe that same-sex and opposite-sex information should contribute differently to the feature embeddings, and we use learnable proportional weights to aggregate this information in an adaptive manner:

$$\mathbf{Z}_{(s)}^{(l+1)} = GraphTransformer\left(\mathbf{Z}^{(l)}, E'_{(s)}\right) \tag{5}$$

$$\mathbf{Z}_{(o)}^{(l+1)} = GraphTransformer\left(\mathbf{Z}^{(l)}, E'_{(o)}\right) \tag{6}$$

$$\mathbf{Z}^{(l+1)} = W_{(s)}^{(l)} \mathbf{Z}_{(s)}^{(l+1)} + W_{(o)}^{(l)} \mathbf{Z}_{(o)}^{(l+1)} \tag{7}$$

where $\mathbf{Z}_{(s)}^{(l+1)}$ and $\mathbf{Z}_{(o)}^{(l+1)}$ denote the feature embeddings learned from the same-sex and opposite-sex neighbours at the $l+1$ layer. $W_{(s)}^{(l)}$ and $W_{(o)}^{(l)}$ are trainable parameters and are updated with a normalisation operation at each layer to make their sum 1. They will do the same at initialisation to prevent large gaps and negative numbers. The predicted labels are finally obtained using MLP :$\hat{y}_i = MLP(||_{l=1}^{L} \mathbf{z}_i{}^{(l)})$. We use the cross-entropy function as a loss function:

$$L_{ce} = \frac{1}{n} \sum_{i=1}^{n} - \left[y_i \cdot \log\left(\hat{y}_i\right) + \left(1 - y_i\right) \log\left(1 - \hat{y}_i\right)\right] \tag{8}$$

3 Dataset and Experimental Evaluation

3.1 Dataset and Experimental Setup

ABIDE dataset [6]:It collected and publicly shared rs-fMRI and phenotypic data (gender, age, cognitive level, DSM-IV, site, etc.) on a total of 1112 subjects from 17 different international research centres. For a fair comparison, we chose data from the same 871 subjects with higher quality that many researchers have used for their experiments [8], including 468 normal subjects and 403 subjects with ASD. These data were preprocessed by the widely used configurable pipeline for the analysis of connectomes (C-PAC) [4].

MDD dataset [27]: In our experiments, we used rs-fMRI and phenotype data containing 282 MDD patients and 251 healthy participants from Southwest University in the publicly available dataset REST-meta-MDD. All data were preprocessed by DPARSF [28], and the main processes were similar to C-PAC, including slice time correction, motion correction, global mean intensity normalisation, and spatial smoothing.

Table 1. Experimental results on ABIDE dataset and MDD dataset.

Type	Method	ABIDE		MDD	
		ACC	AUC	ACC	AUC
M.L.	SVM	66.60±2.47	65.24±3.26	59.41±3.12	56.98±4.02
	RF	60.16±3.89	66.24±3.10	60.73±3.98	60.10±4.21
D.L.	ASD-SAENet(2021) [1]*	70.83	-	-	-
	MFC-PL(2022) [16]*	70.22	73.11	-	-
Ind.GCN	GCN(2017) [12]	65.79±6.24	64.62±5.50	61.04±5.91	60.92±6.84
	BrainGNN(2021) [15]	71.60±4.38	65.08±5.21	68.52±8.01	68.34±7.76
	MVS-GCN(2022) [25]*	68.61	67.38	-	-
Pop.GCN	PopulationGCN(2017) [21]	71.62±4.62	71.98±4.74	96.81±2.34	96.60±1.79
	SGC (2019) [22]	77.78±5.91	77.73±4.20	81.90±5.91	83.97±4.984
	EV-GCN(2020) [8]	84.96±1.77	88.10±2.34	97.56±3.35	98.78±1.83
	Hi-GCN(2020) [10]*	72.81	77.12	-	-
	Cao et al. (2021) [3]	86.45±2.40	91.78±2.66	75.27±4.31	76.31±4.50
	MMGL(2022) [30]	89.88±2.41	89.58±2.80	89.32±2.02	91.20±3.62
	MAMF-GCN(2022) [19]	88.63±2.76	94.41±3.01	99.24±0.02	99.79±0.01
	LG-GNN(2023) [29]	81.97±2.30	87.80±3.1	96.81±1.45	98.16±1.73
	FCP-GNN(ours)	**93.40±1.41**	**96.49±1.58**	**99.81±0.01**	**99.94±0.01**

Note: " *" indicates that the values are from the original literature and not from our reproduction.

In this study, the HarvardOxford (HO) [5] brain atlas was utilized to extract each brain region's average time series and create a functional connectivity matrix. The learning rate for ABIDE was set at 0.01, with a dropout rate of 0.3, $\alpha = 0.65$ and $\beta = 1.5$. Similarly, MDD had the same learning and dropout rates but $\alpha = 0.68$ and $\beta = 0.5$.

3.2 Cross-Validation

We evaluated the effectiveness of the models using tenfold cross-validation with both datasets. The experimental results on the two datasets are presented in Table 1, showing the mean and standard deviation, where bold indicates the best effect. According to the results the following conclusions can be analysed:

1. Deep learning-based (D.L.) models outperform machine learning (M.L.) models that rely on manual feature inputs. Moreover, the binary classification performance of machine learning models should be improved, primarily due to the complex and high-dimensional nature of rs-fMRI data. Extracting distinct features based on these data with manual methods is challenging.

2. The performance of the individual graph (Ind.GCN) models is comparable to that of the DNN-based ASD-SAENet and MFC-PL models, achieving approximately 70% accuracy on the ABIDE dataset. However, graph-structured processing of rs-fMRI data is advantageous for exploring biomarkers through individual graph models. Regarding the predictive performance, population graph (Pop.GCN) methods usually outperform individual graph models. The reason for this is that they account for the connections between subjects based on both image and nonimage data. As a result of this advantage, population graph models are superior to individual graph models. Moreover, population graph models that consider the fusion of multiple modes of information and use a multibrain atlas perform better than those that do not. Examples of such models include MAMF-GCN, MMGL, and EVGCN, all of which have achieved good results on the ABIDE dataset. However, MMGL could have performed better on the MDD dataset. This is due to the lack of modal data in the MDD dataset, which reduces the learning fusion capability of the model.
3. The Hi-GCN and LG-GNN models were both used to analyse functional brain networks on the different datasets, and the FCP-GNN model produced the best results. Our model outperformed Hi-GCN and LG-GNN by 20.59% and 11.43% in terms of the accuracy (ACC) on the ABIDE dataset and showed significant improvement across all metrics. This is because the FCP-GNN model mimics the brain's inner functions and structure, utilizing various types of phenotypic data to extract information. Moreover, HPG is used to classify subjects base on sex, improving its ability to learn comprehensive characteristics.
4. FCP-GNN outperforms all other methods with both datasets, even outperforming MMGL, which uses more modal data, and MAMF-GCN, which uses multibrain mapping data. This suggests that our first-stage method of extracting brain functional connectivity information provides excellent features for the subsequent classification tasks and outperforms features obtained by filtering, as performed by previous population graph models. On the other hand, there is a noticeable effectiveness gap between the HiGCN and LG-GCN models and recent population graph methods.

Table 2. Analysis of the ablation experiments on the ABIDE dataset.

	ACC	AUC	SEN	SPE	F1
w/o IIH-GCN	90.85	92.54	92.41	89.02	91.72
w/o LGP	88.68	91.36	87.48	88.61	89.62
w/o Heterogeneous GNN	91.43	93.72	93.72	88.50	91.88
FCP-GNN	**93.40**	**96.49**	**95.30**	**91.32**	**93.99**

3.3 Ablation Experiment and Parameter Sensitivity Analysis

We conducted ablation experiments on the ABIDE dataset to assess the effectiveness of the IIH-GCN, LGP, and sex-based heterogeneous graph strategies suggested in this paper. Table 2 displays the individual evaluation metrics, showing that the absence of the LGP scheme results in the most significant overall performance degradation. The ACC and AUC metrics decreased by 4.72% and 5.13%, respectively, indicating that the proposed pooling method effectively eliminates noisy connections while preserving distinguishable nodes and edges. In addition, we replaced the IIH-GCN layer with a standard global graph convolutional layer. Our findings show that the ACC and AUC metrics decreased by 2.55% and 3.95%, respectively. Therefore, the IIH-GCN layer aggregates information in a manner that closely resembles the actual information propagation process within the brain. As a result, this layer can more effectively learn the embedded representation of each subject. To evaluate the practical benefits of the heterogeneous classification aggregation layer, we constructed an isomorphic population GNN without taking sex into account. When sex was not considered, the ACC and AUC metrics decreased by 1.97% and 2.77%, respectively, suggesting that it may be clinically meaningful to consider sex when diagnosing mental illness. Overall, the results show that each module contributes substantially to the prediction results of the model, with the optimal outcomes achieved when all the different components are integrated.

In the FCP-GNN model, the significant hyperparameters are the pooling rate for each pooling channel layer in the LGP layer and the number of layers in the IIH-GCN and graph transformer. We conducted parameter sensitivity experiments on the ABIDE dataset to understand the impact of these parameters on the performance. The data shown in Fig. 2(a) clearly demonstrate that the pooling rate of channel 1 significantly impacts classification performance. If too many nodes are discarded, the performance is negatively impacted. Moreover, removing noisy and redundant nodes and edges can improve classification performance. In contrast, channel 2's pooling rate has a more stable effect on the results, and it only significantly reduces performance when there are a large or inadequate number of clusters. In these cases, the clustering results become redundant or completely undifferentiated, which negatively impacts performance. Figure 2(b) shows that the number of convolutional layers has varying effects on the model's performance. The model may become oversmoothed if there are too many layers. The optimal results are achieved by incorporating two layers of the IIH-GCN and four layers of the graph transformer on the ABIDE dataset.

3.4 Interpretability Analysis

We examined the brain regions that have the most prominent roles in the diagnosis of ASD and MDD to better understand how these disorders are diagnosed. Our study focused on the nodes after the LGP operation and their scores. Then, we extracted the top ten ROIs with the highest scores across all subjects in the ten cross-validation folds. These regions were deemed biomarkers that can help with diagnosing these conditions. Figure 3 shows the distribution of the biomarkers.

For ASD classification, biomarkers with high scores were mainly located in the temporal gyrus, parahippocampal gyrus, frontal cortex, accumbens, central cortex, thalamus, brainstem, and caudate nucleus regions. These findings are in line with the findings of some previous studies [18], in which functional abnormalities in the temporal gyrus have been widely reported to be associated with the neural mechanisms of ASD [11]. All these regions are associated with consciousness, arousal, memory, emotion, sensation, and movement [14], and patients with ASD have significant deficits in these functions.

For MDD classification, the highest-scoring ROIs were the pallidum, inferior temporal gyrus, frontal operculum cortex, parahippocampal gyrus, thalamus, temporal fusiform cortex, amygdala, accumbens, and orbital cortex regions. Previous studies have shown that patients with depression have functional abnormalities in neuroanatomical circuits, particularly in the frontal cortex, amygdala, and hippocampus [13]. Symptoms such as apprehension, guilt, inability to think clearly, and indecision are associated with the entire prefrontal cortex. The amygdala is the key brain area that dominates anxiety, fear, and other negative emotions [2]. In MDD patients, the hippocampus is more biased towards negative emotional memories, and the nucleus ambiguus produces impulsive behaviour [17]. Depression is highly correlated with all regions we identified.

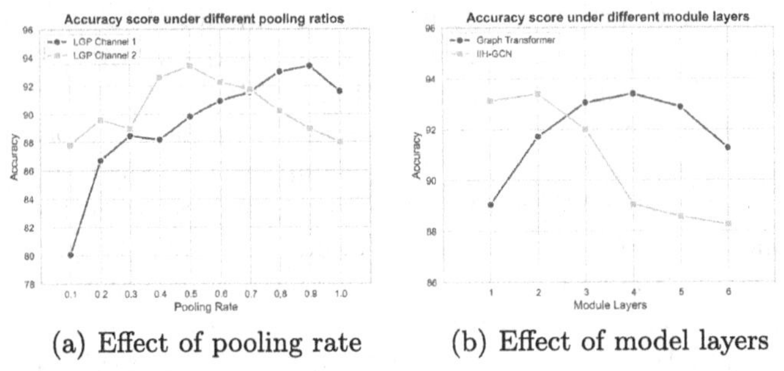

(a) Effect of pooling rate (b) Effect of model layers

Fig. 2. Experimental results for hyperparameter sensitivity on ABIDE dataset.

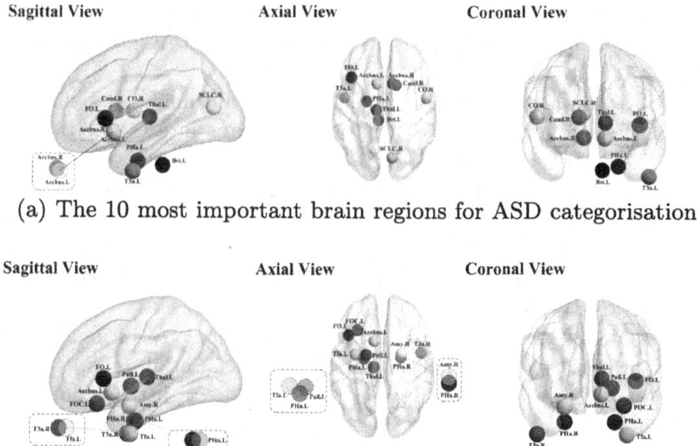

(a) The 10 most important brain regions for ASD categorisation

(b) The 10 most important brain regions for MDD categorization

Fig. 3. The most important brain regions for disease classification.

4 Conclusion

We present FCP-GNN, a comprehensive end-to-end disease prediction model based on functional connectivity-based heterogeneous graph learning. To capture potentially important information regarding the brain's functional connectivity networks, we propose IIH-GCN and LGP, which provide information on intra-hemispheric and interhemispheric interconnections as well as brain regions that can be saliently distinguished. Furthermore, considering the different brain patterns of males and females, we constructed an HPG based on sex and utilized the graph transformer architecture to develop a multilayer information aggregation module to determine the relationship between same-sex and opposite-sex subjects, assign appropriate weights, and complete downstream tasks as efficiently as possible. The FCP-GNN model achieved excellent results based on different public datasets and offers advantages over all existing methods.

References

1. Almuqhim, F., Saeed, F.: ASD-Saenet: a sparse autoencoder, and deep-neural network model for detecting autism spectrum disorder (ASD) using FMRI data. Front. Comput. Neurosci. **15**, 654315 (2021)
2. Bellani, M., Baiano, M., Brambilla, P.: Brain anatomy of major depression ii. focus on amygdala. Epidemiol. Psychiatr. Sci. **20**(1), 33–36 (2011)
3. Cao, M., Yang, M., Qin, C., et al.: Using deepGCN to identify the autism spectrum disorder from multi-site resting-state data. Biomed. Signal Process. Control **70**, 103015 (2021)

4. Craddock, C., Sikka, S., et al.: Towards automated analysis of connectomes: the configurable pipeline for the analysis of connectomes (C-PAC). Front. Neuroinform. **42**(10.3389) (2013)
5. Desikan, R.S., Ségonne, F., Fischl, B., et al.: An automated labeling system for subdividing the human cerebral cortex on MRI scans into GYRAL based regions of interest. Neuroimage **31**(3), 968–980 (2006)
6. Di Martino, A., Yan, C.G., Li, Q., et al.: The autism brain imaging data exchange: towards a large-scale evaluation of the intrinsic brain architecture in autism. Mol. Psychiatr. **19**(6), 659–667 (2014)
7. Gazzaniga, M.S.: Forty-five years of split-brain research and still going strong. Nat. Rev. Neurosci. **6**(8), 653–659 (2005)
8. Huang, Y., Chung, A.C.S.: Edge-variational graph convolutional networks for uncertainty-aware disease prediction. In: Martel, A.L., et al. (eds.) MICCAI 2020. LNCS, vol. 12267, pp. 562–572. Springer, Cham (2020). https://doi.org/10.1007/978-3-030-59728-3_55
9. Ingalhalikar, M., Smith, A., Parker, D., et al.: Sex differences in the structural connectome of the human brain. Proc. Natl. Acad. Sci. **111**(2), 823–828 (2014)
10. Jiang, H., Cao, P., Xu, M., Yang, J., Zaiane, O.: Hi-GCN: a hierarchical graph convolution network for graph embedding learning of brain network and brain disorders prediction. Comput. Biol. Med. **127**, 104096 (2020)
11. Kim, D., Lee, J.Y., Jeong, B.C., et al.: Overconnectivity of the right Heschl's and inferior temporal gyrus correlates with symptom severity in preschoolers with autism spectrum disorder. Autism Res. **14**(11), 2314–2329 (2021)
12. Kipf, T.N., Welling, M.: Semi-supervised classification with graph convolutional networks. arXiv preprint arXiv:1609.02907 (2016)
13. Korgaonkar, M.S., Fornito, A., Williams, L.M., Grieve, S.M.: Abnormal structural networks characterize major depressive disorder: a connectome analysis. Biol. Psychiat. **76**(7), 567–574 (2014)
14. Li, D., Karnath, H.O., Xu, X.: Candidate biomarkers in children with autism spectrum disorder: a review of MRI studies. Neurosci. Bull. **33**, 219–237 (2017)
15. Li, X., Zhou, Y., Dvornek, N., Zhang, M., et al.: BrainGNN: interpretable brain graph neural network for fMRI analysis. Med. Image Anal. **74**, 102233 (2021)
16. Liang, Y., Xu, G.: Multi-level functional connectivity fusion classification framework for brain disease diagnosis. IEEE J. Biomed. Health Inform. **26**(6), 2714–2725 (2022)
17. MacQueen, G., Frodl, T.: The hippocampus in major depression: evidence for the convergence of the bench and bedside in psychiatric research? Mol. Psychiatry **16**(3), 252–264 (2011)
18. Monk, C.S., Peltier, S.J., Wiggins, J.L., et al.: Abnormalities of intrinsic functional connectivity in autism spectrum disorders. Neuroimage **47**(2), 764–772 (2009)
19. Pan, J., Lin, H., Dong, Y., Wang, Y., Ji, Y.: MAMF-GCN: multi-scale adaptive multi-channel fusion deep graph convolutional network for predicting mental disorder. Comput. Biol. Med. **148**, 105823 (2022)
20. Papakostas, G.I.: Managing partial response or nonresponse: switching, augmentation, and combination strategies for major depressive disorder. J. Clin. Psychiatry **70**(suppl 6), 11183 (2009)
21. Parisot, S., et al.: Spectral graph convolutions for population-based disease prediction. In: Descoteaux, M., Maier-Hein, L., Franz, A., Jannin, P., Collins, D.L., Duchesne, S. (eds.) MICCAI 2017. LNCS, vol. 10435, pp. 177–185. Springer, Cham (2017). https://doi.org/10.1007/978-3-319-66179-7_21

22. Rakhimberdina, Z., Murata, T.: Linear graph convolutional model for diagnosing brain disorders. In: Cherifi, H., Gaito, S., Mendes, J., Moro, E., Rocha, L. (eds.) Complex Networks and Their Applications VIII. COMPLEX NETWORKS 2019. Studies in Computational Intelligence, vol. 882, pp. 815–826. Springer, Cham (2020). https://doi.org/10.1007/978-3-030-36683-4_65
23. Shi, Y., Huang, Z., et al.: Masked label prediction: Unified message passing model for semi-supervised classification. arXiv preprint arXiv:2009.03509 (2020)
24. Vigo, D., Thornicroft, G., Atun, R.: Estimating the true global burden of mental illness. The Lancet Psychiatry **3**(2), 171–178 (2016)
25. Wen, G., Cao, P., Bao, H., et al.: MVS-GCN: a prior brain structure learning-guided multi-view graph convolution network for autism spectrum disorder diagnosis. Comput. Biol. Med. **142**, 105239 (2022)
26. Yahata, N., Morimoto, J., Hashimoto, R., Lisi, G., et al.: A small number of abnormal brain connections predicts adult autism spectrum disorder. Nat. Commun. **7**(1), 11254 (2016)
27. Yan, C.G., Chen, X., Li, L., et al.: Reduced default mode network functional connectivity in patients with recurrent major depressive disorder. Proc. Natl. Acad. Sci. **116**(18), 9078–9083 (2019)
28. Yan, C.G., Wang, X.D., Zuo, X.N., Zang, Y.F.: DPABI: data processing & analysis for (resting-state) brain imaging. Neuroinformatics **14**, 339–351 (2016)
29. Zhang, H., et al.: Classification of brain disorders in RS-fMRI via local-to-global graph neural networks. IEEE Trans. Med. Imaging **42**(2), 444–455 (2023)
30. Zheng, S., Zhu, Z., Liu, Z., et al.: Multi-modal graph learning for disease prediction. IEEE Trans. Med. Imaging **41**(9), 2207–2216 (2022)

Weighted Chaos Game Representation for Molecular Sequence Classification

Taslim Murad, Sarwan Ali(✉), and Murray Patterson

Georgia State University, Atlanta, GA, USA
{tmurad2,sali85}@student.gsu.edu, mpatterson30@gsu.edu

Abstract. Molecular sequence analysis is a crucial task in bioinformatics and has several applications in drug discovery and disease diagnosis. However, traditional methods for molecular sequence classification are based on sequence alignment, which can be computationally expensive and lack accuracy. Although alignment-free methods exist, they usually do not take full advantage of deep learning (DL) models since DL models traditionally perform below power on tabular data compared to their effectiveness on image-based data. To address this, we propose a novel approach to classify molecular sequences using a Chaos Game Representation (CGR)-based approach. We utilize k-mers-based frequency chaos game representation (FCGR) to generate 2D images for molecular sequences. Additionally, we incorporate scaling features for the sliding windows, including Kyte and Doolittle (KD) hydropathy scale, Eisenberg hydrophobicity scale, Hydrophilicity scale, Flexibility of the characters, and Hydropathy scale, to assign weights to the k-mers. By selecting multiple features, we aim to improve the accuracy of molecular sequence classification models. The motivations to incorporate weights for the k-mers in the molecular sequence analysis are the fact that different k-mers may have different levels of importance or relevance to the classification task at hand and that incorporating additional information, such as hydropathy scales, could improve the accuracy of classification models. The proposed method shows promising results in molecular sequence classification by outperforming the baseline methods and provides a new direction for analyzing sequences using image classification techniques.

Keywords: Chaos Game Representation · Molecular Sequences · Classification

1 Introduction

Molecular sequence analysis is a fundamental task in bioinformatics with wide-ranging applications in drug discovery, disease diagnosis, and personalized medicine. Understanding the properties, function, structure, and evolution of molecular sequences is crucial for gaining insights into biological processes and developing effective treatments [1].

T. Murad and S. Ali—Joint First Authors.

D.-N. Yang et al. (Eds.): PAKDD 2024, LNAI 14648, pp. 234–245, 2024.
https://doi.org/10.1007/978-981-97-2238-9_18

Traditional methods [2,3] for molecular sequence classification often rely on sequence alignment, which can be computationally expensive [4] and may not always deliver high accuracy. Alignment-free methods [5–7] offer an alternative by involving the concept of k-mers to get the numerical embeddings of molecular sequences. However, they often encounter sparsity and high-dimensionality issues, making them computationally expensive. Additionally, they may not fully exploit the potential of deep learning models, which have shown remarkable performance on image-based data compared to tabular data.

The motivation behind this work is to address the limitations of existing molecular sequence classification methods and leverage the power of deep learning models in analyzing molecular sequences. We introduce a novel strategy based on Chaos Game Representation (CGR) [8] and weighted k-mers for molecular sequence analysis by transforming the sequences into images with the preservation of all the sequence's information. This approach combines the advantages of deep learning models and image classification techniques to unlock the potential of analyzing molecular sequences as images. CGR was originally proposed for creating images of DNA sequences by using fractals. A modification of CGR, known as FCGR (frequency-matrix-based CGR) [9] is proposed to deal with mapping molecular sequences into images. However, it has a one-to-one mapping from characters of a sequence to pixels of the created image, and it also uses a constant value for every character to increment its respective pixel value.

In our approach, we incorporate weighted k-mers with FCGR to enhance the information captured in the generated molecular sequence images. Different k-mers may have varying levels of importance or relevance to the classification task at hand. We introduce scaling features, such as the Kyte and Doolittle (KD) hydropathy scale, Eisenberg hydrophobicity scale, Hydrophilicity scale, Flexibility of the characters, and Hydropathy scale, to assign weights to the k-mers. By considering multiple features and assigning appropriate weights, we aim to improve the accuracy of molecular sequence classification models. The proposed method shows promising results in molecular sequence classification and provides a new direction for analyzing molecular sequences using image classification techniques.

The objectives of this study are threefold. Firstly, we aim to propose an efficient approach for generating images of molecular sequences by combining the power of k-mers with FCGR. This enables the utilization of sophisticated deep learning (DL) vision models on molecular sequence classification tasks. Secondly, we introduce various methods, such as hydropathy scales, to determine the weights of k-mers and incorporate these weights in the image-generation pipeline to create optimal molecular sequence images. Lastly, we evaluate the performance of the proposed approach on the host classification tasks of the rabies virus, demonstrating its superiority over baseline methods. Our contributions to this paper are the following:

1. We propose an efficient approach for generating images of molecular sequences by combining the power of k-mer with FCGR to enable the application of sophisticated DL vision models on molecular sequence classification.

2. We use various methods (Kyte and Doolittle (KD) hydropathy scale, Eisenberg hydrophobicity scale, Hydrophilicity scale, Flexibility of the characters, and Hydropathy scale) to determine the weights of k-mers, and incorporate these weights in the image-generation pipeline to create optimal images.
3. Our results demonstrate that the proposed approach is able to achieve high performance on the rabies virus's host classification task.

The rest of the paper is organized as follows: Sect. 2 discusses the existing methods to perform sequence analysis. Section 3 talks about the details of our proposed work to convert the molecular sequences into images. Furthermore, the dataset information and experimental setup are highlighted in Sect. 4. Section 5 deals with an in-depth discussion of the experimental results, and Sect. 6 highlights the conclusion and future work.

2 Related Work

Molecular sequence analysis is a popular topic in research, and many works already exist in this domain. Most of the techniques aim to extract numerical features from the molecular sequences to apply machine learning algorithms to analyze them. Feature-engineering-based methods are one such example. For instance, OHE [2] deals with transforming molecular sequences into binary vectors for performing machine learning-based classification. However, the computed vectors by OHE are very sparse and have high dimensions. These vectors also lack the ability to capture any local positional information about the sequence. Other feature-engineering-based mechanisms involve the k-mers based alignment-free approaches [5–7] which usually deal with the k-mers frequencies to get the embeddings, but they are usually computationally expensive and generate very sparse numerical embeddings with high dimensions. Similarly, PWM2Vec [3] incorporate the notion of k-mer with position weight matrix to compute the feature vectors, but it requires the sequences to be aligned, and sequence alignment is a very expensive step [4].

Furthermore, employing a neural network to extract the sequence's embeddings is a strategy used by some studies, like [10], but they require large training data for efficient performance, and acquiring medical data is an expensive process. The usage of kernel-matrix to compute the similarity between sequences to get their numerical representations is also an alternate option, like [11]. However, the large size and high dimensions of the kernel matrix limit its application as its space is inefficient.

Another alternate sequence analysis consists of sequence-to-image encoding methods. These techniques work by converting the molecular sequences into images and using these generated images for further analysis. They are based on the CGR [8] concept like FCGR [9] creates images of molecular sequences by constructing n-flakes. However, it uses only one-to-one mapping from characters to the pixels, which can't efficiently capture all the information about the sequence in the image. It also increments the pixel values of all the characters with a constant weight, which indicates that it's unable to consider the relative importance of the characters.

3 Proposed Approach

This section highlights the proposed method to convert molecular sequences into images so that sophisticated DL models can be applied to them to perform the analysis.

An existing popular approach for image encoding of molecular sequences is based on Chaos Game Representation (CGR) [8,12]. It was originally designed for DNA genome sequences, where it maps the nucleotides of a sequence into an image by dividing the image into 4 quadrants recursively, each quadrant representing a unique nucleotide. An example of allocating a pixel to a nucleotide-based sequence using CGR is shown in Fig. 1a. This method computes the k-mers of a given sequence first and then assigns pixels to the k-mers in the corresponding image respectively. k-mers refer to a set of consecutive substrings of length k driven from an input sequence. An example of 3-mers for a given spike sequence is illustrated in Fig. 1b. The CGR-based image encoding works best for nucleotide-based sequences, where $n = 4$ and n is the number of unique nucleotides. However, it causes overlapping in the generated images for molecular sequences because of n being 20, where n is the number of unique characters.

To overcome the overlapping challenge for molecular sequences, a modified version of CGR is presented, known as Frequency Matrix-based Chaos Game Representation (FCGR) [9] (we use the word Chaos and FCGR alternatively from this point onwards). This method generates images based on n-flakes (where n represents the number of unique characters, so for molecular sequences, FCGR generates 20-flakes [13] based images) consisting of multiple icosagons (visual example given in Fig. 1c). An icosagon is a 20-sided polygon. An iterative approach to creating fractals from an n-gon is followed by the n-flake. For instance, a sequence with 20 characters will make the FCGR to produce 20 edges and then 20 icosagons within an icosagon (see Fig. 1c). The fractal construction begins at the center of the image. Then, the following steps are used to determine the pixel's location based on the previous one:

The contraction ratio r between the outer and inner polygon is computed using the equation with $n = 20$,

$$r = \frac{sin(\frac{\pi}{n})}{sin(\frac{\pi}{n}) + sin(\frac{\pi}{n} + \frac{2\pi m}{n})}, \text{for m} = \lfloor \frac{n}{4} \rfloor \tag{1}$$

After that, the ratio of the distance between the current location and the target edge (character) known as the scaling factor (SF) is calculated using the equation,

$$\text{sf} = 1 - r \tag{2}$$

Then, the pixel's coordinates are computed with the given equations,

$$x[i] = r \cdot sin(\frac{2\pi i}{n} + \theta) \tag{3}$$

$$y[i] = r \cdot cos(\frac{2\pi i}{n} + \theta) \tag{4}$$

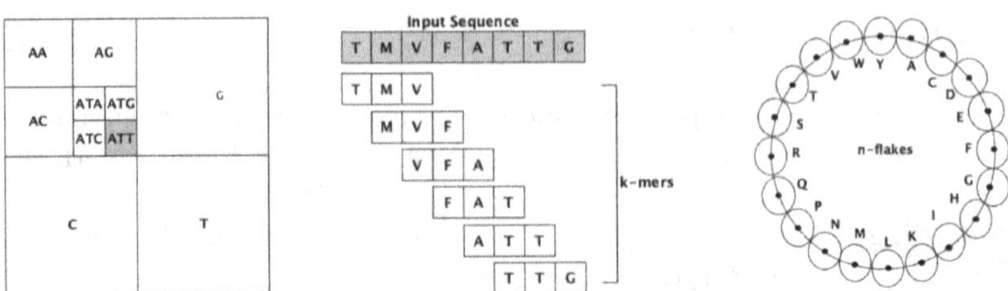

(a) CGR-based allocation for nucleotide sequence.
(b) 3-mers for a protein sequence
(c) 20-flakes for protein sequence.

Fig. 1. (a) illustrates the CGR-based space allocation for a given k-mer in the respective image.(b) shows an example of 3-mers from a given sequence. (c) shows an example of 20-flakes for protein sequences.

where $i \in \{1, 2, \ldots, |Seq|\}$ (where Seq is a sequence and), and θ is the angle of orientation. Given a molecular sequence, FCGR works by getting the pixel location of each character using Eq. 3 & Eq. 4, and incrementing the pixel value by 1 and yields a greyscale image as output.

Although FCGR [9] works well in eliminating the overlapping in the images for sequences with $n < 4$, however, it only operates on 1-mers of a sequence. This single-pixel value assignment for a character is not an effective image representation of the sequence. Therefore, we proposed to combine the power of k-mers (with $k > 1$) with FCGR to get an optimal image representation, as k-mers are well known to be the effective representation of molecular sequences. With a higher value of k, a single character of the original sequence can be mapped to multiple pixels in the corresponding image, which can enable the capturing of more information about the given sequence in its respective image encoding. Moreover, FCGR [9] uses a constant value of 1 to increment the pixel values for all the characters, which ignores the relative importance of the characters. Therefore, rather than using a constant value of 1, we propose to assign weights to each character of the k-mer and utilize the cumulative weight to increase the corresponding pixel value of the k-mer. In this way, we are able to capture the relative importance of the characters, which can improve our image encoding in terms of performing classification using those generated images. We are using the following 5 methods to assign weights to the characters of a k-mer:

Kyte and Doolittle (KD) Hydropathy Scale: This technique has devised a hydropathy scale that considers the hydrophilic and hydrophobic properties of each character of a molecular sequence to assign it weight [14]. With advances in the molecular sequence, this method continuously determines the average hydropathy of the characters. The KD score for a k-mer is an average of all the scores of its respective characters, and it can be expressed as follows,

$$KD_{score} = \frac{1}{m} \sum_{j=1}^{m} KD_j \tag{5}$$

where m refers to the total number of characters in a k-mer, KD_{score} represents the final KD score of the k-mer, and KD_j is the KD score of the jth characters.

Eisenberg Hydrophobicity Scale: This method determines the free energy change of transfer of characters from water to a non-polar solvent in the form of hydrophobicity scale [15]. The partitioning behavior of a character between water and a hydrophobic solvent is used to give it a hydrophobicity score. For a k-mer, this weight is determined by averaging the scores of its characters.

Hydrophilicity Scale: Hydrophilicity scale [16] is a well-known measure of hydrophilicity of character based on its partitioning behavior between water and 1-octanol. Hydrophilicity refers to the ability of a character to interact with water molecules, and based on the tendency of the character to bond formulation with water molecules or other polar groups, the hydrophilicity score is assigned to it. The scores of all the corresponding characters are averaged to get the respective k-mers hydrophilicity score.

Flexibility of the Characters: The tendency of a character to adapt to various conformations determines its flexibility score [17]. The scores of all the respective characters are averages to get the corresponding k-mers's score.

Hydropathy Scale: The tendency of a character to interact with water is determined using the hydropathy scale, and a numerical score is assigned to the character based on this interaction ability [18]. The hydropathy score of a k-mer is determined by averaging its character scores.

We have illustrated the workflow of our proposed approach in Fig. 2. Given a sequence, its k-mers are extracted in step (2), followed by getting the weights of each k-mer using different techniques in step (3). Then, each k-mer, along with its weights, is passed on to step (4) to construct the image. In step (4) (a), the coordinates of every character of the k-mers are determined using Eqs. 3, and 4, and these coordinates are further utilized to locate the pixel in the corresponding created image. The pixel value is incremented with the respective weight value, and the FCGR- and weighted k-mers-based image is generated.

3.1 Vision DL Models

We use three different popular deep-learning (DL) models for classification tasks using our generated image datasets. These models are Vision Transformers (ViT) model [19], the Convolution Neural Network (CNN) model [20], and the pre-trained EfficientNetB0 model [21]. As CNN is known to achieve state-of-the-art performance for image-based classification and ViT utilizes the power of transformer architecture for vision to yield the best performance, therefore, we use these models. Moreover, to investigate the capability of transfer learning for image-based classifications, we employ the pre-trained EfficientNetB0 model, as it is one of the most efficient models with high levels of accuracy and known to perform better for complex tasks [22].

The CNN model has 4 hidden convolution layers followed by a fully connected (FC) layer and a softmax classification layer. Each convolution layer has a 2D

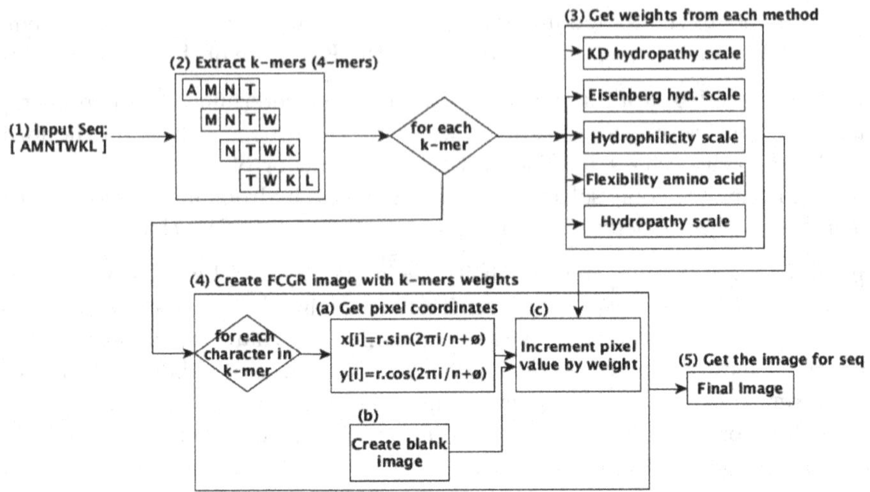

Fig. 2. Workflow of the proposed method for creating an image of a sequence.

convolution operation with a ReLU activation function and a 2D max pooling operation. The architecture diagram of this model is given in Fig. 3.

The ViT model architecture is shown in Fig. 4. For a given input image, its patches are extracted (we use 20 as patch size). Then, the patches are fed to the linear embedding module to get corresponding vectors (we use 8 as vector dimension), and a positional embedding is added to them. These vectors are then passed to the transformer encoder block, which consists of a normalization layer followed by a multi-head self-attention layer and a residual connection. Then, there is a second normalization layer, a multi-layer perceptron (MLP), and, again, a residual connection. These blocks are connected back-to-back, and our model uses 2 transformer blocks. The result from the transformer encoder is given to the softmax classification module to get the final label of the image.

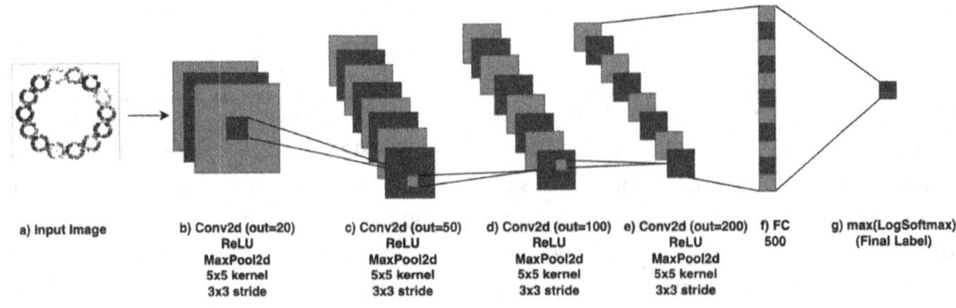

Fig. 3. The architecture of the CNN model with 4 hidden convolution layers and softmax classification layer.

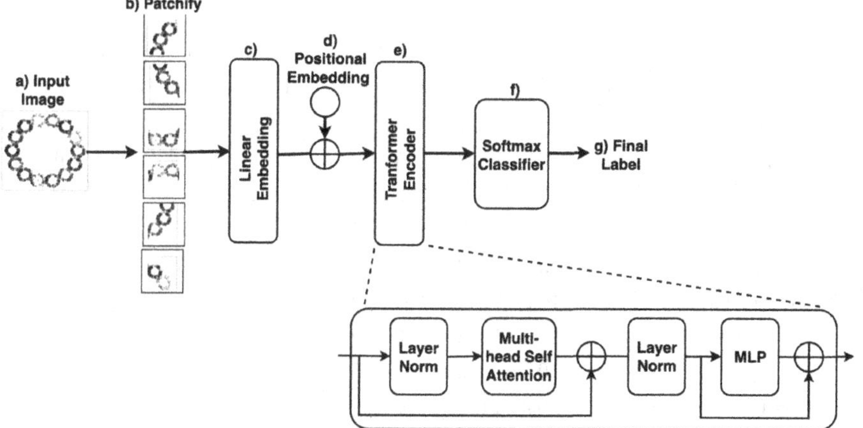

Fig. 4. The architecture of the ViT model.

4 Experimental Evaluation

This section discusses the details of the datasets along with the baseline models used to evaluate the system based on the mentioned evaluation metrics. We evaluated the classifiers using the following performance metrics: accuracy, precision, recall, weighted F1, F1 macro, and ROC AUC macro. The reason for reporting many metrics is to get more insight into the classifiers' performance, especially in the class imbalance scenario where reporting only accuracy does not provide sufficient performance information. All experiments are conducted using an Intel(R) Xeon(R) CPU E7-4850 v4 @ 2.10 GHz having Ubuntu 64-bit OS (16.04.7 LTS Xenial Xerus) with 3023 GB memory. The dataset that we use to evaluate our method consists of 20051 sequences of rabies virus corresponding to various infected hosts extracted from RABV-GLUE [23] website. These sequences belong to 12 unique host categories, which are used as labels for the classification task. The further distribution detail of the dataset is shown in Table 1.

4.1 Baselines Models:

To evaluate our proposed method, we are using the following 2 different techniques as baselines. In the first technique, we transform the molecular sequence data into numerical features without converting the data to images. We then use 2 tabular deep learning (i.e., CNN) models along with the classical Naive Bayes classifier to perform the classification based on these numerical forms. This comparison can give an overview of the classification performance using vision data versus original sequence data. The methods used to get numerical features are One-hot Encoding (OHE) [2], Wasserstein Distance Guided Representation Learning (WDGRL) [10], and position weight matrix-based approach, called

Table 1. Dataset Statistics for Rabies data.

Host Name	Count	Rabies Sequence Length			Number of Sequences		
		Min	Max	Average	Training	Validation	Testing
Canis Familiaris	9065	90	11928	1600.50	5802	1450	1813
Bos Taurus	2497	117	11928	995.29	1599	399	499
Vulpes Vulpes	2221	133	11930	2923.77	1422	355	444
Felis Catus	1125	90	11928	1634.43	720	180	225
Procyon Lotor	884	291	11926	6763.80	567	141	176
Desmodus Rotundus	875	164	11923	1051.50	560	140	175
Mephitis Mephitis	864	220	11929	1266.59	554	138	172
Homo Sapiens	838	101	11928	1537.85	537	134	167
Eptesicus Fuscus	718	264	11924	1144.35	460	115	143
Skunk	492	211	11928	6183.26	316	78	98
Tadarida Brasiliensis	270	264	11923	1175.67	173	43	54
Equus Caballus	202	163	11924	1376.74	130	32	40
Total	20051	-	-	-	-	-	-

PSSM2Vec [24]. The WDGRL is an unsupervised domain adoption technique that extracts the numerical embeddings of the sequences by utilizing a neural network. The optimization of this network is done by minimizing the Wasserstein distance between the source and target encoded networks. The PSSM2Vec approach combines the k-mers and position weight matrix (PWM) concepts to get numerical embeddings of molecular sequences.

In the second baseline technique, we compare the results of the proposed method with the original chaos [9] method to see how our method has improved the prediction performance for molecular sequence-based image data. This method produces an n-flakes-based image representation, which consists of an image with multiple icosagons. The Chaos method takes a molecular sequence as input and yields an image as output by considering the characters of the sequence one by one, following Eq. 3 and Eq. 4 to get the coordinates of every character in the image.

The OHE-, PSSM2Vec-, and WDGRL-based classifications are performed by two tabular deep learning models, namely 3-Layer Tab CNN and 4-Layer Tab CNN. The 3-Layer Tab CNN network consists of 3 hidden linear layers while 4-Layer Tab CNN has 4 layers. Both of them have a final classification linear layer, respectively. To train these models, $80-20\%$ train-test splits, 10 epochs, 0.003 learning rate, ADAM optimizer, and 64 batch size are used. Moreover, the cross entropy loss function is employed as a training loss function. The WDGRL-based input vectors are of dimensions 10 for both datasets. The features of OHE are made to a fixed length using the zero padding approach.

To train the vision models (CNN, ViT, EfficientNetB0), we use almost $80-20\%$ train-test splits based on stratified sampling, as it preserves the proportions between the classes. Moreover, the generated input images are of size 480x480 for both datasets. To train the models, we use 10 epochs, ADAM optimizer, 64 batch size, and 0.001 learning rate. Furthermore, as we are doing multi-class classification, therefore we employ cross-entropy loss as the training loss function.

5 Results and Discussion

The classification results are reported in Table 2 for the Rabies virus dataset. The results of the rabies virus dataset's host classification task demonstrate that our proposed imaged-based encodings, Eisenberg & hydrophilicity, are able to achieve the highest performance for almost all the evaluation metrics using the CNN classifier. The feature-engineering-based classifiers (NB, Tab 3 CNN, Tab 4 CNN) illustrate lower performance than the image-based custom ones (CNN & ViT), which indicates that generating images from sequences is a more efficient strategy than mapping the sequences to vectors in terms of classification performance as images can capture more relevant information about the sequences for classification. Moreover, the pre-trained models depict very low performance for

Table 2. Classification results for different methods on Rabies Virus datasets. The top 2 best values for each evaluation metric are shown in bold.

	Method	Acc. ↑	Prec. ↑	Recall ↑	F1 (Weig.) ↑	F1 (Macro) ↑	ROC AUC ↑	Train Time (Sec.) ↓
NB	OHE	0.124	0.447	0.124	0.134	0.195	0.585	979.44
	WDGRL	0.514	0.441	0.514	0.410	0.184	0.575	**0.01**
	PSSM2Vec	0.125	0.296	0.125	0.072	0.105	0.58	**0.04**
3 Layer Tab CNN	OHE	0.451	0.203	0.451	0.280	0.050	0.500	4191.34
	WDGRL	0.450	0.202	0.450	0.279	0.049	0.500	1737.65
	PSSM2Vec	0.452	0.204	0.452	0.281	0.051	0.500	2040.81
4 Layer Tab CNN	OHE	0.452	0.204	0.452	0.281	0.051	0.500	5974.26
	WDGRL	0.535	0.318	0.535	0.395	0.103	0.500	964.97
	PSSM2Vec	0.450	0.204	0.450	0.282	0.052	0.500	3790.09
ViT	Chaos	0.448	0.201	0.448	0.277	0.051	0.500	2943.45
	KD	0.440	0.194	0.440	0.269	0.050	0.500	3593.00
	Eisen	0.465	0.216	0.465	0.295	0.052	0.500	3474.12
	Flex	0.441	0.194	0.441	0.270	0.051	0.500	3035.72
	Hydrophil	0.455	0.207	0.455	0.285	0.052	0.500	2829.95
	Hydropathy	0.449	0.201	0.449	0.278	0.051	0.500	3029.90
CNN	Chaos	0.780	0.763	0.780	0.767	**0.662**	**0.813**	12505.91
	KD	0.771	0.757	0.771	0.756	0.647	0.807	13331.11
	Eisen	**0.787**	**0.779**	**0.787**	**0.773**	**0.668**	0.810	14127.47
	Flex	0.775	0.763	0.775	0.758	0.647	0.807	13068.88
	Hydrophil	**0.785**	**0.770**	**0.785**	**0.774**	0.659	**0.817**	14286.38
	Hydropathy	0.773	0.766	0.773	0.765	0.653	0.809	13115.00
Pretrain	Chaos	0.202	0.365	0.202	0.230	0.081	0.500	146831.05
	KD	0.210	0.370	0.210	0.229	0.079	0.510	147221.45
	Eisen	0.284	0.451	0.284	0.364	0.095	0.530	161828.01
	Flex	0.274	0.441	0.274	0.387	0.087	0.500	144477.50
	Hydrophil	0.283	0.431	0.283	0.363	0.093	0.521	150921.41
	Hydropathy	0.252	0.331	0.252	0.323	0.073	0.500	142441.85

all the metrics as compared to the other classifiers. As these models are trained originally using different kinds (foreground, background, content) of images, they are unable to generalize well to the CGR-based images. We can also observe that our proposed image encodings can outperform the Chaos baseline using the CNN model. Note that the CNN model is achieving higher performance than the ViT model. A reason for this can be that ViT works by using a global attention mechanism that focuses on the entire image, but in our case, the CGR-based images have relevant content in very specific pixels and the rest of the image is blank. Therefore, the global attention mechanism may not be very efficient for it as compared to the local operation-based CNN model. Additionally, ViT requires large training data to outperform the CNN model, and our dataset is very limited. Overall, the results show that our proposed methods have better performance than the baselines for the rabies host classification task.

6 Conclusion

In this study, we proposed a novel approach for molecular sequence classification using chaos game representation by incorporating scaling features for the sliding windows, which assigns weights to the k-mers and aims to improve the accuracy of models. The results demonstrate that our proposed approach outperforms traditional methods, indicating that incorporating additional information, such as hydropathy scales and utilizing deep learning models, can improve classification accuracy. Our findings show promising results and provide a new direction for analyzing molecular sequences using image classification techniques. For our future work, we can incorporate a 3D structure formulation of proteins in the images, along with evaluating the robustness of the existing models.

References

1. Whisstock, J.C., Lesk, A.M.: Prediction of protein function from protein sequence and structure. Q. Rev. Biophys. **36**(3), 307–340 (2003)
2. Kuzmin, K., et al.: Machine learning methods accurately predict host specificity of coronaviruses based on spike sequences alone. Biochem. Biophys. Res. Commun. **533**(3), 553–558 (2020)
3. Ali, S., Bello, B., Chourasia, P., Punathil, R.T., Zhou, Y., Patterson, M.: PWM2Vec: an efficient embedding approach for viral host specification from coronavirus spike sequences. Biology. **11**(3), 418 (2022)
4. Chowdhury, B., Garai, G.: A review on multiple sequence alignment from the perspective of genetic algorithm. Genomics **109**(5–6), 419–431 (2017)
5. Ma, Y., Yu, Z., Tang, R., Xie, X., Han, G., Anh, V.V.: Phylogenetic analysis of HIV-1 genomes based on the position-weighted K-mers method. Entropy **22**(2), 255 (2020)
6. Zhang, J., Bi, C., Wang, Y., Zeng, T., Liao, B., Chen, L.: Efficient mining closed K-mers from DNA and protein sequences. In: International Conference on Big Data and Smart Computing, pp. 342–349 (2020)

7. Ali, S., Patterson, M.: Spike2vec: an efficient and scalable embedding approach for COVID-19 spike sequences. In: IEEE Big Data, pp. 1533–1540 (2021)
8. Jeffrey, H.J.: Chaos game representation of gene structure. Nucleic Acids Res. **18**(8), 2163–2170 (1990)
9. Löchel, H.F., Eger, D., Sperlea, T., Heider, D.: Deep learning on chaos game representation for proteins. Bioinformatics **36**(1), 272–279 (2020)
10. Shen, J., Qu, Y., Zhang, W., Yu, Y.: Wasserstein distance guided representation learning for domain adaptation. In: AAAI Conference (2018)
11. Farhan, M., et al.: Efficient approximation algorithms for strings kernel based sequence classification. In: NeurIPS, pp. 6935–6945 (2017)
12. Barnsley, M.F.: Fractals everywhere: New edition (2012)
13. Tzanov, V.: Strictly self-similar fractals composed of star-polygons that are attractors of iterated function systems. arXiv preprint arXiv:1502.01384 (2015)
14. Kyte, J., Doolittle, R.F.: A simple method for displaying the hydropathic character of a protein. J. Mol. Bio. **157**(1), 105–132 (1982)
15. Eisenberg, D.: Three-dimensional structure of membrane and surface proteins. Annu. Rev. Biochem. **53**(1), 595–623 (1984)
16. Hopp, T.P., Woods, K.R.: Prediction of protein antigenic determinants from amino acid sequences. PNAS **78**(6), 3824–3828 (1981)
17. Kabsch, W., Sander, C.: Dictionary of protein secondary structure: pattern recognition of hydrogen-bonded and geometrical features. Biopolym. Orig. Res. Biomol. **22**(12), 2577–2637 (1983)
18. MacCallum, J.L., Tieleman, D.P.: Hydrophobicity scales: a thermodynamic looking glass into lipid-protein interactions. Trends Biochem. Sci. **36**(12), 653–662 (2011)
19. Dosovitskiy, A., et al.: An image is worth 16x16 words: Transformers for image recognition at scale. arXiv preprint arXiv:2010.11929 (2020)
20. O'Shea, K., Nash, R.: An introduction to convolutional neural networks. arXiv preprint arXiv:1511.08458 (2015)
21. Tan, M., Le, Q.: Efficientnet: rethinking model scaling for convolutional neural networks. In: International Conference on Machine Learning, pp. 6105–6114 (2019)
22. Hassan, Z.: 3 pre-trained image classification models (2022). https://www.folio3.ai/blog/image-classification-models/
23. Campbell, K., et al.: Making genomic surveillance deliver: A lineage classification and nomenclature system to inform rabies elimination. PLoS Pathog. **18**(5), e1010023 (2022)
24. Ali, S., Murad, T., Patterson, M.: PSSM2Vec: a compact alignment-free embedding approach for coronavirus spike sequence classification. In: Neural Information Processing (ICONIP), pp. 420–432 (2023)

Robust Influence-Based Training Methods for Noisy Brain MRI

Minh-Hao Van, Alycia N. Carey, and Xintao Wu(✉)

University of Arkansas, Fayetteville, AR 72701, USA
{haovan,ancarey,xintaowu}@uark.edu

Abstract. Correctly classifying brain tumors is imperative to the prompt and accurate treatment of a patient. While several classification algorithms based on classical image processing or deep learning methods have been proposed to rapidly classify tumors in MR images, most assume the unrealistic setting of noise-free training data. In this work, we study a difficult but realistic setting of training a deep learning model on noisy MR images to classify brain tumors. We propose two training methods that are robust to noisy MRI training data, Influence-based Sample Reweighing (ISR) and Influence-based Sample Perturbation (ISP), which are based on influence functions from robust statistics. Using the influence functions, in ISR, we adaptively reweigh training examples according to how helpful/harmful they are to the training process, while in ISP, we craft and inject helpful perturbation proportional to the influence score. Both ISR and ISP harden the classification model against noisy training data without significantly affecting the generalization ability of the model on test data. We conduct empirical evaluations over a common brain tumor dataset and compare ISR and ISP to three baselines. Our empirical results show that ISR and ISP can efficiently train deep learning models robust against noisy training data.

Keywords: Noisy images · Brain tumor classification · Robust training

1 Introduction

Accurate brain tumor detection and diagnosis is a crucial step in providing appropriate medical treatment and improving the prognosis of individuals with brain cancer and other brain ailments. However, the human brain is complex, and it can be difficult to correctly determine both the type and stage of a brain tumor. Often, magnetic resonance imaging (MRI) is utilized to create 3-D scans of the brain, which a specialist can use to plan a course of treatment based on the determined tumor type and stage. Unfortunately, analyzing MR images is a non-trivial task, and relying on human perception alone can result in misdiagnosis and delayed treatment. More and more, domain experts are relying on computer software that can analyze large-scale MR images in rapid time. Many image processing methods [4–6,8,13,22] have been proposed to classify brain tumors based on type by extracting useful features from an MR image to feed into the classification algorithm. More recently, techniques from deep learning have also been used to improve the performance of medical image classification systems [1,2,10].

D.-N. Yang et al. (Eds.): PAKDD 2024, LNAI 14648, pp. 246–257, 2024.
https://doi.org/10.1007/978-981-97-2238-9_19

Despite the advances in classifying MR images, many of the proposed works assume an unrealistic setting of noise-free data. Obtaining clean and high-quality MR images unperturbed by noise is challenging and time-consuming. Noise can be generated by a variety of different sources during the process of an MR scan (e.g., the patient moving, the sampling procedure used, or hardware issues), and in most cases collecting noisy MRI data is unavoidable. When the MR images are noisy, the performance ability of a classifier significantly decreases in terms of accuracy. Therefore, it is important to consider the setting where noisy data is assumed. In this work, we simulate the challenging but realistic scenario of the training and test dataset containing noisy MR images. Specifically, we propose two training methods, named Influence-based Sample Reweighing (ISR) and Influence-based Sample Perturbation (ISP), using the influence function from robust statistics to improve the generalization ability of a classification model trained on noisy brain tumor data.

2 Related Work

Improving the generalization ability of models trained to classify medical images has been a significant area of research over the past several years. While the field is broad, many of the proposed works utilize classical image processing techniques to extract features from the medical images to use in the training of a classification model. For example, the wavelet transform [13] and the Gray Level Co-occurrence Matrix (GLCM) [13,22] are two commonly used methods from image processing that can be used to construct intensity and texture features of the MR images. However, processing techniques from outside classical image processing have been proposed as well. For example, Bag-of-Words (BoW) method, which was originally proposed in the text-mining field to build a representation vector for each word in a document, has been shown to be capable of representing the complex features of a medical image [4,6,8]. Some approaches combining traditional and non-traditional image processing methods have also been proposed. For instance, [5] combines classical feature extractors (HoG and DSUFT) and a Support Vector Machine (SVM) to improve the performance and efficiency of the classifier in a multi-class setting. Additionally, [8] showed that when tumor regions are available, they can serve as a region of interest (ROI) and be augmented using image dilation to help boost the accuracy of BoW, the intensity histogram, and the GLCM.

With the growing popularity of deep learning, many works have proposed to adopt deep neural networks (DNNs) in the medical imaging field. In [1], the authors evaluated the effectiveness of different Convolutional Neural Network (CNN) architectures on a brain tumor classification task, and they found that a simple CNN architecture can achieve good performance without any prior knowledge of the segmented tumor regions. Further, transfer learning has been proven to be an effective method in helping a model to be more generalizable when there is limited training data across several different settings, and [10] showed that transfer learning is also useful to boost generalizability when working with medical images. [2] ventured beyond classic DNNs and showed that using more advanced architectures like Capsule Networks [12] can help to improve the accuracy of brain tumor classification, and they additionally explored why CapsNet tends to over-fit the data more than classic CNN models do.

Most of the works discussed above are proposed (and shown to be successful) based on the assumption that both the training and test datasets are free of noise. However, noisy data, in general, is unavoidable due to randomness and error during data collection. There are many robust training methods proposed to alleviate the potential harm of noisy data to a classification model [11,18]. Specifically in the medical imaging domain, [19,20] showed that adversarial training methods are effective when adapted to medical image analysis. Further [15,16] showed that image denoising methods can be utilized to remove noise and reconstruct clean medical images without requiring prior knowledge of clean data.

3 Preliminaries

3.1 Influence Function on Single Validation Point

Influence functions, first proposed in the field of robust statistics [9], have recently been found useful for explaining machine learning models. In [14], the authors proposed a method for estimating the influence that a single training point $z = (x, y)$ has on both the parameters of a machine learning model and the loss of a single test point. Here $x \in X$ represents a feature in the domain of features X and $y \in Y$ represents a label in the domain of labels Y. Let f_θ be a classification model parameterized by θ, $\hat{\theta}$ be the optimal parameters of f, and $D_{trn}/D_{val}/D_{tst}$ be the training/validation/test datasets. Let $l(\cdot, \theta)$ represent the loss and $L(D_*, \theta) = \frac{1}{|D_*|}\sum_{z_i \in D_*} l(z_i, \theta)$ be the empirical loss function to be minimized, where D_* denotes "any dataset". To see how a training point z affects the model's parameters, we write the empirical risk minimization as:

$$\hat{\theta}_{\epsilon,z} = \arg\min_{\theta \in \Theta} \frac{1}{|D_{trn}|} \sum_{z_i \in D_{trn}} l(z_i, \theta) + \epsilon l(z, \theta) \tag{1}$$

In Eq. 1, we simulate the removal of z from the training set by upweighting it by a small weight ϵ (usually on the order of $-\frac{1}{n}$ where n is the number of training points). Rather than calculating Eq. 1 through training, [14] shows that the influence of training point z on the model parameters can be estimated as:

$$\mathcal{I}_{up,param}(z) = \left.\frac{d\hat{\theta}_{\epsilon,z}}{d\epsilon}\right|_{\epsilon=0} = -H_{\hat{\theta}}^{-1} \nabla_\theta l(z, \hat{\theta}) \tag{2}$$

and the resulting change in parameters can be calculated as $\hat{\theta}_{\epsilon,z} \approx \hat{\theta} - \frac{1}{n}\mathcal{I}_{up,params}(z)$. In addition to showing the effect a training point z has on the model parameters, [14] extends Eq. 2 to calculate the influence that a training point z has on a test point z_{test}.

$$\mathcal{I}_{up,loss}(z, z_{test}) = -\nabla_\theta l(z_{test}, \hat{\theta})^\top H_{\hat{\theta}}^{-1} \nabla_\theta l(z, \hat{\theta}) \tag{3}$$

Both Eq. 2 and Eq. 3 estimate the effect of a training point z being completely removed from the training dataset. However, in [14], the authors additionally show that influence functions can be used to estimate the effect of perturbing $z = (x, y)$

to $\hat{z} = (x + \delta, y)$ on the model parameters. The empirical risk minimization can be written as:

$$\hat{\theta}_{\epsilon,\hat{z},-z} = \arg\min_{\theta \in \Theta} \frac{1}{|D_{trn}|} \sum_{z_i \in D_{trn}} l(z_i, \theta) - \epsilon l(\hat{z}, \theta) + \epsilon l(z, \theta) \tag{4}$$

and Eq. 4 can be estimated as $\hat{\theta}_{\epsilon,\hat{z},-z} \approx \hat{\theta} - \frac{1}{n}\mathcal{I}_{pert,param}(z)$ where:

$$\mathcal{I}_{pert,param}(z) = -H_{\hat{\theta}}^{-1}\left(\nabla_\theta l(\hat{z}, \hat{\theta}) - \nabla_\theta l(z, \hat{\theta})\right) \tag{5}$$

As in the case with Eq. 2, [14] extends Eq. 5 to show how perturbing $z \rightarrow \hat{z}$ would affect the loss of a test point z_{test}:

$$\mathcal{I}_{pert,loss}(z, z_{test}) = -\nabla_\theta l(z_{test}, \hat{\theta})^\top H_{\hat{\theta}}^{-1} \nabla_x \nabla_\theta l(z, \hat{\theta}) \tag{6}$$

The main difference between Eq. 3 and Eq. 6 is that in Eq. 6, the gradient of $\nabla_\theta L(z, \hat{\theta})$ w.r.t x is additionally calculated. This additional gradient computation captures how changing z along each dimension of x affects the loss of a test point.

3.2 Influence Function on Validation Group

It is essential to note that Eqs. 3 and 6 consider the influence that a single training point has on a single test point. However, calculating the influence score with respect to a single test point may not produce a good estimation when the training data is noisy. Further, it is computationally expensive to calculate the influence score for each pair of training and test points individually because each calculation requires the inverse Hessian matrix to be computed (either directly or via estimation). Therefore, we extend the influence functions of Eqs. 3 and 6 to estimate the impact that a single training point has on a group of test (or validation) points. Since influence is additive [14], we can extend both equations to consider how the loss of a group of validation points changes when z is either removed or perturbed:

$$\mathcal{I}_{up,loss}(z, D_{val}) = -\nabla_\theta L\left(D_{val}, \hat{\theta}\right)^\top H_{\hat{\theta}}^{-1} \nabla_\theta l\left(z, \hat{\theta}\right) \tag{7}$$

$$\mathcal{I}_{pert,loss}(z, D_{val}) = -\nabla_\theta L\left(D_{val}, \hat{\theta}\right)^\top H_{\hat{\theta}}^{-1} \nabla_x \nabla_\theta l\left(z, \hat{\theta}\right) \tag{8}$$

With deep learning models, the top layers generally serve as classification layers while the lower layers work as feature extractors. Therefore, instead of computing the influence score over all the deep learning model's layers, we can simply use the top-most layers. However, computing the inverse Hessian matrix ($H_{\hat{\theta}}^{-1}$) is still computationally intensive even if we only consider the top-most layers. We leverage the inverse Hessian-vector product (IHVP) method to approximate $H_\theta^{-1} \nabla_\theta L(D_{val}, \hat{\theta})$ in Eq. 7 and Eq. 8 using the Linear time Stochastic Second-Order Algorithm (LiSSA) [3].

4 Influence-Based Sample Reweighing

We propose Influence-based Sample Reweighing (ISR) – a training procedure robust to noisy MR image data that uses influence scores to reweigh the training examples.

4.1 Framework

Most robust training approaches either manipulate the entire training dataset, or a portion of it, to train a model robust to noisy training points. One popular approach is to introduce a weight for each training point, which forces the model to pay more attention to important samples. We formulate our ISR robust training procedure by modifying the empirical loss function introduced in Sect. 3 to incorporate sample weights. Let $w \in \mathbb{R}^{|D_{trn}|}$ be the vector of weights for the training data where weight w_i corresponds to training point z_i. The per-sample weighted loss function can be defined as:

$$L_w(D_{trn}, \theta, w) = \frac{1}{|D_{trn}|} \sum_{z_i \in D_{trn}} w_i l(z_i, \theta) \tag{9}$$

The loss function $L_w(D_{trn}, \theta, w)$ in Eq. 9 is a generalized formulation of an empirical loss function. Recall from Eq. 7 that the influence score can tell us the effect that a training point has on the validation loss. For ISR, we construct the weight vector w based on the influence score of each training point. Since $\mathcal{I}_{up,loss}(z, D_{val})$ reflects the influence that training point z has on the validation loss, we can determine the weight of each training sample based on if they have a negative (helpful) or positive (harmful) score. We introduce our ISR in Algorithm 1.

Algorithm 1. Training with Influence-based Sample Reweighing

Input: Training data D_{trn}, validation data D_{val}, train epochs T, pre-train epochs T_{pre}, learning rate η, maximum weight m
Output: Trained model $\hat{\theta}$

```
1: Initialize θ^0
2: for t = 1...T_pre do
3:     θ^t ← θ^t − η∇L(D_trn, θ^t)
4: Initialize w as an array of size |D_trn|        ▷ Calculating sample weights in lines 4-7
5: for i = 1...|D_trn| do
6:     w_i ← −I_up,loss(z_i, D_val) using Eq. 7
7: Scale w to range [1, m] using MinMax scaler
8: for t = T_pre + 1...T do
9:     θ^t ← θ^t − η∇L_w(D_trn, θ^t, w)
```

In Algorithm 1, the model is randomly initialized with θ^0 and then trained for T_{pre} epochs in lines 1–3. Normally, calculating influence scores requires a fully trained model. However, we find that allowing the model to complete several pre-training steps sufficiently warms the θ parameters to generate good influence scores. In lines 4–7, ISR calculates the weights for training samples. Finally, we continue the training procedure using the per-sample weighted loss function for the remaining epochs in lines 8–9.

4.2 Calculating Sample Weights

Lines 4–7 of Algorithm 1 shows how to calculate the weights w_i's of the training points based on their influence score. After initializing w in line 4, we go through each training sample, estimate its influence score using Eq. 7, and assign the score to the vector w in lines 5–6. Since the influence score values are usually close to zero and can be either negative or positive, the loss function will be deactivated if we directly use the influence scores as weights. Therefore, in line 7, we scale the values in w to be in the range from 1 to m using the MinMax scaler. Specifically, samples with negative (positive) influences should have weights close to 1 (m). By doing so, helpful samples keep their original effect on the model, while harmful samples are penalized and weighted higher such that the model will pay more attention to them. The weighted loss in the ISR training procedure is therefore bounded based on the empirical loss:

$$L(D_*, \theta) \leq L_w(D_*, \theta, w) \leq mL(D_*, \theta) \tag{10}$$

where equality only holds when $m = 1$. When m approaches 1, the weighted loss approximates the empirical loss, and the addition of weights has no impact on the learning of the model. On the other hand, when $m \gg 1$, it takes longer for the model to converge as the loss is sufficiently larger. For this reason, in our experiments, we choose $m = 2$.

5 Influence-Based Sample Perturbation

We counteract the noise in a training image by adding perturbation proportional to the image's influence score – a process we call Influence-based Sample Perturbation (ISP).

5.1 Framework

ISP works as follows: (1) ISP selects a subset of training points D_s which have the most impact on the model by calculating the influence of each training point on the model loss and selecting those with the highest influence score; and (2) ISP generates (helpful) perturbation for every example in D_s based on the influence score to fortify the model against noisy training examples. More specifically, we construct a subset $D_s \subset D_{trn}$ by choosing the most influential examples in D_{trn} and let $D_u = D_{trn} \setminus D_s$. Let δ_i be the influence-based perturbation for training point $z_i = (x_i, y_i)$ and let $\hat{z}_i = (x_i + \delta_i, y_i)$ be the influence-based perturbed version of z_i. We define the ISP training set as $\hat{D}_{trn} = \hat{D}_s \cup D_u$, where $\hat{D}_s$ is the set D_s *after* influence-based perturbation is added. Under this setting, we define the empirical loss function of the robust model as $L(\hat{D}_{trn}, \theta) = L(\hat{D}_s, \theta) + L(D_u, \theta)$. We define our objective as:

$$\min_{\delta \in \Delta} L(D_{val}, \theta_\delta) \text{ s.t. } \theta_\delta = \arg\min_{\theta \in \Theta} L(\hat{D}_{trn}, \theta) \tag{11}$$

The naïve approach to the above problem would be to train/optimize several different models with different perturbation values δ_i, which is intractable. By using influence functions, we avoid the requirement of costly retraining. Specifically, we use the influence function of Eq. 7 to estimate the change in the model's loss on D_{val} when a particular training point is perturbed to train a model robust to noisy input data efficiently. We fully present the robust training procedure ISP in Algorithm 2.

Algorithm 2. Training with Influence-based Sample Perturbation

Input: Training data D_{trn}, validation data D_{val}, train epochs T, pre-train epochs T_{pre}, scaling factor γ, ratio of selected examples r, learning rate η
Output: Trained model $\hat{\theta}$
1: Initialize θ^0
2: **for** $t = 1 \ldots T_{pre}$ **do**
3: $\quad \theta^t \leftarrow \theta^t - \eta \nabla L(D_{trn}, \theta^t)$
4: $D_s \leftarrow \emptyset$ ▷ Selecting influential samples in lines 4-9
5: **for** $z_i \in D_{trn}$ **do**
6: $\quad$ Compute $\mathcal{I}_{up,loss}(z_i, D_{val})$ using Eq. 7
7: Sort data points in D_{trn} in ascending order of influence scores
8: Assign the first $\lceil r \times |D_{trn}| \rceil$ examples of D_{trn} to D_s
9: $D_u \leftarrow D_{trn} \setminus D_s$
10: $\hat{D}_s \leftarrow D_s$ ▷ Adding influence-based perturbation in lines 10-15
11: **for** $z_i \in D_s$ **do**
12: $\quad \delta_i \leftarrow -\gamma \mathcal{I}_{pert,loss}(\hat{z}_i, D_{val})$
13: $\quad \hat{x}_i \leftarrow \text{Clip}(x_i + \delta_i)$
14: $\quad \hat{z}_i \leftarrow (\hat{x}_i, y_i)$
15: $\quad$ Update new $\hat{z}_i$ in $\hat{D}_s$
16: $\hat{D}_{trn} \leftarrow \hat{D}_s \cup D_u$
17: **for** $t = T_{pre} + 1 \ldots T$ **do**
18: $\quad \theta^t \leftarrow \theta^t - \eta \nabla L(\hat{D}_{trn}, \theta^t)$

Similar to ISR, Algorithm 2 begins by pre-training the model parameterized by θ for T_{pre} epochs (lines 2–3). The parameters in early epochs of training can be unstable, and it is crucial to avoid adding perturbation to the training examples based on unstable parameters. In lines 4–9, we select the most influential training points D_s (and thereby the points not in D_s, D_u) (discussed in Sect. 5.2). In lines 10–16, we first add perturbation to selected training examples (lines 10–15), and then update $\hat{D}_{trn}$ (line 16). The complete details of how the influence-based perturbation is generated will be given in Sect. 5.3. The model then undergoes additional training on the updated training set $\hat{D}_{trn}$ (lines 17 and 18) to construct a model robust to noisy examples.

5.2 Selecting Influential Samples

The data distribution of a (clean) training dataset, and the same training dataset with added noisy samples, are not the same. In other words, the noisy samples shift the data distribution and, therefore, will shift the model's decision boundary away from the one generated when the model is trained on the clean data. Further, training samples that are highly affected by noise will tend to stand apart in the loss space [21]. Intuitively, we can increase the generalization ability of the model by focusing on the training samples which have a harmful impact on the model.

Recall from Sect. 3 that the influence function tells how the model parameters (or loss) would change if a data point z was removed from the training set. We can utilize this idea to select the training examples which cause the increase in the model's loss when they are included in the training phase. We present our subset selection method based on influence scores. For each sample in the training data, we calculate the influence score using the upweighting approach of Eq. 7 in lines 5–6. We then build the subset D_s (and D_u) by selecting the $\lceil r \times |D_{trn}| \rceil$ examples, which have the highest impact in lines 7–9.

5.3 Adding Influence-Based Perturbation

In Eq. 8, $\mathcal{I}_{pert,loss}(z, D_{val})$ tells us the contribution of each input pixel to the loss of the whole validation set. While the gradient of the loss of the validation set, $\nabla_\theta L(D_{val}, \hat{\theta})$, provides information on how the trained model performs on unseen validation data, the term $\nabla_x \nabla_\theta l(z, \hat{\theta})$ tells how each input pixel contributes to the loss. Pixels with either significantly positive or negative influence scores highlight where the model's attention is focused. Since the influence score represents the change in loss, a pixel that has a positive (negative) score will cause an increase (decrease) in the loss. Therefore, by crafting perturbation in the opposite direction of the influence score, i.e., $\delta = -\mathcal{I}_{pert,loss}(z, D_{val})$, we can perturb the original image in a way that further strengthens helpful pixels and weakens harmful pixels.

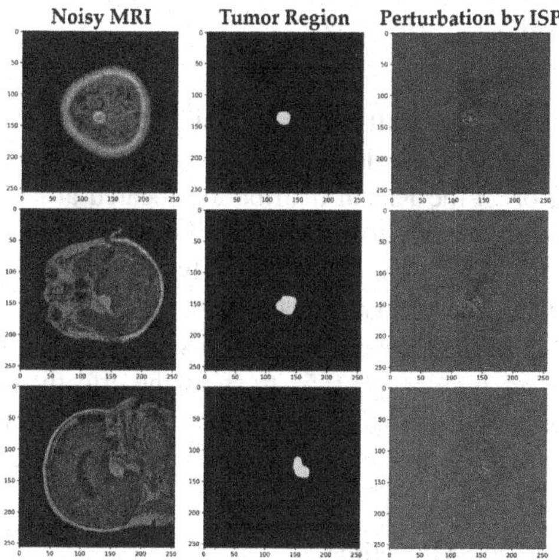

Fig. 1. Visualization of perturbation by our ISP.

Let $\delta_i \in \Delta$ be the influence-based perturbation corresponding to training input x_i, i.e., $\hat{x}_i = x_i + \delta_i$ (line 12). Adding δ_i to x_i could potentially result in $\hat{x}_i$ being a non-feasible image. Therefore, we clip the perturbed image $\hat{x}_i$ (line 13) before setting the perturbed training point as $\hat{z}_i = (\hat{x}_i, y)$ (line 14). Finally, we add the newly perturbed example $\hat{z}_i$ to $\hat{D}_s$ in line 15. In Fig. 1, we visualize training MR images and influence-based perturbations to show how our ISP can make the model more robust by adding friendly perturbation. The first column shows the noisy MR image. The second column is the information on the tumor region provided by the dataset. The last column is the influence-based perturbation generated by our ISP. Looking at the second and last columns, we can see that the brighter region in the perturbation aligns with the tumor region. ISP leads the attention of the classifier to the critical regions of an image by adding more perturbation to those areas.

6 Experiments

6.1 Evaluation Setup

Dataset and Model. For all experiments conducted, we use the Brain Tumor Dataset [7]. The dataset contains 3,064 brain T1-weighted CE-MRI slices with a size of 512×512 from 233 patients. There are three kinds of tumors: 708 meningiomas, 1426 gliomas, and 930 pituitary images. We allocate 75% of the patients to be in the training set, 5% to be in the validation set, and 20% to be in the test set. We use a Convolutional Neural Network (CNN) as our classification model. The architecture of the CNN model consists of five convolutional layers (32, 32, 64, 64, and 64). The last layer is a fully connected layer. We use Batch Normalization, ReLU activation, and Max Pooling after each convolutional layer to improve the performance and efficiency of the model. Data augmentation (e.g., random horizontal flipping) is incorporated to increase the diversity of the training set. The model is trained for 40 epochs using the Adam optimizer with a learning rate of 0.01 and batch size of 32.

Simulating Noisy Images. We craft the noisy brain tumor MR images by adding random noise from either: 1) the Gaussian distribution centered at $\mu = 0$ with a standard deviation of $\sigma = 32$, or 2) the Rician distribution located at $\nu = 1$ with a scale of $\sigma = 16$. In our evaluation, we aim to simulate a difficult operating scenario where the majority of the labeled data is noisy. Consequently, we add noise to every image in the training data and leave the validation data clean. For the test data, we add random noise to a subset of the original images where the size of the subset is determined by a parameter ρ_{test}. We set $\rho_{test} = \{0, 0.5\}$ to show the effectiveness of our methods. Our evaluation setting is realistic and mirrors the reality that only a small number of collected MR images are noise-free, while the majority have some degree of noise.

6.2 Proposed Methods and Baselines

We compare our proposed methods with the following three baselines:

Naïve Training: In this setting, the CNN model is trained on the original brain tumor dataset, and no noise correction or modified training procedure is applied.

Adversarial Training (AT): This method hardens the base CNN model by crafting adversarial examples and optimizing the model over the generated adversarial data. We use FGSM with uniform initialization [18] to generate the adversarial perturbations. I.e., $\delta = \alpha \cdot \text{sign}(\nabla_x l(f_\theta(x), y))$, where in our experiments we set $\alpha = 0.03$ and $\epsilon = 0.1$.

Noise2Void (N2V): N2V aims to reconstruct clean images from the given noisy data with a U-Net [17] backbone architecture and uses a blind-spot network to train the denoiser. We train the N2V model using the default settings from [15], and we train a

CNN model on the denoised images using settings defined in Sect. 6.1. In the results, we denote a combined method of N2V denoising and CNN training as N2V-Training to make it distinguishable from N2V, which only involves image denoising.

ISR and ISP: To efficiently compute the influence scores, we only use the weights from the fully connected layer of the CNN model (see Sect. 3.2 for more details). We choose T_{pre} as 10, $r = 0.1$, $\gamma = 0.01$ and use the CNN architecture.

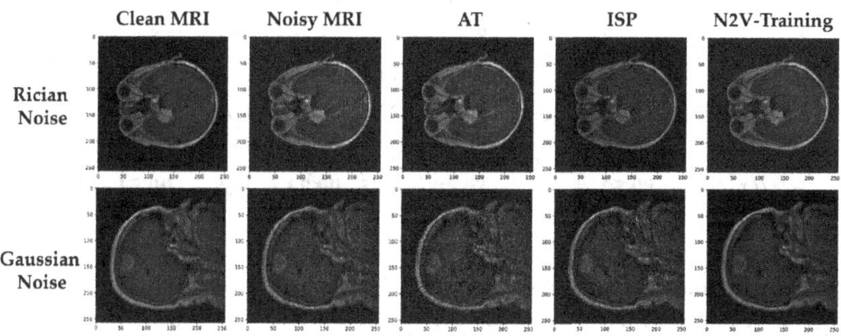

Fig. 2. Training samples crafted by each training method.

Metrics: For each experiment, we report the average and standard deviation of the test accuracy over five trials. We use a GPU Tesla V100 (32GB RAM) and CPU Xeon 6258R 2.7 GHz to conduct all experiments.

6.3 Results

Intervened Training Data by Methods. While ISR puts the model's attention on important samples via reweighing the training loss, training methods like AT, our ISP, and N2V-Training intervene on the noisy training images. In Fig. 2, we illustrate the training images generated by the AT, ISP, and N2V-Training. Note that since our ISR only reweighs the images, the noisy training image is not perturbed, and hence we do not provide a column for ISR in Fig. 2.

AT and ISP craft additional perturbation to inject into the training images to increase the generalization ability of the model. By comparing images perturbed by AT and our ISP (third and fourth columns) to clean and noisy images (first and second columns), we can see that ISP alleviates the harmful noise and makes the generated images look closer to the clean ones. While ISP can significantly reduce the noise in the background (darker background), AT, on the contrary, makes the value of background noise larger (brighter background). In other words, our ISP is more successful than AT in reverting harmful noise. ISP also makes the tumor regions, which are the key regions of the classification task, more detailed. For example, in the first row, the tumor region of the ISP image has a clearer boundary and is more prominent than the surrounding regions. This observation is clearer when looking at the example of Gaussian Noise on the second row.

N2V-Training is able to generate images with less noise than AT and ISP because it denoises the training images before training the classifier. However, generated images can lose important fine-grained information in the tumor image. Further, when considering Rician noise, N2V-Training generates an image where the brain itself is free from noise, but the background noise becomes stronger.

Table 1. Test accuracy (%) of training methods on noisy Brain Tumor Dataset

Method	$\rho_{test} = 0$		$\rho_{test} = 0.5$	
	Gaussian Noise	Rician Noise	Gaussian Noise	Rician Noise
Naïve	87.77 ± 0.84	86.50 ± 3.38	88.89 ± 0.88	88.44 ± 2.43
AT	88.38 ± 1.47	86.85 ± 1.62	88.95 ± 0.94	89.04 ± 0.54
N2V-Training	88.12 ± 1.69	83.92 ± 2.52	89.08 ± 1.70	86.78 ± 1.79
ISR (Our)	88.79 ± 1.75	88.12 ± 2.28	89.17 ± 1.09	89.49 ± 1.82
ISP (Our)	$\mathbf{89.52 \pm 2.61}$	$\mathbf{89.49 \pm 1.45}$	$\mathbf{90.73 \pm 1.58}$	$\mathbf{90.54 \pm 1.19}$

Overall Performance. Table 1 shows the comparisons of our ISR and ISP methods with other baselines in terms of accuracy. We do the evaluation under the noise generated by two different distributions as discussed previously. Each column represents a dataset affected by Gaussian noise or Rician noise. Each row shows the results of one method. In most cases, both ISR and ISP outperform other baselines, showing that our proposed methods effectively improve the robustness of the trained classifier. ISP achieves the best performance with a descent gap in test accuracy compared to the worst performer (N2V-Training). When we compare the results under Gaussian noise and Rician noise, one can observe that Rician noise impacts the trained models more than Gaussian noise does. This is due to the Gaussian distribution producing zero-mean noise, while the Rician distribution produces non-zero mean noise. Out of all the methods, N2V-Training performs the worst with a significant degradation in accuracy under Rician noise. Recall our previous observation in Sect. 6.3 that reconstructed images from Rician noise have larger background noise than the original noisy images, so it is understandable that N2V-Training loses its performance significantly in this setting.

7 Conclusion

We have presented two effective and efficient training methods, Influence-based Sample Reweighing and Influence-based Sample Perturbation, to improve the robustness of a brain tumor classification model against noisy training MR image data. Our comprehensive experiments showed that models trained with our methods are stable and robust against the effect of different noise distributions (Gaussian and Rician). Further, we showed that our methods could be directly applied to noisy training data without the need for additional clean images or denoising steps. Futher, our method can be applicable to other types of medical images such as X-ray or CT images.

Acknowledgements. This work was supported in part by National Science Foundation under awards 1946391, the National Institute of General Medical Sciences of National Institutes of Health under award P20GM139768, and the Arkansas Integrative Metabolic Research Center at University of Arkansas.

References

1. Abiwinanda, N., Hanif, M., Hesaputra, S.T., Handayani, A., Mengko, T.R.: Brain tumor classification using convolutional neural network. In: World Congress on Medical Physics and Biomedical Engineering (2019)
2. Afshar, P., Mohammadi, A., Plataniotis, K.N.: Brain tumor type classification via capsule networks. In: ICIP (2018)
3. Agarwal, N., Bullins, B., Hazan, E.: Second-order stochastic optimization for machine learning in linear time. J. Mach. Learn. Res. **18**, 1–40 (2017)
4. Avni, U., Greenspan, H., Konen, E., Sharon, M., Goldberger, J.: X-ray categorization and retrieval on the organ and pathology level, using patch-based visual words. IEEE Trans. Med. Imaging. **30**, 733–746 (2010)
5. Ayadi, W., Charfi, I., Elhamzi, W., Atri, M.: Brain tumor classification based on hybrid approach. Vis. Comput. **38**, 107–117 (2022)
6. Bosch, A., Munoz, X., Oliver, A., Marti, J.: Modeling and classifying breast tissue density in mammograms. In: CVPR (2006)
7. Cheng, J.: Brain tumor dataset (2017). https://doi.org/10.6084/m9.figshare.1512427.v5
8. Cheng, J., et al.: Enhanced performance of brain tumor classification via tumor region augmentation and partition. PloS one (2015)
9. Cook, R.D., Weisberg, S.: Residuals and Influence in Regression. Chapman and Hall, New York (1982)
10. Deepak, S., Ameer, P.: Brain tumor classification using deep CNN features via transfer learning. Comput. Biol. Med. **111**, 103345 (2019)
11. Goodfellow, I.J., Shlens, J., Szegedy, C.: Explaining and harnessing adversarial examples. arXiv preprint arXiv:1412.6572 (2014)
12. Hinton, G.E., Krizhevsky, A., Wang, S.D.: Transforming auto-encoders. In: ICANN (2011)
13. John, P., et al.: Brain tumor classification using wavelet and texture based neural network. Int. J. Sci. Eng. Res. **3**, 1–7 (2012)
14. Koh, P.W., Liang, P.: Understanding black-box predictions via influence functions. In: ICML (2017)
15. Krull, A., Buchholz, T.O., Jug, F.: Noise2void-learning denoising from single noisy images. In: CVPR (2019)
16. Lehtinen, J., et al.: Noise2noise: learning image restoration without clean data. arXiv preprint arXiv:1803.04189 (2018)
17. Ronneberger, O., Fischer, P., Brox, T.: U-net: convolutional networks for biomedical image segmentation. In: Navab, N., Hornegger, J., Wells, W.M., Frangi, A.F. (eds.) MICCAI 2015. LNCS, vol. 9351, pp. 234–241. Springer, Cham (2015). https://doi.org/10.1007/978-3-319-24574-4_28
18. Wong, E., Rice, L., Kolter, J.Z.: Fast is better than free: Revisiting adversarial training. arXiv preprint arXiv:2001.03994 (2020)
19. Wu, D., Liu, S., Ban, J.: Classification of diabetic retinopathy using adversarial training. In: IOP Conference Series: Materials Science and Engineering (2020)
20. Xie, Y., Fetit, A.E.: How effective is adversarial training of CNNs in medical image analysis? In: MIUA (2022)
21. Yang, Y., Liu, T.Y., Mirzasoleiman, B.: Not all poisons are created equal: robust training against data poisoning. In: ICML (2022)
22. Zulpe, N., Pawar, V.: GLCM textural features for brain tumor classification. IJCSI **9**, 354 (2012)

Co-ReaSON: EEG-based Onset Detection of Focal Epileptic Seizures with Multimodal Feature Representations

Uttam Kumar[1,2](✉), Ran Yu[1,2], Michael Wenzel[3], and Elena Demidova[1,2]

[1] Data Science and Intelligent Systems Group (DSIS), University of Bonn, Bonn, Germany

[2] Lamarr Institute for Machine Learning and Artificial Intelligence, Bonn, Germany
{uttam.kumar,elena.demidova}@cs.uni-bonn.de, ran.yu@uni-bonn.de

[3] Department of Epileptology, University Hospital Bonn, Bonn, Germany
michael.wenzel@ukbonn.de

https://lamarr-institute.org

Abstract. Early detection of an epileptic seizure's onset is crucial to reduce the impact of seizures on the patient's health. The Electroencephalogram (EEG) has been widely used in clinical epileptology for continuous, long-term measurement of electrical activity in the brain. Despite numerous EEG-based approaches employing diverse models and feature extraction methods for seizure detection, these methods rarely tackle the more challenging task of early detection of the seizure onset, especially as only a few EEG channels are impacted at the onset, and the seizure evidence is minimal. Furthermore, EEG-based seizure onset detection remains challenging due to the sparse, imbalanced, and noisy data, as well as the complexity posed by the diverse nature of epileptic seizures in patients. In this paper, we propose *Co-ReaSON* – a novel approach towards early detection of focal seizure onsets by considering the onset-specific increase in spatio-temporal correlations across the EEG channels observed over a range of multimodal EEG feature representations, combined in a ResNet18-based model architecture. Evaluation on a real-world dataset demonstrates that *Co-ReaSON* outperforms the state-of-the-art baselines in focal seizure onset detection by at least 5 percent points regarding the macro-average F1-score.

Keywords: Multivariate EEG Time Series · Seizure Onset Detection

1 Introduction

Early detection of an epileptic seizure onset can facilitate timely intervention to prevent a seizure from progressing into a more severe and potentially life-threatening event and support appropriate medication management. Such detection helps to improve the quality of life for individuals with epilepsy by minimizing the impact and unpredictability of seizures, allowing patients to engage more

D.-N. Yang et al. (Eds.): PAKDD 2024, LNAI 14648, pp. 258–270, 2024.
https://doi.org/10.1007/978-981-97-2238-9_20

fully in daily activities in their social and professional environments. Epileptic seizures can be detected by continuously monitoring scalp electroencephalogram (EEG) data. Surface EEG is a non-invasive technique that measures the electrical activity of the brain by placing electrodes (channels) on the scalp to record the electrical signals generated by brain cells. Developing precise detection methods for seizure onsets based on EEG remains an important research topic.

Early seizure onset detection from multivariate EEG data is a significant challenge. Firstly, EEG data is highly imbalanced, as seizures represent rare events. Further, seizure patterns are different across individuals and etiologies of epilepsy, making it nearly impossible to find a one-size-fits-all solution for seizure detection across all types of epilepsy. Additionally, EEG data is noisy. The variety of brain activities recorded by the sensors and the inherent sensor inaccuracy make pattern extraction from EEG data particularly challenging. In this paper, we focus on detecting seizure onsets in focal epilepsies for feasibility reasons. Focal epilepsies represent one major subset of epilepsy etiologies, and comprise, among many other types, temporal lobe epilepsy, the most common type of epilepsy. Still, even within the class of focal epilepsies, a large variety of seizure EEG patterns can be found, whose accurate detection represents an immense challenge. As per ILAE [1], focal seizures originate within networks limited to one hemisphere, from where they can subsequently spread across larger portions of the brain and across both hemispheres. Notably, the initial seizure initiation site can be miniscule, limited to a volume less than 1 mm^3 [16,17]. Thus, focal seizures may initially be undetectable at the anatomical macroscale, or restricted to a single electrode. Thus, seizure onset may be detectable only in one of the EEG channels, while other channels continue to show unspecific patterns. This underscores the challenges related to early seizure detection.

When data storage and processing capacity were limited, seizure detection had many algorithmic approaches based on extracting statistical features [7] from EEG data. The next approaches relied on features extracted using time-frequency analysis and signal-processing techniques and then applying the extracted features to machine learning models such as random forests and support vector machines. More recently, state-of-the-art deep learning architectures [5,20] have also been applied. Most of the work relies on taking full or longer segments of EEG for feature extraction and training, which is unsuitable for early seizure onset detection. Existing studies often do not distinguish between seizure types and fail to achieve high accuracy in focal seizure onset detection.

In this paper, we propose *Co-ReaSON* – a novel approach for early focal seizure onset detection based on spatio-temporal correlations across the EEG channels, adopting rich multimodal EEG-based feature representations. In particular, we base our approach on the idea that the spread of a focal epileptic seizure in the brain during the seizure onset can be observed as an increased correlation of the neighboring EEG channels. We capture such correlations in various features, including novel EEG graph features, Discrete Wavelet Transform (DWT)-based features and EEG-based images. Capturing such features in the state-of-the-art ResNET18 [3]-based architecture enables *Co-ReaSON* to obtain stable patterns clearly indicating seizure onset and substantially increases the precision of the early onset prediction.

Our contributions are as follows.

- We propose *Co-ReaSON* – a novel approach to early focal seizure onset detection based on learning the onset-specific spatio-temporal correlations across the EEG channels in a ResNet18-based model architecture.
- We propose a rich set of EEG-based features and their multimodal representations and demonstrate their impact on the *Co-ReaSON* performance. To the best of our knowledge, we are the first to propose a novel visual representation of spatio-temporal correlations in the EEG graph as a feature and demonstrate its effectiveness.
- We evaluate our model on a real-world seizure dataset and demonstrate that our approach is effective and outperforms the state-of-the-art baselines by at least 5 percent points regarding the macro average F1-score.

2 Problem Definition

As input for seizure onset detection, we consider a multivariate time series with c channels representing an EEG signal. We discretize an EEG signal into a sequence of chronological time intervals of a specific duration. Each channel in the time interval is associated with a set of features. Such features can be extracted from the time series and its contextual information. Specific time intervals within the EEG channels can be labeled as seizures. Typically, such labeling is conducted manually by experts.

Definition 1 (Multivariate EEG Time Series). *A multivariate EEG time series is a collection of chronological observations* $X_t = [X_{1,t}, ..., X_{|C|,t}]$, *where* t *is the time interval index,* C *is the set of the observed variables (EEG channels). Each EEG channel* $c \in C$ *is observed at a discrete time interval* t, *and* $X_{c,t}$ *represents the value of the channel* c *at the time interval* t. *The signal* $X_{c,t}$ *is associated with the feature vector* $F_{c,t}$.

The variables (EEG channels) within the multivariate EEG time series reflect spatial relations based on the physical placement of the sensors on the scalp and their proximity. To capture these relations, we represent EEG data as a graph, where the nodes correspond to the channels and the edges to the connections of nodes according to the international 10–20 sensor placement system for EEG [4].

Definition 2 (EEG Graph). *An EEG graph* $G = (V, E)$ *is an undirected graph, where the nodes* V *represent the EEG channels (sensors) and the edges* E *represent the connections between the sensors within the international 10–20 system of electrode placement on the scalp for recording EEG signals.*

A seizure onset is a specific pattern in a multivariate EEG time series that indicates the beginning of an epileptic seizure.

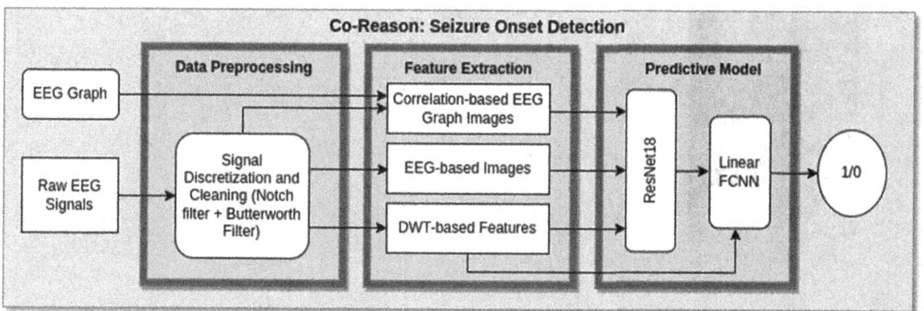

Fig. 1. The *Co-ReaSON* architecture includes three main components: data preprocessing, feature extraction, and a predictive model.

Definition 3 (Seizure Onset). *Seizure onset at time interval* t *indicates that the seizure not observed at time interval* $t-1$ *is observed in any channel* c *at time interval* t.

The task of seizure onset detection is defined as follows.

Definition 4 (Seizure Onset Detection). *Given a multivariate EEG time series* X_t, *and the sequence of observations at previous* k *time intervals* $X_{t-k}, \ldots,$ X_{t-1}, *identify if* X_t *represents a seizure onset.*

3 The *Co-ReaSON* Approach

The proposed *Co-ReaSON* approach includes three main components, namely, data preprocessing, feature extraction, and a predictive model to detect seizure onsets. An overview of the approach is illustrated in Fig. 1.

3.1 Data Preprocessing

In this step, we process the raw multivariate EEG signal to extract features. We obtain EEG data from all channels using a unipolar Average Reference (AR) montage. The reference channel is the average potential of all channels. Given that the raw EEG data is susceptible to power line interference at 50 Hz, we first apply a notch filter to remove this noisy frequency. Subsequently, we apply a Butterworth bandpass filter spanning from 0.1 Hz to 64 Hz to obtain epileptic features from this low sub-band frequency range, aligning with established practices in prior research [8,19,20]. We then split filtered, discretized EEG signals into time intervals. We label the time intervals containing a seizure onset as positive and others as negative. In addition, we remove seizure onsets, where a seizure starts within three seconds of the end of an earlier seizure, representing a so-called stop-start EEG pattern within the same seizure rather than a new seizure. Finally, as a clear empirical cut-off is lacking to this day, the demarcation line between interictal (between seizures) and ictal (seizure) epileptic activity in the

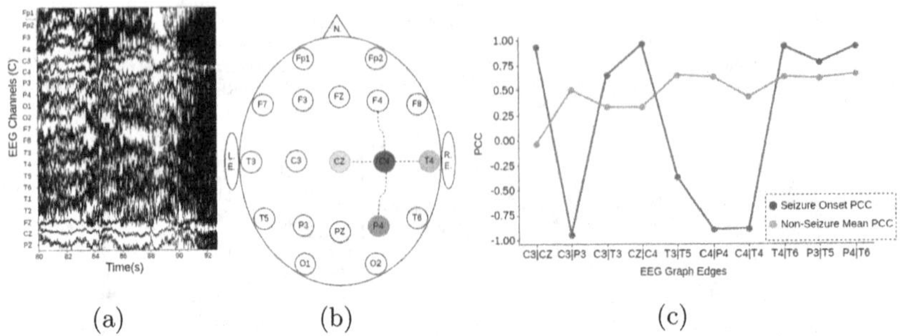

(a) (b) (c)

Fig. 2. (a) A multivariate EEG time series X_t with a focal seizure onset at the 83rd sec in the node C4 for a patient. (b) 1-hop neighbors of node C4 connected by dotted lines (edges) in the EEG Graph; N.-Nose, L.E.-Left Ear, R.E.-Right Ear (c) PCC of selected EEG Graph edges for a patient at the time interval $80s - 84s$ during the focal seizure onset. The PCC at the seizure onset (blue) and the average PCC of the EEG Graph edges of the non-seizure intervals (orange). (Color figure online)

EEG is typically set arbitrarily. In line with the common notion of spike-wave paroxysms in genetic epilepsies, we consider rhythmic epileptic EEG patterns shorter than 3 s as interictal activity, not seizures.

3.2 Feature Extraction

We extract three types of features: Correlation-based EEG Graph Images, EEG-based Images, and Discrete Wavelet Transform (DWT)-based features. We create a specific image representation for each feature type and provide this image as input into the ResNet18 model.

EEG Graph. To capture spatio-temporal correlations among neighboring EEG channels, we represent the EEG channels as graph nodes according to the Definition 2. We take the number of graph nodes $V = 19$, placed as per the standard 10–20 system depicted in Fig. 2(b). The number of node connections, i.e., the edges in this EEG graph, is $E = 32$. For each time interval t, we compute the 32 edge weights corresponding to the Pearson Correlation Coefficient (PCC) of the EEG channels within the 1-hop distance in the EEG graph.

Correlation-Based EEG Graph Images. Most focal seizures start in a circumscribed local region of the brain, from where they may propagate to other regions or generalize across larger portions of the brain, including both hemispheres, i.e., the nodes of the EEG graph start showing similar EEG patterns. For example, in Fig. 2(a), a focal seizure onset occurs at the 83rd s. Subsequently, around the 91st s, we can observe almost all the nodes demonstrating a similar signal pattern. Thus, to detect focal seizures before they propagate to all nodes, we base our correlation features on the intuition that the correlation between neighboring EEG channels increases during focal seizure onset near the affected brain region. To capture this intuition statistically, for a multivariate EEG time series X_t at the time interval t, we compute the PCC of EEG graph nodes. For example, if $P4$ and $C4$ are neighboring nodes, with w observations each in t, i.e. $X_{C4,t} = [m_1, .., m_w]$ and $X_{P4,t} = [n_1, .., n_w]$, then the PCC between them at the time interval t is computed as:

$$X_{C4,t}|X_{P4,t} = \frac{\sum_{i=1}^{w}(m_i - \bar{m})(n_i - \bar{n})}{\sqrt{\sum_{i=1}^{w}(m_i - \bar{m})^2}\sqrt{\sum_{i=1}^{w}(n_i - \bar{n})^2}}, \quad (1)$$

where $\bar{m}$ is the mean of all observations of $X_{C4,t}$ and $\bar{n}$ is the mean of all observations of $X_{P4,t}$. We compute the PCC for each pair of nodes in the graph.

Figure 2(c) illustrates the PCCs computed with Eq. 1 for the pairs of EEG graph nodes for a time interval for a patient. For each time interval t, we graphically represent the PCC values of all EEG graph edges (i.e., node pairs) and their overall non-seizure mean PCCs. On the x-axis, we represent the edges E of the EEG graph, and the y-axis represents the PCC value for each edge. As we can observe, the correlations between some neighboring nodes increase drastically during the seizure onset (blue line) compared to their respective mean correlations in the non-seizure intervals (orange line). We adopt such correlation-based images as model input. For new patients, we use the non-seizure mean computed over all the time intervals of all patients. We observe that for focal seizures, it is likely that the difference in correlations between an edge current correlation value and its non-seizure mean are high when its neighboring nodes reside near the focal seizure onset brain region. For example, in Fig. 2(b) and (c), when correlation-based features are computed for a patient, we can observe that in a focal seizure initiated at node C4, it propagates to three out of its four one-hop neighbors, P4, T4, and CZ, and the difference between the above-discussed correlations is high for P4, T4, and CZ. These correlations, combined with other feature representations, help our model learn the seizure onset pattern.

EEG-Based Images. Clinical neurologists identify seizures in EEG recordings by reviewing a sequence of recorded EEG images. Thus, for the multivariate EEG time series X_t at the time interval t, we map the raw EEG signal to an image that contains observations of all channels in this interval. To extract significant features, reduce the impact of noise, and improve the convergence of the model, the input images are normalized by subtracting the mean and standard deviation of all image tensors from the current image tensor. Figure 2(a) illustrates an example EEG image that can be segmented into time intervals.

DWT-Based Features. We extract DWT-based images and statistical features. Previous research [12,19] demonstrates that the seizure patterns can be more accurately identified in the multi-resolution analysis of the EEG provided by the Discrete Wavelet Transform. The wavelet decomposition of discretized EEG aids in localizing features in the frequency and time domains. Figure 3

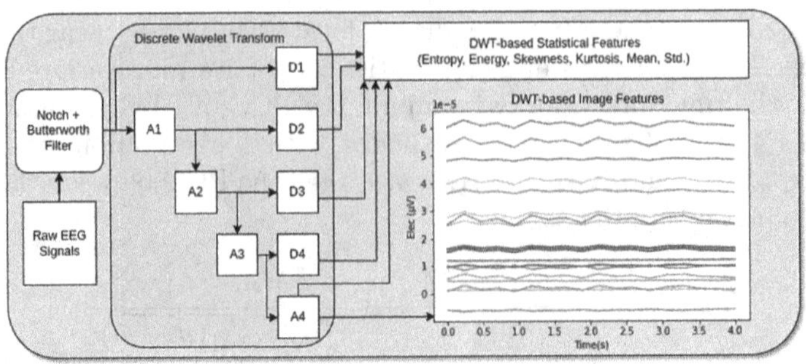

Fig. 3. Feature extraction using DWT to obtain detailed and approximate coefficients in frequency sub-bands, namely, D1:32–64 Hz, D2:16–32 Hz, D3:8–16 Hz, D4:4–8 Hz, A4:0.1–4 Hz, A1-A2-A3: previous approximate of DWT.

illustrates the DWT-based feature extraction process. DWT [6] decomposes the discretized signal by taking the inner products of the EEG signal and a Basis Wavelet Function (BWF) by simultaneously applying a low pass and a high pass filter to obtain the approximate and the detailed coefficients, respectively, and then further decomposing the approximate coefficients iteratively to obtain the next frequency sub-band's approximate and detailed coefficients until a determined level is reached. To obtain the best BWF for applying DWT in our task, as adapted from [15] we choose Daubechies 4 (DB4) with level 4 for extracting shannon entropy, energy, kurtosis, skewness, mean, and standard deviation features, and Daubechies 10 (DB10) with level 4 for extracting permutation entropy features, as shown to be efficient in [2]. By choosing level 4 for a particular Daubechies, upon decomposition of EEG signals bandpassed in our case to be in the frequency range of 0.1–64 Hz, we obtain the coefficients in 5 sub-bands of frequency (D1:32–64 Hz, D2:16–32 Hz, D3:8–16 Hz, D4:4–8 Hz, A4:0.1–4 Hz), where D1-D4 are the detailed coefficients and A4 contains the approximate coefficients.

DWT-based Image Features. Per time interval, we map the last approximate coefficients A4 into an image so that the representation of even small oscillations of EEG observable only in the lower frequency sub-bands can be learned by our model. The approximates provide dimensionality reduction and are computationally efficient for further feature extraction.

DWT-based Statistical Features. Using coefficients D1, D2, D3, D4, and A4, we extract features like entropy, energy, skewness, kurtosis, mean, and standard deviation as proposed in [14] to finally obtain 595 features per time interval to form the feature vector $F_{c,t}$.

3.3 *Co-ReaSON* Predictive Model

We adopt the state-of-the-art ResNet18 [3] model pre-trained on the ImageNet-1K dataset to process the correlation-, EEG-, and DWT-based images created based on the three image feature types we proposed. ResNet18 extracts the features from the input RGB images using its multiple convolutional layers with skip connections and then uses batch normalization followed by ReLu activation and max-pooling. We fuse the output from ResNet18 with the feature vector $F_{c,t}$ having DWT-based statistical features and feed them into a linear, fully connected layer for predicting seizure onset. This layer outputs 1 for a seizure onset prediction and 0 otherwise.

4 Evaluation Setup

To evaluate the performance of *Co-ReaSON* for seizure onset detection, we conducted experiments on a real-world dataset.

Dataset. We conduct experiments on the TUH EEG Seizure Corpus v2.0.0 [10], which provides anonymous EEG data with annotations for seizure onset and offset times. After the data preprocessing described in Sect. 3.1, we use data from 139 individuals from 628 EEG data recording sessions. We segment the recordings into fixed time intervals to create the multivariate EEG time series as defined in Definition 1. We then randomly sample 80% of the segments to form the training set and the rest to form the test set. This results in 991 seizure onsets and 15,867 non-onset samples in the training set. The test set contains 239 seizure onsets and 3959 non-onset samples. To prevent the model from bias toward the majority class, we over-sample the minority class in the training set.

Metrics. To evaluate the performance on the binary classification task of detecting focal seizure onset, we adopt the standard metrics, such as precision (P), recall (R), and F1 score on the minority class (i.e., seizure onset) and macro averages across two classes (Macro Avg P, Macro Avg R, and Macro Avg F1).

Baselines. We compare *Co-ReaSON* with state-of-the-art baselines. GCN Teacher [20] uses a Graph Convolutional Networks (GCN) model to learn the features of nodes and their neighbors, considering all 19 EEG channels as nodes. ResNet Short+LSTM [5] utilizes raw EEG images and their downsampled versions to create features for seizure onset detection. RUSBoost [12] is an ensemble tree model that uses statistical features extracted using DWT from 10 and 30-s EEG time-frame windows to detect seizures. TDNN+LSTM [13] works by first convolving the EEG signal across time using 1D convolutional layers and then using LSTM to model the spatial dependency. Chrononet [9] uses multiple filters of exponentially varying lengths in the 1D convolution layers and dense connections within the GRU layers to extract features from multivariate EEG input and classify abnormal brain activity. We adapt Chrononet to seizure onset detection, considering a seizure onset as an abnormal activity.

Environment and Parameter Settings. The experiments were conducted on a 64-bit machine with Nvidia GPU (NVIDIA A40, 7251 MHz, 48 GB memory). We set the time interval to 8 s according to the evaluation results in Sect. 5. We train the pre-trained ResNet18 for ten epochs with batch size 8, learning

Table 1. Evaluation results of seizure onset detection. Precision (P), Recall (R), and F1 scores are reported for the minority class (seizure onset).

Model	P	R	F1	Macro Avg P	Macro Avg R	Macro Avg F1
Chrononet	0.62	0.11	0.18	0.66	0.37	0.47
RUSBoost	0.15	0.58	0.23	0.55	0.63	0.51
TDNN+LSTM	0.62	0.28	0.39	0.69	0.41	0.51
ResNet_Short+LSTM	0.65	0.47	0.55	0.73	0.54	0.62
GCN Teacher	0.94	0.48	0.63	0.96	0.74	0.81
Co-ReaSON	0.99	0.58	0.73	0.98	0.79	0.86

rate 0.001, momentum 0.9, and weight decay 0.003. We use a cross-entropy loss function, giving a higher penalty to incorrect high-confidence predictions. All results average the model performance over the last three epochs.

5 Evaluation Results

In this section, we report and discuss our evaluation results.

Effectiveness of Seizure Onset Detection. The overall evaluation result of our model and the baselines on the seizure onset detection task are illustrated in Table 1. Our model achieved at least five p.p. higher Macro Avg. F1-score than all the baselines. More specifically, Chrononet results in the lowest recall compared to all other models, RUSBoost achieved the same recall as *Co-ReaSON* but with low precision. TDNN+LSTM has a low F1 score, possibly because, in the original paper, it took longer seizure segments of the 20 s, 30 s, and 40 s, whereas our dataset has shorter EEG signal segments, more suitable for early seizure onset detection. *Co-ReaSON* outperforms ResNet_Short + LSTM model, demonstrating that our additional representations of correlation and DWT-based features help in learning the seizure onset patterns. The GCN model has a precision close to *Co-ReaSON*, which may be because graph-based networks learn some node-neighbor correlations; however, this baseline underperforms in terms of recall. This result underlines the effectiveness of multimodal EEG feature representations performed by *Co-ReaSON*.

Feature Analysis. To demonstrate the impact of different features, we remove one feature type at a time from our approach and evaluate the model's performance. The results are presented in Table 2. We observe that removing any feature types results in overall lower model performance in terms of F1 and macro average F1 scores, demonstrating the effectiveness of our features in seizure onset detection. Removing the raw EEG image feature (*Co-ReaSON* $-$ *Raw*) has the largest impact on the overall model performance, with a reduction of 0.34 and 0.18 p.p. in F1 and macro average F1 scores, respectively. Removing the correlation features (*Co-ReaSON* $-$ *ECorr*) has a strong negative impact on the

Table 2. Feature importance in seizure onset detection with an 8-sec interval. Precision (P), Recall (R), and F1 scores for the minority class (seizure onset).

Feature	P	R	F1	Macro Avg P	Macro Avg R	Macro Avg F1
Co-ReaSON	0.99	0.58	0.73	0.98	0.79	0.86
Co-ReaSON - E_Corr	0.48	0.67	0.56	0.73	0.80	0.76
Co-ReaSON - DWT	0.99	0.45	0.62	0.98	0.73	0.80
Co-ReaSON - Raw	0.56	0.30	0.39	0.76	0.64	0.68

Table 3. Impact of removing features corresponding to specific brain regions. Precision (P), Recall (R), and F1 score for the minority class (seizure onset).

Brain Region	P	R	F1	Macro Avg P	Macro Avg R	Macro Avg F1
Co-ReaSON	0.99	0.58	0.73	0.98	0.79	0.86
Co-ReaSON- Temporal	0.90	0.56	0.69	0.94	0.78	0.84
Co-ReaSON- Parietal	0.95	0.52	0.67	0.96	0.76	0.83
Co-ReaSON- Central	0.95	0.48	0.63	0.96	0.74	0.81
Co-ReaSON- Frontal	0.94	0.56	0.70	0.95	0.78	0.84
Co-ReaSON- Occipital	0.96	0.59	0.73	0.97	0.80	0.86

Table 4. Impact of the time interval lengths (Interval Length). Precision (P), Recall (R), and F1 scores are reported for the minority class (seizure onset).

Interval Length	Avg. Lag	P	R	F1	Macro Avg P	Macro Avg R	Macro Avg F1
4 sec	1.08 sec	0.87	0.55	0.68	0.92	0.77	0.83
8 sec	1.19 sec	0.99	0.58	0.73	0.98	0.79	0.86
12 sec	1.36 sec	0.84	0.67	0.74	0.91	0.83	0.87
16 sec	1.52 sec	0.85	0.64	0.72	0.92	0.82	0.86

precision of the model performance, confirming that the correlation between channels provides important indicators of seizure onset. Removing DWT features (*Co-ReaSON – DWT*) has the least model performance impact but also reduces recall and F1, indicating that DWT features can capture meaningful patterns.

Influence of Brain Regions. To analyze the impact of signals from different brain regions on the model's performance, we remove features corresponding to the nodes in specific brain regions one at a time. The evaluation results are presented in Table 3. We observe that the model using features from all channels results in the highest performance. As compared to all other regions, the signals from the temporal brain region have the strongest impact on precision. This can be because most of the focal epilepsies are temporal lobe epilepsies, though the seizures strongly depend on etiology. We also notice that the features corresponding to the central and parietal regions strongly impact recall.

Time Interval Length Impact. We evaluate *Co-ReaSON* with different time interval lengths to better understand its ability to detect seizure onsets early. The results using time intervals of 4, 8, 12, and 16 s are presented in Table 4. We also report the average time (Avg. Lag) the model takes to predict. We observe that *Co-ReaSON* achieves better performance on the 8-s than the 4-s intervals, which may be because the 8-s intervals contain a longer seizure period after the onset, and the extracted features are more informative. Compared to 8 s, the model achieved only a slightly improved F1 score on longer intervals due to increased recall, whereas precision dropped. Overall, we consider 8 s an optimal time interval for *Co-ReaSON*.

6 Related Work

Numerous studies focused on the seizure detection task, but there are few baselines for detecting focal seizure onset. Working on multivariate time series data, Roy et al. [9] applied a convolutional GRU model on EEG data to detect abnormal brain activity. You et al. [18] used unsupervised learning through modified anomaly Generative Adversarial Networks (anoGAN) to learn the time-frequency variance in the spectral images of behind-the-ear normal EEG recordings and then apply the models to detect seizures. Regarding feature representations, the Fast Fourier Transform (FFT) on sequenced sliding windows of EEG signals has been applied for feature extraction [11,13]. Glory et al. [2] identify a suitable basis wavelet function and use the minimum entropy of EEG signals as evaluation criteria, showing that Daubechies Wavelet DB10 is most efficient for seizure detection. Shen et al. [12] utilizes DWT features over the RUSBoost ensemble tree model and detects seizure onsets within a 30-s interval.

The above-stated works train their models either on very few patients' data (CHB-MIT or Bonn Epilepsy dataset) or use full-length EEG signals to train their models. Instead of detecting seizure onsets, they focus on detecting if the full segments of EEG recordings have a seizure period. Thus, generalizing their model performance to the early detection of seizure onsets is challenging. Recent work, such as [20], uses GCN to explore EEG channel connectivity information. They take 5-s time segments of seizure and non-seizure periods for early detection of seizures, but they consider all types of seizures as one class and detect seizures as a whole instead of detecting the focal seizure onset, which is more challenging. Lee et al. [5] study early seizure onset detection using ResNet convolutions with skip connections over raw EEG signals and then learning the sequenced features using LSTM. In contrast to the above-mentioned works, we focus on the more challenging task of focal seizure onset detection and incorporate multimodal feature representations to capture specific patterns.

7 Conclusion

In this paper, we proposed *Co-ReaSON* - a novel approach for early detection of focal epileptic seizure onsets. *Co-ReaSON* is based on learning the onset-specific spatio-temporal correlations across the EEG channels in a ResNet18-based model architecture. *Co-ReaSON* relies on various rich graph-, image- and DWT-based EEG features to capture such correlations effectively. Our evaluation on a real-world seizure dataset demonstrates that *Co-ReaSON* outperforms state-of-the-art baselines by at least five p.p. regarding the macro average F1-score.

Acknowledgements. This work was partially funded by the Ministry of Culture and Science of the State of North Rhine-Westphalia, Germany ("iBehave").

References

1. Berg, A.T., Berkovic, S.F., Brodie, M.J., Buchhalter, J., et al.: Revised terminology and concepts for organization of seizures and epilepsies: report of the ilae commission on classification and terminology, 2005–2009 (2010)
2. Glory, A., et al.: Identification of suitable basis wavelet function for epileptic seizure detection using EEG signals. In: ICTSCI (2019)
3. He, K., Zhang, X., Ren, S., Sun, J.: Deep residual learning for image recognition. In: IEEE CVPR (2016)
4. Jasper, H.H.: Ten-twenty electrode system of the international federation. Electroencephalogr. Clin. Neurophysiol. **10**, 371–375 (1958)
5. Lee, K., et al.: Real-time seizure detection using EEG: a comprehensive comparison of recent approaches under a realistic setting. arXiv:2201.08780 (2022)
6. Mallat, S.G.: A theory for multiresolution signal decomposition: the wavelet representation. IEEE Trans. Pattern Anal. Mach. Intell. **11**(7), 674–693 (1989)
7. Mormann, F., Andrzejak, R.G., Elger, C.E., Lehnertz, K.: Seizure prediction: the long and winding road. Brain **130**(2), 314–333 (2007)
8. Raghu, S., Sriraam, N., et al.: Performance evaluation of dwt based sigmoid entropy in time and frequency domains for automated detection of epileptic seizures using SVM classifier. Comput. Biol. Med. **110**(C), 127-143 (2019)
9. Roy, S., Kiral-Kornek, I., Harrer, S.: Chrononet: a deep recurrent neural network for abnormal EEG identification. In: AIME (2019)
10. Shah, V., et al.: The temple university hospital seizure detection corpus. Front. Neuroinform. **12**, 83 (2018)
11. Shawki, N., Elseify, T., Cap, T., Shah, V., Obeid, I., Picone, J.: A deep learning-based real-time seizure detection system. In: IEEE SPMB (2020)
12. Shen, M., Wen, P., Song, B., Li, Y.: An EEG based real-time epilepsy seizure detection approach using discrete wavelet transform and machine learning methods. Biomed. Signal Process. Control **77**, 103820 (2022)
13. Thyagachandran, A., Kumar, M., Sur, M., Aghoram, R., Murthy, H.: Seizure detection using time delay neural networks and LSTMS. In: IEEE SPMB (2020)
14. Wagh, K.P., Vasanth, K.: Performance evaluation of multi-channel electroencephalogram signal (EEG) based time frequency analysis for human emotion recognition. Biomed. Signal Process. Control **78**, 103966 (2022)
15. Wang, D., Miao, D., Xie, C.: Best basis-based wavelet packet entropy feature extraction and hierarchical EEG classification for epileptic detection. Expert Syst. Appl. **38**(11), 14314–14320 (2011)
16. Wenzel, M., Hamm, J.P., Peterka, D.S., Yuste, R.: Acute focal seizures start as local synchronizations of neuronal ensembles. J. Neurosci. **39**(43), 8562–8573 (2019)

17. Worrell, G.A., Gardner, A.B., Stead, S.M., Hu, S., Goerss, S., et al.: High-frequency oscillations in human temporal lobe: simultaneous microwire and clinical macroelectrode recordings. Brain **131**(4), 928–937 (2008)
18. You, S., et al.: Unsupervised automatic seizure detection for focal-onset seizures recorded with behind-the-ear EEG using an anomaly-detecting generative adversarial network. Comput. Methods Programs Biomed. **193**, 105472 (2020)
19. Zarei, A., Asl, B.M.: Automatic seizure detection using orthogonal matching pursuit, discrete wavelet transform, and entropy based features of EEG signals. Comput. Biol. Med. **131**, 104250 (2021)
20. Zheng, Q., Venkitaraman, A., Petravic, S., Frossard, P.: Knowledge distillation with graph neural networks for epileptic seizure detection. In: ECML PKDD (2023)

A Data-Driven Approach for Building a Cardiovascular Disease Risk Prediction System

Hongkuan Wang[1(✉)], Raymond K. Wong[1], and Kwok Leung Ong[2]

[1] University of New South Wales, Kensington Sydney, Australia
hongkuan.wang@unsw.edu.au

[2] The University of Sydney, Camperdown Sydney, Australia

Abstract. Cardiovascular disease is a leading cause of mortality worldwide. The disease can develop without showing apparent symptoms at an early stage, making it difficult for domain experts to provide intervention. Using machine learning techniques in chronic disease prediction is becoming popular because of their ability in processing a large amount of data and analysing the patterns buried in the datasets. To increase the accessibility for healthcare professionals to ready-to-use machine learning prediction pipelines, we introduce an automated machine learning system called Auto-Imblearn that can process and analyse the imbalanced clinical data; automatically compare different classification algorithms and apply the best algorithm for prediction. Using a real patient dataset, the prediction of our proposed system achieves the best performance against the state-of-the-art baselines while saving significant computations from the exhaustive approach.

Keywords: Cardiovascular Disease · Automated Machine Learning

1 Introduction

Cardiovascular disease (CVD) is a leading cause of mortality worldwide, with an estimated 17.9 million deaths per year [44]. Machine learning techniques are widely used in life science because of their ability to analyse and predict data [39]. It is beneficial for some diseases, such as coronary heart disease [23,25], type 2 diabetes [28], or Alzheimer's disease [22], especially for accurate screening and early detection, saving medical resources and making medical treatment sooner for patients.

CVD Risk Prediction: Machine learning techniques have shown promise in CVD risk prediction. Ganguly et al. [23] proposed a novel framework that is based on convolutional neural networks (CNN). The proposed approach achieved high accuracy in the classification of different types of CVD, including arrhythmias, myocardial infarction, and heart failure. Wada et al. [25] did a comparative study of various machine learning algorithms, which were presented for the classification of CVD using clinical data. The results show that the SVM-based and logistic regression approach outperforms other algorithms in terms of classification accuracy. Amal et al. [2] proposed a multi-modal

D.-N. Yang et al. (Eds.): PAKDD 2024, LNAI 14648, pp. 271–283, 2024.
https://doi.org/10.1007/978-981-97-2238-9_21

approach for the classification of CVD using multiple clinical data sources, including ECG signals and echocardiography images. The proposed approach achieved high accuracy in the classification of different types of CVD.

Although machine learning is becoming popular in CVD risk predictions, two of the most significant challenges that still need to be addressed for building a machine learning pipeline. 1) Missing values are prevalent in clinical data. In reality, more than half of the data may be missing for some important features [46]. 2) Most patient/clinical data are highly imbalanced. In practice, the ratio of positive cases to negative cases can be close to 1/100. Some regular methods, such as support vector machines or random forests, can achieve about 99% accuracy by predicting everyone as negative [3,9,37], which make them not very viable for actual clinical uses. Because of these challenges, selecting the appropriate methods is needed to handle imbalanced clinical data, provide accurate screening and make early detection.

Imputation: Unlike experimental data, clinical data are collected from clinicians to take care of their patients better. Therefore, the missing value problem wouldn't be an issue for them until the data were decided to be used for analysis purposes afterwards [6]. Based on the research of [41], if the missing rate of the datasets is small, the missing data can be simply removed. However, in the healthcare domain, the missing rate can reach more than 50% [13]. In this context, imputation methods can be used to address the challenge of missing values. Imputation methods include simple methods such as filling with median or mean values, and some complex models. For example, k-NN imputation method uses the k nearest neighbours of the missing data to compute the filling values [5]; the iterative imputation method treats the imputation as a regression problem by sequentially imputing each feature using other features [12]. Besides those traditional methods, some state-of-the-art models have been proposed recently. [27] proposed MIRACLE imputation method, which simultaneously models the missingness generating mechanism and encourages imputation to be consistent with the causal structure of the data; [45] proposed GAIN imputation method, which adapts the Generative Adversarial Nets framework into the imputation process; [31] proposed MIWAE imputation method, which is based on the importance-weighted autoencoder [11].

Imbalanced Classification: As for solving the imbalanced data classification problem, data-level methods have become more popular in recent years [9]. The theory behind the data-level methods is applying modification directly to the datasets to make them balanced before feeding them to any classification methods. This process involves changing the data distribution, and the new distribution can still be representative enough for each class. There are studies showing that resampling can improve the performance of some classifiers [30]. Another benefit of data-level methods is that the modified datasets can be applied to almost any existing classifiers [9,35].

There are some simple yet good-performing resampling models, such as Random Over Sampler (ROS) [32] and Random Under Sampler (RUS) [8]. During the training with ROS, minority class instances are replicated randomly to relieve the imbalance problem. In contrast, data from the majority class are removed randomly by RUS. One of the benefits of those two resampling methods is that can greatly preserve the data distribution of all features.

SMOTE is one of the most classical oversampling methods [15]. It synthetic new instances of the minority class in the following process: 1) find the differences between the instances and some or all of their k nearest neighbours; 2) multiply the differences with a random number between 0 and 1; 3) add the multiplied results to the original instances to form new synthetic instances.

Motivated by the success of SMOTE, SMOTE-based oversampling approaches attract more attention. For example, the process of MWMOTE [4] is: i) find the minority class instances that are not surrounded by majority class instances only; ii) calculate a weight for each selected instance; iii) generate new instances based on the weights. The search space of AutoSMOTE [47] covers all areas for generating samples with SMOTE and most of SMOTE extension. Besides, AutoSMOTE used hierarchical reinforcement learning [26] to optimize the generalization process.

Automated Machine Learning (AutoML): As the complexity and scale of machine learning models emerge and evolve, the demand for people in other fields to efficiently build accessible and ready-to-use machine learning pipelines has increased, too. AutoML, as an emerging field, focuses on filling this gap [24]. It's beneficial in the healthcare domain, as it has the potential to reduce the burden of the labour-intensive model selection and early diagnosis of some diseases [34]. Besides, for designing machine learning pipelines in the healthcare domain, collaborations between healthcare experts and machine learning experts are needed [1,43,48]. Healthcare experts are familiar with the data they collect, but machine learning experts are often needed for searching and applying the appropriate machine learning models. The designing process of a machine learning pipeline is a labour-intensive process for both parties. Therefore, there is a clear need for an AutoML system for imputation and class imbalance while being efficient in the training process.

The topic of AutoML in the healthcare domain has been thoroughly surveyed [43]. The first AutoML system that combines both algorithm selection and hyperparameter optimization (CASH) is Auto-WEKA [42]. Auto-WEKA is a single hierarchical optimization-based algorithm/model selection pipeline system, and it treats the choice of model as a hyperparameter to optimize.

Another approach for building AutoML systems is meta-learning [10]. The core concept of the meta-learning strategy is that the models are more likely to perform better than others on a new dataset if they performed better on similar datasets before. Auto-Sklearn [19] also attempts to solve the CASH problem defined by [42]. It uses meta-learning to warmstart the Bayesian optimization procedure to boost efficiency. Auto-Sklearn first trains the meta-learning process with 140 datasets to find good instantiation results of machine learning pipelines. When training with new data, Auto-Sklearn selects the $k = 25$ nearest datasets to the new dataset in meta-feature space before starting Bayesian optimization with their results. Because Auto-Sklearn lacks support for resampling methods, to deal with imbalanced classification problems, it requires machine learning knowledge to manually tweak some classification models in its selection pool by assigning different weights to different classes.

Several methods have been introduced to the AutoML system topic in recent years. TPOT [36] is a popular AutoML system. It treats classification models, feature preprocessing models and feature selection models as features in a normal tree-based classification method. However, it sometimes overfits the training data and generalizes poorly to held-out data [43]. AlphaD3M [18] brings reinforcement learning [26] into the topic, and it requires less computation power for similar prediction accuracy than TPOT. Besides training with individual classification methods, H2O AutoML [29] attempts different ways to combine the classification models in the selection pool with H2O's stacked ensemble method.

Because of the class imbalance problem can greatly affect the performance of some classification models [3,9,37]. In the last 2 years, two AutoML systems, ICLL [14] and ATOMIC [33], were proposed with a focus on imbalanced classification. ICLL [14] uses a layered learning [40] method to bring the imbalanced classification task into simpler sub-tasks. It first perform a hierarchical clustering procedure to separate the datasets into 3 different clusters: pure majority, pure minority and mix of both. The first layer only solve the classification of pure majority and the rest of data; The second layer only deal with the binary classification task with less imbalanced class distribution. ATOMIC [33] also meta-learning approach to build the AutoML system. It extract the meta-features from the datasets used for development phase and the pipeline configuration descriptions. Then they are fed into eXtreme Gradient Boosting model [16] to learn the relationship between the dataset descriptions and pipeline configuration descriptions. In the prediction phase, the meta-features of the new data are used to predict the rank of different pipeline configurations. The whole meta-learning process borrows the idea from recommendation system.

Whilst AutoML systems may be able to determine a good machine learning pipeline for normal datasets, their performances are usually not good for datasets with lots of missing values and high imbalance ratios, such as CVD data, as shown in the experiment section of this paper. To tackle this challenge, we propose an AutoML system called Auto-Imblearn, an automated model selection pipeline system that performs iterative and level-wise optimization to improve the performances on those datasets. It can handle both missing values and class imbalance problems and automatically finds ideal models for a given dataset while at the same time reducing computation cost. The contributions of our work can be summarized as follows.

1. It provides an AutoML system with high accuracy for CVD risk prediction;
2. It has shown to be effective for a dataset with a high percentage of missing values as well as a high imbalance ratio;
3. It selects the best imputation and classification methods for a given dataset without exploring all possible combinations of different methods and models.
4. It can further reduce the computation cost by using a portion of the data and yet still achieves high prediction accuracy.

Using real patient data, our proposed system has been shown to achieve good accuracy. The source code of Auto-Imblearn is available at Github[1].

2 The Proposed AutoML System

In this paper, we propose the Auto-Imblearn pipeline system. The objective is to automate the development of pipelines to tackle both imputation and imbalanced

[1] https://github.com/AutoImblearn/auto-imblearn.

classification tasks in CVD risk prediction. Our primary motivation is to relieve the burden of labour-intensive and computation-intensive model selection process in CVD risk prediction. The Auto-Imblearn is an iterative optimization approach for AutoML system. In each iteration, the optimization is sequentially performed along the pipeline for each level. Unlike gradient descent, which take a small step in each iteration using all features, Auto-Imblearn further divides the step into sub-steps. The number of sub-steps is dependent on the number of levels in the Auto-Imblearn. Because the positions of the models in each level are irrelevant to the optimization process, so in each sub-step the optimization process need to use all models within the level.

There are three levels in Auto-Imblearn: Imputation level, Resampling level, and Classification level. We will further explain them in the following subsections. Algorithm 1 provides details of Auto-Imblearn.

Imputation Level. The first thing of our proposed pipeline system is to deal with missing values in a given dataset. For example, NHANES dataset consists real-world recordings from real people, there are missing values in their lifestyle survey and also in their lab test results. In NHANES, there are important features with more than 50% of missing values for its lab test results, such as the insulin resistance index, fasting insulin level, LDL-cholesterol, fasting glucose level, among others [13]. Finding a right method to perform imputation is crucial before feeding the data to the classification methods. Currently, Auto-Implearn automatically identifies a right method from a selection pool containing median values, iterative imputer, k-NN imputer, GAIN, MIRACLE and MIWAE for the imputation level. More methods will be added later.

Resampling Level. The NHANES dataset has an imbalance ratio of 14:1, so before feeding the modified data into the classification level, it needs to be passed through a resampling process. The selection pool in this level consists of RUS, ROS, SMOTE, MWMOTE and AutoSMOTE.

Classification Level. After the previous two levels, the modified NHANES data can be trained by almost any classification method. At the classification level, the system will test the accuracy variations based on different classifiers (CLF for short). Currently, there are four classifiers supported by the classification level: LR [17], SVM [7], AdaBoost (ADA for short) [21] and MLP [38].

The AutoML System. As show in Algorithm 1, at the initial iteration of the system, Auto-Imblearn initialize a pipeline as the temporary result with all component randomly selected from different levels. Then in the imputation level, the imputation method in the pipeline will be replaced by the best imputer found by function *FindBest*. It is a similar process for both resampling level and classification level. Auto-Imblearn stops when two adjacent iterations produces same pipeline, and outputs the pipeline found from the last iteration. To further reduce computation costs and to address the situation when insufficient labelled data is available, Auto-Imblearn can also be trained with a portion of a given dataset.

Algorithm 1: Auto-Imblearn pipeline system

```
Result: The best pipeline best_pipeline and it's prediction result
imps ← All available imputation methods, rss ← All available resampling methods,
  clfs ← All available classification methods;
best_imp ← Randomly select one element from imps, best_rsp ← Randomly select
  one element from rss, best_clf ← Randomly select one element from clfs;
current_pipeline ← ∅;
best_pipeline ← {best_imp, best_rsp, best_clf};
best_result ← 0 ;
pipelines ← 2-D empty array;
while true do
    current_pipeline ← ∅;
    level ← 1 // Imputation level;
    best_imp, best_result ← FindBest (imps, best_rsp, best_clf, level) ;
    current_pipeline.Insert(best_imp);
    level ← 2 // Resampling level;
    best_rsp, best_result ← FindBest (rss, best_imp, best_clf, level) ;
    current_pipeline.Insert(best_rsp);
    level ← 3 // Classification level;
    best_clf, best_result ← FindBest (clfs, best_imp, best_rsp, level) ;
    current_pipeline.Insert(best_clf);
    if best_pipeline ≠ current_pipeline then
        best_pipeline ← current_pipeline;
    else
        return best_pipeline, best_result;

Function FindBest (pool, model_1, model_2, level) :
    // Find the best model from selection pool;
    tmp_pipeline ← ∅ ;
    tmp_pipeline.append(model_1);
    tmp_pipeline.append(model_2);
    tmp_pipeline.insert(level, 0)  // Insert 0 at position level of tmp_pipeline;
    tmp_result, result ← 0;
    for tmp_model ∈ pool do
        // Find the best performing model in current level;
        tmp_pipeline[level] ← tmp_model;
        tmp_result ← RunModel (tmp_pipeline) ;
        if tmp_result > result then
            selected_model, result ← tmp_model, tmp_result;

    tmp_pipeline[level] ← selected_model;
    return selected_model, result;
Function RunModel (tmp_pipeline) :
    // Train the tmp_pipeline and save the result to pipelines;
    if tmp_pipeline ∈ pipelines then
        return ← pipelines[tmp_pipeline];
    else
        result ← train the tmp_pipeline with a portion of the data;
        pipelines[tmp_pipeline] ← result;
        return result;
```

3 Experiment

3.1 The Dataset

The National Health and Nutrition Examination Survey (NHANES) [13], is a crucial initiative assessing the health and nutritional well-being of the American population. This program employs a dual approach, conducting interviews and physical examinations, to comprehensively collect data.

The experiments in this paper are based on the data from the NHANES conducted between 1999 and 2008. There were a total of 51623 participants. Among them, 24693 participants aged $\geq$ 20 years were both interviewed and examined in the mobile examination center. The dataset includes issues such as missing values and high imbalance ratio. For example, after excluding pregnant women and subjects with missing data on mortality status there are 23513 participants, and 1576 of them died because of CVD in the tracking period, so the imbalance ratio is approximately 14:1.

3.2 Evaluation Metrics

To measure the quality of the results, we use the Area Under a Receiver Operating Characteristic curve (AUROC) and Macro-F1 (m-F1 for short) as evaluation metrics.

3.3 Pipelines

Imputation Method Selection. Table 1(a) shows a snapshot of imputation method selection in our system. As mentioned above, currently the selection pool consists of median values, iterative imputation, k-NN imputation, GAIN, MIRACLE and MIWAE. For this snapshot, the resampling model is ROS with the classification method being MLP.

Table 1. The optimization process experiment results of one iteration

Imputation	AUROC	m-F1
Median	0.8997	0.6488
Iterative	0.8997	0.6458
k-NN	0.8993	0.6492
GAIN	0.8999	0.6451
MIRACLE	0.8993	0.6498
MIWAE	0.9000	0.6391

(a) Different imputation methods

Method	AUROC	m-F1
RUS	0.8960	0.6628
ROS	0.9000	0.6391
SMOTE	0.8985	0.6612
MWMOTE	0.8980	0.6543
AutoSMOTE	0.7408	0.4963

(b) Different resamplers

Classifier	AUROC	m-F1
MLP	0.9000	0.6391
LR	0.8918	0.6393
SVM	0.8947	0.6554
ADA	0.8695	0.6480

(c) Best classifier results

Resampler Selection. Auto-Imblearn evaluates the performance of different resampling methods. The selection pool currently consists of RUS, ROS, SMOTE, MWMOTE and AutoSMOTE. A snapshot of their performances are shown in Table 1(b). In this snapshot, the imputation method is MIWAE, and the classifier is MLP.

Classifier Selection. Finally, Auto-Imblearn will benchmark different classification methods including MLP, LR, SVM and ADA. Their performances can be found in Table 1(c). The imputation method used in this iteration is MIWAE, and the resampling method is ROS.

Based on the results above, the overall best combination comes from the pipeline which consists of the MIWAE imputation model, ROS and MLP classification. The pipeline reaches the best AUROC score of 0.90. For your reference, the performances of all combinations of different models in Auto-Imblearn using the NHANES dataset are included and shown in the Appendix.

3.4 Computation Cost

To evaluate the computation cost of Auto-Imblearn, we run Auto-Imblearn until it achieves the best pipeline that could be found by an exhaustive approach as shown in the Appendix. The amount of pipelines evaluated by Auto-Imblearn is recorded. We repeat this experiment 100 times and plot a histogram for both evaluation metrics. Their results are shown in Fig. 1. On average Auto-Imblearn saves about 75.16% of computation when compared with the exhaustive approach using AUROC as the evaluation metric, and saves about 78.03% when compared with the exhaustive approach using Macro-F1.

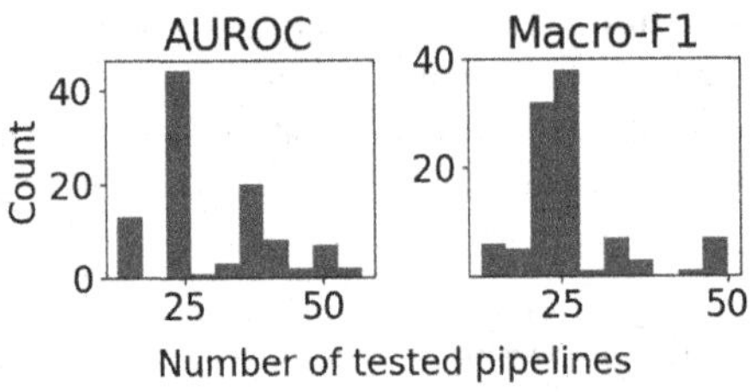

Fig. 1. Histograms showing the distribution of number of pipelines evaluated in order to find the best performing pipeline for each metric.

3.5 Comparison with Other AutoML Systems

The above experiments have proved that our Auto-Imblearn reduces computation cost by reducing a significant amount of pipelines as mentioned in the previous subsection. The performance of Auto-Imblearn is compared with other baseline AutoML systems and the result is shown in Table 2. Using its iterative and level-wise optimization, Auto-Imblearn outperforms all other AutoML systems using all the data from the dataset NHANES (i.e., 100% NHANES).

In particular, AlphaD3M failed to find a pipeline when the evaluation metirc is AUROC. H2O and ICLL don't support optimizing with evaluation metirc Macro-F1. Because TPOT provides a more flexible machine learning pipeline, it sometimes overfit the data [43]. Therefore, the generated pipelines from TPOT shown in Table 2 have been validated with the whole dataset using 10-fold cross validation.

To further reduce computation costs and to address some applications that do not have sufficient labelled data, Auto-Imblearn also supports training with a small portion of a given dataset. For example, Table 2 shows the average performances of Auto-Imblearn on 10% and 20% of the NHANES dataset (i.e., 10% NHANES and 20% NHANES respectively); and they still outperforms all other AutoML systems. Currently, we are extending Auto-Imblearn to automatically determine the minimum amount of data needed to identify an ideal pipeline, using approaches such as [20].

Table 2. Comparison with other baseline AutoML systems

System	AUROC	Macro-F1
Auto-Sklearn	0.5027	0.4887
TPOT	0.8754	0.5910
AlphaD3M	N/A	0.5798
H2O	0.7686	N/A
ICLL	0.8884	N/A
Auto-Imblearn (10% NHANES)	0.8913	0.6565
Auto-Imblearn (20% NHANES)	0.8951	0.6469
Auto-Imblearn (100% NHANES)	**0.9000**	**0.6669**

4 Conclusion

We have presented an effective AutoML system for CVD risk prediction. The proposed system utilises various machine learning techniques and achieves high accuracy in CVD risk prediction. It tolerates a high percentage of missing values and can deal with imbalanced data. Our proposed system can also determine the best classification method that is appropriate for a given dataset, utilizing a portion of its data or a full dataset. This system has been applied and tested with a real-world patient dataset, NHANES. Based on NHANES, it has shown that our proposed system achieves the best prediction accuracy with respect to other AutoML systems. We are currently working with medical professionals from hospitals and medical research institutes and believe that the proposed system can play an important role in the actual diagnosis and treatment of CVD diseases.

A Appendix

An Exhaustive Approach Without using our proposed AutoML system to automatically to determine the best methods/models, one would need to conduct an exhaustive approach to compare the performances of different combinations of imputation, resampling

Table 3. Exhaustive search

Resampling	Imputation	Classification method							
		MLP		LR		SVM		ADA	
		AUROC	m-F1	AUROC	m-F1	AUROC	m-F1	AUROC	m-F1
RUS	Median	0.8960	0.6645	0.8900	0.6456	0.8927	0.6600	0.8633	0.6381
	Iterative	0.8961	0.6638	0.8899	0.6610	0.8928	0.6615	0.8657	0.6430
	k-NN	0.8956	0.6612	0.8891	0.6586	0.8921	0.6595	0.8655	0.6437
	GAIN	0.8963	**0.6669**	0.8909	0.6618	0.8932	0.6618	0.8709	0.6169
	MIRACLE	0.8956	0.6662	0.8901	0.6608	0.8936	0.6596	0.8655	0.6460
	MIWAE	0.8960	0.6628	0.8900	0.6458	0.8923	0.6582	0.8693	0.6424
ROS	Median	0.8997	0.6488	0.8918	0.6387	0.8947	0.6598	0.8673	0.6448
	Iterative	0.8997	0.6458	0.8919	0.6395	0.8950	0.6617	0.8687	0.6199
	k-NN	0.8993	0.6492	0.8911	0.6398	0.8939	0.6588	0.8689	0.6211
	GAIN	0.8999	0.6451	0.8928	0.6400	0.8947	0.6541	0.8730	0.6199
	MIRACLE	0.8993	0.6498	0.8924	0.6414	0.8952	0.6593	0.8693	0.6280
	MIWAE	**0.9000**	0.6391	0.8918	0.6393	0.8947	0.6554	0.8695	0.6480
SMOTE	Median	0.8986	0.6459	0.8894	0.6458	0.8922	0.6501	0.8638	0.6058
	Iterative	0.8995	0.6526	0.8896	0.6459	0.8922	0.6498	0.8645	0.5976
	k-NN	0.8981	0.6608	0.8887	0.6467	0.8909	0.6528	0.8659	0.6013
	GAIN	0.8983	0.6615	0.8899	0.6427	0.8915	0.6505	0.8694	0.6087
	MIRACLE	0.8990	0.6524	0.8902	0.6467	0.8927	0.6577	0.8640	0.5933
	MIWAE	0.8985	0.6612	0.8895	0.6461	0.8922	0.6500	0.8612	0.6391
MWMOTE	Median	0.8988	0.6574	0.8924	0.6498	0.8956	0.6582	0.8539	0.6146
	Iterative	0.8980	0.6605	0.8926	0.6480	0.8959	0.6573	0.8576	0.6178
	k-NN	0.8970	0.6602	0.8916	0.6529	0.8948	0.6564	0.8529	0.6215
	GAIN	0.8982	0.6535	0.8935	0.6454	0.8958	0.6565	0.8623	0.6320
	MIRACLE	0.8980	0.6583	0.8927	0.6497	0.8956	0.6536	0.8570	0.6174
	MIWAE	0.8980	0.6543	0.8925	0.6475	0.8956	0.6578	0.8590	0.6234
AutoSMOTE	Median	0.7493	0.4968	0.8927	0.5392	0.7707	0.5543	0.8592	0.5384
	Iterative	0.7585	0.4964	0.8972	0.5585	0.7776	0.5391	0.8682	0.5182
	k-NN	0.7467	0.5173	0.8919	0.5366	0.7856	0.5394	0.8539	0.5783
	GAIN	0.7582	0.5107	0.8976	0.5826	0.7755	0.5519	0.8506	0.5427
	MIRACLE	0.7159	0.4951	0.8958	0.5656	0.7657	0.5253	0.8685	0.5352
	MIWAE	0.7408	0.4963	0.8937	0.5639	0.7900	0.5584	0.8547	0.5240

and classification methods, which will require far more computations. In this paper, the imputation methods include median values, k-NN imputation, iterative imputation, GAIN, MIRACLE and MIWAE. The resampling methods include RUS, ROS, SMOTE, MWMOTE and AutoSMOTE. The classification methods include LR, SVM, ADA and MLP. Their combination performances are shown in Table 3.

References

1. Romero, A., et al.: Benchmarking automl frameworks for disease prediction using medical claims. BioData Mining **15**(1), 15 (2022)
2. Amal, S., Safarnejad, L., Omiye, J.A., Ghanzouri, I., Cabot, J.H., Ross, E.G.: Use of multi-modal data and machine learning to improve cardiovascular disease care. Front. Cardiovascular Med. **9**, 840262 (2022)
3. Barbara, P.: Learning from high-dimensional biomedical datasets: The issue of class imbalance. IEEE Access **8**, 13527–13540 (2020). https://doi.org/10.1109/ACCESS.2020.2966296
4. Barua, S., Islam, M.M., Yao, X., Murase, K.: Mwmote-majority weighted minority oversampling technique for imbalanced data set learning. IEEE Trans. Knowl. Data Eng. **26**(2), 405–425 (2012)
5. Batista, G.E., Monard, M.C., et al.: A study of k-nearest neighbour as an imputation method. His **87**(251–260), 48 (2002)
6. Bell, M.L., Fiero, M., Horton, N.J., Hsu, C.H.: Handling missing data in rcts; a review of the top medical journals. BMC Med. Res. Methodol. **14**(1), 118 (2014)
7. Boser, B.E., Guyon, I.M., Vapnik, V.N.: A training algorithm for optimal margin classifiers. In: Proceedings of the Fifth Annual Workshop on Computational Learning Theory, pp. 144–152 (1992)
8. Branco, P., Torgo, L., Ribeiro, R.P.: A survey of predictive modeling on imbalanced domains. ACM Comput. Surv. (CSUR) **49**(2), 1–50 (2016)
9. Branco, P., Torgo, L., Ribeiro, R.P.: A survey of predictive modeling on imbalanced domains. ACM Comput. Surv. **49**(2) (2016).https://doi.org/10.1145/2907070
10. Brazdil, P., Carrier, C.G., Soares, C., Vilalta, R.: Metalearning: Applications to data mining. Springer Science & Business Media (2008)
11. Burda, Y., Grosse, R., Salakhutdinov, R.: Importance weighted autoencoders. arXiv preprint arXiv:1509.00519 (2015)
12. van Buuren, S., Groothuis-Oudshoorn, K.: mice: Multivariate imputation by chained equations in r. J. Stat. Softw. **45**(3), 1–67 (2011)
13. CDC: National health and nutrition examination survey (2022). http://www.cdc.gov/nchs/nhanes/about_nhanes.html
14. Cerqueira, V., Torgo, L., Branco, P., Bellinger, C.: Automated imbalanced classification via layered learning. Mach. Learn. **112**(6), 2083–2104 (2023)
15. Chawla, N.V., Bowyer, K.W., Hall, L.O., Kegelmeyer, W.P.: Smote: synthetic minority oversampling technique. J. Artifi. Intell. Res. **16**, 321–357 (2002)
16. Chen, T., Guestrin, C.: Xgboost: a scalable tree boosting system. In: Proceedings of the 22nd ACM SIGKDD International Conference on Knowledge Discovery and Data Mining, pp. 785–794 (2016)
17. Cox, D.R.: The regression analysis of binary sequences. J. R. Stat. Soc. Ser. B Stat Methodol. **20**(2), 215–232 (1958)
18. Drori, I., et al.: Alphad3m: machine learning pipeline synthesis. arXiv preprint arXiv:2111.02508 (2021)
19. Feurer, M., Klein, A., Eggensperger, K., Springenberg, J., Blum, M., Hutter, F.: Efficient and robust automated machine learning. In: Advances in Neural Information Processing Systems 28, pp. 2962–2970 (2015)
20. Figueroa, R.L., Zeng-Treitler, Q., Kandula, S., Ngo, L.H.: Predicting sample size required for classification performance. BMC Med. Inform. Decis. Mak. **12**, 1–10 (2012)
21. Freund, Y., Schapire, R.E.: A decision-theoretic generalization of on-line learning and an application to boosting. J. Comput. Syst. Sci. **55**(1), 119–139 (1997)

22. Fuse, H., Oishi, K., Maikusa, N., Fukami, T., Initiative, J.A.D.N.: Detection of alzheimer's disease with shape analysis of mri images, pp. 1031–1034 (2018)
23. Ganguly, B., Ghosal, A., Das, A., Das, D., Chatterjee, D., Rakshit, D.: Automated detection and classification of arrhythmia from ecg signals using feature-induced long short-term memory network. IEEE Sensors Lett. **4**(8), 1–4 (2020)
24. Hutter, F., Kotthoff, L., Vanschoren, J.: Automated machine learning: methods, systems, challenges. Springer Nature (2019)
25. Jinjri, W.M., Keikhosrokiani, P., Abdullah, N.L.: Machine learning algorithms for the classification of cardiovascular disease- a comparative study. In: 2021 International Conference on Information Technology (ICIT), pp. 132–138 (2021)
26. Kaelbling, L.P., Littman, M.L., Moore, A.W.: Reinforcement learning: a survey. J. Artifi. Intell. Res. **4**, 237–285 (1996)
27. Kyono, T., Zhang, Y., Bellot, A., van der Schaar, M.: Miracle: causally-aware imputation via learning missing data mechanisms. Adv. Neural. Inf. Process. Syst. **34**, 23806–23817 (2021)
28. Le, T.M., Vo, T.M., Pham, T.N., Dao, S.V.T.: A novel wrapper-based feature selection for early diabetes prediction enhanced with a metaheuristic. IEEE Access **9**, 7869–7884 (2021)
29. LeDell, E., Poirier, S.: H2o automl: Scalable automatic machine learning. In: Proceedings of the AutoML Workshop at ICML, vol. 2020. ICML (2020)
30. Lee, P.H.: Resampling methods improve the predictive power of modeling in class-imbalanced datasets. Int. J. Environ. Res. Public Health **11**(9), 9776–9789 (2014)
31. Mattei, P.A., Frellsen, J.: Miwae: deep generative modelling and imputation of incomplete data sets. In: International Conference on Machine Learning, pp. 4413–4423. PMLR (2019)
32. Menardi, G., Torelli, N.: Training and assessing classification rules with imbalanced data. Data Min. Knowl. Disc. **28**, 92–122 (2014)
33. Moniz, N., Cerqueira, V.: Automated imbalanced classification via meta-learning. Expert Syst. Appl. **178**, 115011 (2021)
34. Mustafa, A., Rahimi Azghadi, M.: Automated machine learning for healthcare and clinical notes analysis. Computers **10**(2), 24 (2021). https://www.mdpi.com/2073-431X/10/2/24
35. Nguyen, H.M., Cooper, E.W., Kamei, K.: A comparative study on sampling techniques for handling class imbalance in streaming data. In: The 6th International Conference on Soft Computing and Intelligent Systems, and The 13th International Symposium on Advanced Intelligence Systems, pp. 1762–1767. IEEE (2012)
36. Olson, R.S., Urbanowicz, R.J., Andrews, P.C., Lavender, N.A., Kidd, L.C., Moore, J.H.: Automating biomedical data science through tree-based pipeline optimization. In: Squillero, G., Burelli, P. (eds.) EvoApplications 2016. LNCS, vol. 9597, pp. 123–137. Springer, Cham (2016). https://doi.org/10.1007/978-3-319-31204-0_9
37. Pes, B.: Handling class imbalance in high-dimensional biomedical datasets, pp. 150–155 (2019). https://doi.org/10.1109/WETICE.2019.00040
38. Rumelhart, D.E., Hinton, G.E., Williams, R.J., et al.: Learning internal representations by error propagation (1985)
39. Shastry, K.A., Sanjay, H.A.: Machine Learning for Bioinformatics, pp. 25–39. Springer Singapore, Singapore (2020). https://doi.org/10.1007/978-981-15-2445-5_3
40. Stone, P., Veloso, M.: Layered learning. In: López de Mántaras, R., Plaza, E. (eds.) ECML 2000. LNCS (LNAI), vol. 1810, pp. 369–381. Springer, Heidelberg (2000). https://doi.org/10.1007/3-540-45164-1_38
41. Strike, K., El Emam, K., Madhavji, N.: Software cost estimation with incomplete data. IEEE Trans. Software Eng. **27**(10), 890–908 (2001)
42. Thornton, C., Hutter, F., Hoos, H.H., Leyton-Brown, K.: Auto-weka: combined selection and hyperparameter optimization of classification algorithms. In: Proceedings of the 19th ACM SIGKDD International Conference on Knowledge Discovery and Data Mining, pp. 847–855 (2013)

43. Waring, J., Lindvall, C., Umeton, R.: Automated machine learning: review of the state-of-the-art and opportunities for healthcare. Artif. Intell. Med. **104**, 101822 (2020). https://doi.org/10.1016/j.artmed.2020.101822
44. WHO: (2021). https://www.who.int/news-room/fact-sheets/detail/cardiovascular-diseases-(cvds)
45. Yoon, J., Jordon, J., van der Schaar, M.: GAIN: missing data imputation using generative adversarial nets. In: Dy, J., Krause, A. (eds.) Proceedings of the 35th International Conference on Machine Learning. Proceedings of Machine Learning Research, vol. 80, pp. 5689–5698. PMLR (10–15 Jul 2018). https://proceedings.mlr.press/v80/yoon18a.html
46. Yuan, L.: Evaluating the state of the art in missing data imputation for clinical data. Briefings Bioinform. **23**(1), bbab489 (2022). https://doi.org/10.1093/bib/bbab489
47. Zha, D., Lai, K.H., Tan, Q., Ding, S., Zou, N., Hu, X.B.: Towards automated imbalanced learning with deep hierarchical reinforcement learning. In: Proceedings of the 31st ACM International Conference on Information & Knowledge Management, pp. 2476–2485 (2022)
48. Zöller, M.A., Huber, M.F.: Benchmark and survey of automated machine learning frameworks. J. Artifi. Intell. Res. **70**, 409–472 (2021)

TFAugment: A Key Frequency-Driven Data Augmentation Method for Human Activity Recognition

Hao Zhang[1,2], Bixiao Zeng[2,3], Mei Kuang[1], Xiaodong Yang[2], and Hongfang Gong[1(✉)]

[1] Changsha University of Science and Technology, Chang Sha 410114, China
{zhanghao,kuangmei}@stu.csust.edu.cn, ghongfang@126.com

[2] Institute of Computing Technology, Chinese Academy of Sciences, Beijing 100090, China
{zengbixiao19b,yangxiaodong}@ict.ac.cn

[3] University of Chinese Academy of Sciences, Beijing 101408, China

Abstract. Data augmentation enhances Human Activity Recognition (HAR) models by diversifying training data through transformations, improving their robustness. However, traditional techniques with random masking pose challenges by introducing randomness that can obscure critical information. This randomness may lead the model to learn incorrect patterns, yielding variable results across datasets and models and diminishing reliability and generalizability in real-world scenarios. To address this issue, this paper introduces Time-Frequency Augmentation (TFAugment), an adaptive method improving generalizability by selectively enhancing key frequencies across diverse datasets in HAR. The proposed method incorporates a FreqMasking module into the network to extract an importance distribution from incoming frequency channels. This distribution serves as a parameter in a Bernoulli distribution for independent sampling of each frequency channel, thereby generating enriched training data. Experiments on DSADS, MHEALTH, PAMAP2, and RealWorld-HAR datasets demonstrate TFAugment's superior adaptability and significant performance enhancement compared to state-of-the-art techniques.

Keywords: Human Activity Recognition · Data Augmentation · Deep Learning

1 Introduction

Human Activity Recognition (HAR) systems are crucial in healthcare, surveillance, entertainment, and sports analytics. They recognize and interpret human activities, providing context-aware services. This improves user interfaces and supports intelligent decision-making [5].

Various device-based activity recognition methods, such as video [11], radar [14], and ambient sensors [15], have their limitations, like obstructions and

D.-N. Yang et al. (Eds.): PAKDD 2024, LNAI 14648, pp. 284–296, 2024.
https://doi.org/10.1007/978-981-97-2238-9_22

electromagnetic interferences. In contrast, sensor-based HAR systems, utilizing embedded sensors in smartphones or wearables, are more advantageous due to their portability, non-intrusiveness, and robustness, providing real-time activity insights [5]. Traditional HAR methods, based on machine learning techniques like Logistic Regression [22], Decision Trees [18], Random Forests [4], ELM [9], and Naive Bayes [20], have shown notable results but depend on manually crafted features, presenting challenges in activity recognition.

Recently, the widespread application of deep learning in Human Activity Recognition (HAR) has underscored the importance of data augmentation, especially when existing datasets are imbalanced or insufficient. However, current data augmentation methods, particularly random masking, often excessively alter semantics, impeding the model's ability to accurately capture activity patterns. However, current data augmentation methods, particularly random masking, often excessively alter semantics, impeding the model's ability to accurately capture activity patterns. For example, random masking can obscure key frequency differences in activities like walking and running, leading to blurred distinctions in patterns. Some research experiments indicate that randomness leads to significant performance variability across datasets and models [8,17]. This underscores the importance of adopting adaptive data augmentation methods over solely random techniques.

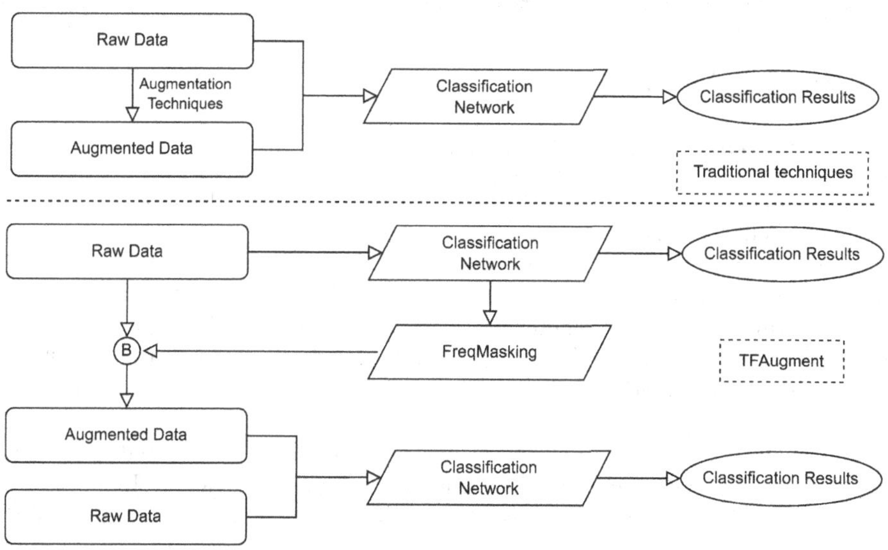

Fig. 1. Training paradigms with traditional data augmentation versus TFAugment.

In our work, as shown in Fig. 1, we transform original sensor signal features into time-frequency domain features and, based on this, propose a novel adaptive data augmentation method: Time-Frequency Augmentation (TFAugment). TFAugment primarily aims to adaptively augment critical frequencies across various datasets in the HAR domain, guided by a data-driven approach. The construction of TFAugment involves two stages. Unlike traditional methods that augment data before training, we adopt a new approach: train -> augment -> retrain. In Stage 1, we insert a FreqMasking module into the network, which first extracts an importance distribution from the input frequency channels. In Stage 2, the frequency importance distribution extracted from the FreqMasking module is used as a parameter for a Bernoulli distribution, independently sampling each frequency channel. The sampling results from the Bernoulli distribution are used as the mask for the frequency channels, with mask values of 0 or 1 determining whether the corresponding frequency channel is included in the enhanced data.

Through this process, we can randomly generate new, augmented data based on the importance of frequency channels, thereby enriching our model training. The main contributions of this paper are as follows:

- We innovatively employ a two-stage training approach, extracting the frequency importance distribution with virtually no impact on model speed.
- We introduce a novel data augmentation technique, Time-Frequency Augmentation (TFAugment), which enhances data based on importance distribution while minimally altering the semantics.
- Extensive evaluation of our method across multiple datasets shows its superior performance over established augmentation techniques.

2 Related Work

2.1 Data Augmentation Approaches

Data augmentation has been a critical task when existing datasets are imbalanced or insufficient. In domains like image recognition, techniques such as cropping, scaling, mirroring, color enhancement, and translation [6] have been effectively employed and achieved notable success. However, due to the inherent nature of time-series data, where the relevance of values depends on their temporal rather than spatial positioning, traditional data augmentation methods either cannot be directly applied or fail to yield desirable outcomes in time-series contexts [10,12].

In the time domain, Gao et al. [8] illustrated several conventional data augmentation strategies, including Jittering, Rotation, and Scaling. In addition, Qian et al. [17] discussed methods like negate, permute, and resample. These approaches demonstrate varying degrees of effectiveness across different datasets and models, highlighting the significant variance in the impact of data augmentation methods.

In the frequency domain, Qian et al. [17] introduced techniques such as hfc and lfc, which respectively preserve the high-frequency and low-frequency components of Human Activity Recognition (HAR) signals. The study's findings underscore the significant variance of many data augmentation methods across different datasets and models.

In the time-frequency domain, Park et al. [16] proposed SpecAugment, an augmentation scheme that includes feature warping, masking frequency channel blocks, and masking time steps. Experimental results indicate that SpecAugment significantly enhances the performance of neural networks for speech recognition.

Data augmentation techniques, despite their progress, face limitations due to their reliance on random processes like masking, distortion, and shuffling. This randomness hinders their ability to adapt to the specific characteristics of various datasets, often failing to fully utilize unique patterns and features. Therefore, there's a pressing need for more adaptive, data-driven augmentation methods that efficiently leverage dataset-specific features.

2.2 Deep Learning Approaches

Machine learning methods have historically been utilized for Human Activity Recognition (HAR), with deep learning algorithms like Convolutional Neural Networks (CNNs) [23] and Recurrent Neural Networks (RNNs) further advancing the field [5]. CNNs, adept at capturing local patterns in data, have been successfully applied in HAR tasks, as demonstrated by Yang et al. [24] and Laput et al. [13], achieving impressive results like a 95.2% classification accuracy. On the other hand, RNNs, especially their evolved forms like Long Short-Term Memory (LSTM) units [7], capture temporal correlations in signal data. However, due to their computational demands, simpler yet efficient alternatives like Gated Recurrent Units (GRUs) have been introduced, as used by Yao et al. [25] to achieve high accuracy with reduced resource consumption.

In this study, we have chosen to employ Dilated Convolutional Neural Networks (DCNNs) and Gated Recurrent Unit (GRU) networks as our backbone architectures. We aim to provide a comprehensive evaluation of our augmentation methods, ensuring that the observed improvements are not confined to a single model type but are generalizable across different neural network architectures.

3 Methodology

This chapter provides a detailed description of TFAugment. It is divided into three sections: (1) Time-Frequency Transformation (2) Time-Frequency Augment(TFAugment) (3) TFAugment Based HAR method.

3.1 Time-Frequency Transformation

Our method begins by transforming sensor data $X = \{x_k | x_k \in R^{T \times S}, k = 0, 1, ..., K-1\}$ into the time-frequency domain using the short time Fourier transform (STFT), where T is the time dimension, S is the sensor dimension, and $(x_k)_{pq}$ represents the signal at time p and sensor q. The transformed data $V = \{v_k | v_k \in R^{T' \times (\lfloor \frac{T'}{2} \rfloor + 1) \times S}, k = 0, 1, ..., K-1\}$, with T' as the new time dimension and $(v_k)mnq$ indicating the signal magnitude at time index m and frequency index n, is obtained using a Hamming window function ω. Finally, as shown in (1), we obtain v_k, the time-frequency representation of x_k.

$$v_k = f_{STFT}(x_k) = \left[(v_k)_{mnq}\right]_{m=0,T'-1;n=0,\lfloor \frac{T'}{2} \rfloor;q=0,S-1} \quad (1)$$

3.2 Time-Frequency Augmentation

The TFAugment technique, as illustrated in Fig. 2, utilizes the FreqMasking module to extract an importance distribution vector from input data's frequency channels, where each element, ranging from 0 to 1, indicates the corresponding frequency channel's importance.

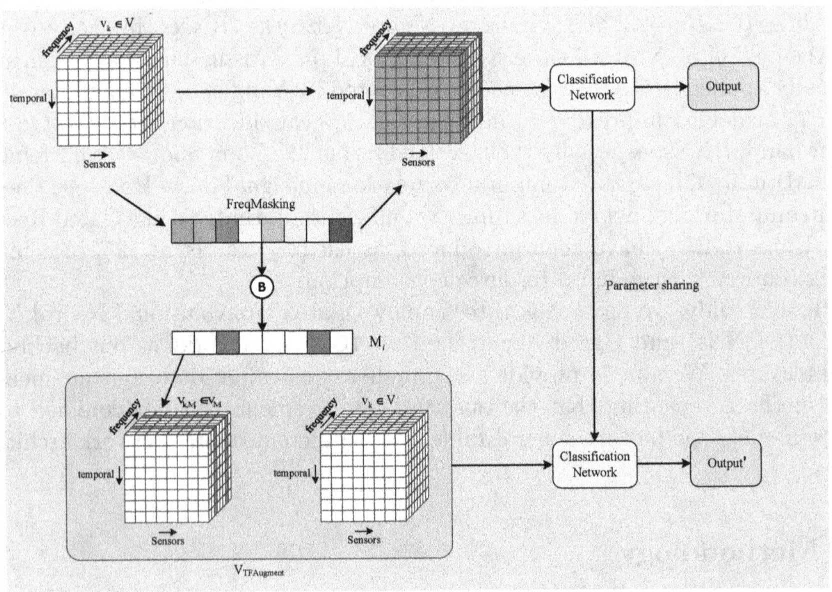

Fig. 2. The structure of TFAugment.

FreqMasking Module. We calculate z_k, which embodies a frequency-centric signal depiction, as the mean of signal magnitudes over all time points and sensor dimensions, as illustrated in Eq. (2).

$$z_k = \left[\frac{1}{T' \times S} \sum_{m=0}^{T'-1} \sum_{q=0}^{S-1} (v_k)_{mnq} \right]_{n=0, \lfloor \frac{T'}{2} \rfloor} \quad (2)$$

Subsequently, the FreqMasking module forms an importance distribution W_k from frequency-based signals via a neural network, involving a linear transformation and a sigmoid activation, as described in Eq. (3).

$$W_k = f_W(z_k, W') = Sigmoid(\tau(W' Z_k)) \quad (3)$$

By applying the importance distribution, we weight each frequency channel in v_k to obtain V_k', as detailed in Eq. (4), enabling integration with the model and extraction of critical frequencies via backpropagation.

$$V_k' = f_{V_k'}(v_k, W_k) = [(v_k)_{mnq}(W_k)_n]_{m=0,T'-1; n=0, \lfloor \frac{T'}{2} \rfloor; q=0,S-1} \quad (4)$$

After generating binary masks W_k for each input data window, we compute an aggregate mask W, representing the average frequency channel importance, as shown in Eq. (5).

$$W = \left[\frac{1}{K} \sum_{k=0}^{K-1} (W_k)_n \right]_{n=0, \lfloor \frac{T'}{2} \rfloor} \quad (5)$$

Data Augmentation. The next phase generates binary masks M_i by independently sampling each frequency channel using a Bernoulli distribution parameterized by the average mask W, as outlined in Eq. (6).

$$(M_i)_n = f'_{Bernoulli}(W_n) = \begin{cases} 1, & \text{With prob } W_n \\ 0, & \text{With prob } 1 - W_n \end{cases} \quad (6)$$

where M_i is defined as:

$$M_i = f_{Bernoulli}(W) = [(M_i)_n]_{n=0, \lfloor \frac{T'}{2} \rfloor} \quad (7)$$

And the set M of all possible masks M_i is described by:

$$M = \{M_i | M_i \in \{0, 1\}, i = 0, 1, ..., I - 1\} \quad (8)$$

Each M_i uniquely corresponds to v_k through non-replacement sampling, which can be mathematically represented as:

$$v_{kM_i} = f_{v_{kM_i}}(v_k, M_i) = [(v_k)_{mnq}(M_i)_n]_{m=0,T'-1;n=0,\lfloor \frac{T'}{2} \rfloor;q=0,S-1} \tag{9}$$

Following the application of M, the masked signal set is denoted by V_M, as illustrated below:

$$V_M = \{v_{k,M_i}\}_{i=0,I-1} \tag{10}$$

In conclusion, the stage 2 dataset, symbolized by $V_{TFAugment}$, comprises both the original and augmented data, represented in Eq. (11).

$$V_{TFAugment} = V_M \cup V \tag{11}$$

3.3 TFAugment Based HAR Method

The primary objective of this paper is to demonstrate an effective and adaptive data augmentation method. To this end, we have selected the Dilated Convolutional Neural Network (DilatedConvNet) and the Gated Recurrent Unit (GRU) from the classic families of CNN and RNN respectively, as our backbone architectures. The detailed design of these networks is presented in Table 1.

Table 1. Detailed design of the DilatedConvNet and GRUNet architectures.

Network	Layer	Parameters
DilatedConvNet	FreqMasking	only stage1
	Conv	(1,S); None; None; 64
	Conv	(3,1); (2,1); (2, 0); 128
	Conv	(3,1); (4,1); (4, 0); 256
	Adaptive Pool	(1, S)
	Linear	(256,Number of Classification)
GRUNet	FreqMasking	only stage1
	GRU	256
	GRU	256
	Linear	(256,Number of Classification)

4 Experiments

4.1 Datasets

To assess our method, we tested it on four wearable HAR datasets: DSADS [2], MHEALTH [3], PAMAP2 [19], and RealWorld-HAR [21], with details on categories and user division in Table 2. We used overlapping sliding windows to prepare samples and transformed the data into the time-frequency domain using a 16-length Hamming window with a stride of 4 and Fourier transform.

Table 2. Overview of Operations Performed on Each HAR Dataset.

Operation	Dataset			
	DSADS	MHEALTH	PAMAP2	RealWorld-HAR
Sampling rate	25 Hz	50 Hz	100 Hz	50 Hz
No. of activity	9	13	12	7
No. of subject	8	10	9	10
Test subject ID	1, 2	1, 2	1, 2	1, 10

4.2 Methods for Comparison

Before presenting our results, we briefly review the methods chosen for comparison in our experiments. For time domain data augmentation, we compared Jittering, Scaling, and Time Warping (TW) methods as discussed by Gao et al. [8], along with Moving Average (MA) and Exponential Moving Average (EMA) techniques proposed by Alawneh et al. [1]. In the time-frequency domain, our comparison included Frequency Masking (FM), Time Masking (TM), and Random Masking (RM) as explored by Jeong et al. [12]. To ensure the validity of the comparisons, all the comparative methods in this paper have adopted the parameter settings as mentioned in the referenced literature.

Furthermore, given the specificity of the behavior recognition field, we excluded methods that could significantly distort the data structure. For instance, the use of Rotation, as demonstrated in Iglesias et al. [10], can lead the model to learn incorrect patterns. Additionally, we did not compare our method with generative model-based approaches due to their typically longer training cycles, slower data generation times, and the challenge in objectively assessing the correctness of their generated data patterns.

4.3 Experimental Results

As demonstrated in Table 3 and 4, our experiments across four diverse datasets and two different backbones demonstrate that for time domain data augmentation, methods like Jittering and Scaling generally yielded positive results. They enable the model to focus more on the patterns intrinsic to the behavior rather than on noise and variations in magnitude. Time Warping (TW) performed well across all datasets and architectures, suggesting that it can effectively simulate the nuances of action duration and amplitude variations, aligning with the original labels and leading to a maximum accuracy increase of 2.46% and 3.50% in the two models respectively.

However, Moving Average (MA) and Exponential Moving Average (EMA) generally diminished model performance, indicating that some key patterns critical for HAR may be compromised by the smoothing process. Notably, on the

Table 3. Experimental results (Acc%) based on DilatedConvNet, with values in parentheses indicating the difference relative to Vanilla.

Methods	Dataset			
	DSADS	MHEALTH	PAMAP2	RealWorld-HAR
Vanilla	91.13	91.74	90.16	88.97
Jittering	91.89(+0.76)	91.94(+0.2)	90.22(+0.06)	89.75(+0.78)
Scaling	91.01(−0.12)	92.45(+0.71)	91.79(+1.63)	91.14(+2.17)
TW	92.97(+1.84)	93.43(+1.69)	92.62(+2.46)	90.85(+1.88)
MA	83.00(−8.13)	87.07(−4.67)	91.51(+1.35)	87.82(−1.15)
EMA	85.47(−5.66)	91.27(−0.47)	91.67(+1.51)	88.91(−0.06)
FM	91.03(−0.10)	92.94(+1.20)	91.88(+1.72)	87.76(−1.21)
TM	91.25(+0.12)	91.36(−0.38)	88.94(−1.22)	89.07(+0.10)
RM	91.24(+0.11)	91.55(−0.19)	90.27(+0.11)	88.91(+0.06)
TFAugment	92.91(+1.78)	94.78(+3.04)	93.67(+3.51)	89.12(+0.15)

Table 4. Experimental results (Acc%) based on GRU.

Methods	Dataset			
	DSADS	MHEALTH	PAMAP2	RealWorld-HAR
Vanilla	90.04	93.03	89.97	87.57
Jittering	91.51(+1.47)	93.07(+0.04)	89.91(−0.06)	87.92(+0.35)
Scaling	91.16(+1.12)	93.46(+0.43)	91.35(+1.38)	88.86(+1.29)
TW	93.03(+2.99)	93.83(+0.80)	91.76(+1.79)	91.07(+3.50)
MA	83.27(−6.77)	88.16(−4.87)	90.09(+0.12)	88.14(+0.57)
EMA	84.48(−5.56)	89.93(−3.10)	91.20(+1.23)	88.35(+0.78)
FM	90.27(+0.23)	93.69(+0.66)	91.86(+1.89)	87.07(−0.50)
TM	89.90(−0.14)	93.18(+0.15)	88.49(−1.48)	87.13(−0.44)
RM	90.18(+0.14)	92.83(−0.20)	89.74(−0.23)	88.73(+1.16)
TFAugment	92.89(+2.85)	94.22(+1.19)	93.79(+3.82)	89.65(+2.08)

PAMAP2 dataset, which has a higher sampling rate, these methods still managed to retain essential pattern information, resulting in positive effects.

In the time-frequency domain, Frequency Masking (FM) and Time Masking (TM) showed inconsistent results across all datasets, with about half of the outcomes harming model performance. This inconsistency may be attributed to their randomness masking critical frequencies and time segments, leading to erroneous pattern learning by the model. Random Masking (RM) had a negligible impact, which we consider within the normal range of training error, without significantly altering the learned patterns.

TFAugment, proposed in this paper, selectively augments data by leveraging the importance distribution across frequencies. This targeted approach minimizes the masking of crucial frequencies compared to random frequency masking, resulting in adaptive enhancements that lead to stable performance gains. With this method, we observed a maximum improvement of 3.51% for DilatedConvNet and up to 3.82% for GRU networks across the four datasets.

Overall, our method demonstrates a clear advantage over traditional augmentation techniques, offering stable improvements in complex HAR tasks and compatibility with multiple mainstream networks across different architectures.

4.4 Discussion

Why Has TFAugment's Performance Improved? To understand TFAugment's performance boost, we analyzed its effect on the MHEALTH dataset's test results. The confusion matrices in Fig. 3 compare DilatedConvNet with and without TFAugment. We note a 10% accuracy improvement for Knees bending (crouching) with TFAugment. It also more accurately distinguishes Jogging from Running by focusing on the frequency of movement-increasing Jogging accuracy by 18% and Running by 6%, although some misclassifications as Jump front & back remain. By implementing TFAugment, we adaptively masked the less important aspects between the two actions, enabling the model to better capture the patterns of significant activity frequencies.

Can Combined Use of TFAugment with Other Methods Enhance Performance Further? We conducted further experiments on the MHEALTH dataset. After performing time-frequency transformation on the data, we expanded a portion of the data using Time Warping (TW) and performed an equal level of expansion using TFAugment. These augmentations were implemented on the baseline network DilatedConvNet. The final accuracy achieved was 95.22%, which is 0.44% higher than using TFAugment alone and 1.79% higher than using TW alone. These results suggest that the classification network's accuracy can be further enhanced by compound use of effective data augmentation methods.

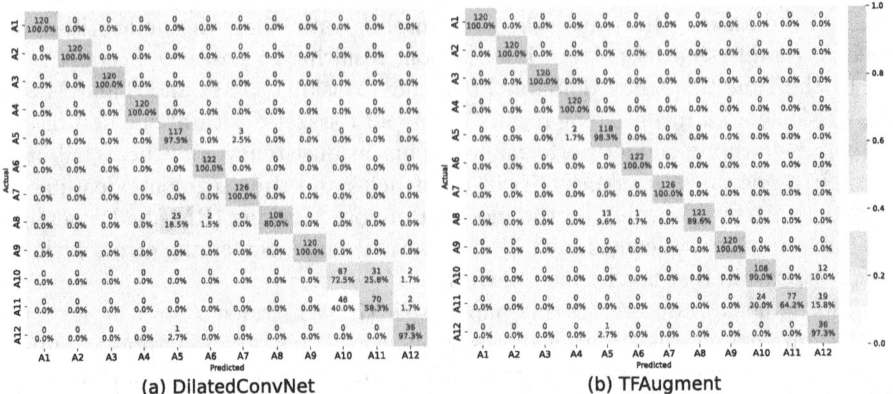

Fig. 3. Comparison of confusion matrices. Subfigure (a) shows the confusion matrix for the baseline DilatedConvNet model, while subfigure (b) illustrates the confusion matrix results of applying our proposed TFAugment method on top of the baseline. The horizontal axis represents the predicted labels, while the vertical axis corresponds to the true labels. The color intensity represents the proportion of each classification outcome.

5 Conclusion

In this paper, we introduce our data augmentation method, Time-Frequency Augmentation (TFAugment). By extracting the importance distribution in the frequency domain after time-frequency transformation, we innovatively adopt a two-stage training approach, enabling TFAugment to enhance data with minimal semantic alteration. Experimental results on four datasets demonstrate that our proposed method exhibits superior performance, achieving steady improvements compared to baseline models and other augmentation techniques. Furthermore, our subsequent analysis validates that our approach can be integrated with other data augmentation methods to further enhance performance, demonstrating robustness and adaptability.

Acknowledgements. This work was supported in part by the National Natural Science Foundation of China under Grant 61972055, in part by the Natural Science Foundation of Hunan Province under Grant 2021JJ30734, and in part by the Postgraduate Scientific Research Innovation Project of Hunan Province under Grant CX20220956.

References

1. Alawneh, L., Alsarhan, T., Al-Zinati, M., Al-Ayyoub, M., Jararweh, Y., Lu, H.: Enhancing human activity recognition using deep learning and time series augmented data. J. Ambient Intell. Humaniz. Comput. 1–16 (2021)
2. Altun, K., Barshan, B., Tunçel, O.: Comparative study on classifying human activities with miniature inertial and magnetic sensors. Pattern Recogn. **43**(10), 3605–3620 (2010)

3. Banos, O., et al.: mHealthDroid: a novel framework for agile development of mobile health applications. In: Pecchia, L., Chen, L.L., Nugent, C., Bravo, J. (eds.) IWAAL 2014. LNCS, vol. 8868, pp. 91–98. Springer, Cham (2014). https://doi.org/10.1007/978-3-319-13105-4_14
4. Breiman, L.: Random forests. Mach. Learn. **45**, 5–32 (2001)
5. Chen, K., Zhang, D., Yao, L., Guo, B., Yu, Z., Liu, Y.: Deep learning for sensor-based human activity recognition: overview, challenges, and opportunities. ACM Comput. Surv. (CSUR) **54**(4), 1–40 (2021)
6. Chen, S., Yang, X., Chen, Y., Yu, H., Cai, H.: Uncertainty-based fusion network for automatic skin lesion diagnosis. In: 2022 IEEE International Conference on Bioinformatics and Biomedicine (BIBM), pp. 1487–1492. IEEE (2022)
7. Chen, Y., Zhong, K., Zhang, J., Sun, Q., Zhao, X.: LSTM networks for mobile human activity recognition. In: 2016 International Conference on Artificial Intelligence: Technologies and Applications, pp. 50–53. Atlantis Press (2016)
8. Gao, Z., Li, L., Xu, T.: Data augmentation for time-series classification: An extensive empirical study and comprehensive survey. arXiv preprint arXiv:2310.10060 (2023)
9. Huang, G.B., Zhu, Q.Y., Siew, C.K.: Extreme learning machine: theory and applications. Neurocomputing **70**(1–3), 489–501 (2006)
10. Iglesias, G., Talavera, E., González-Prieto, Á., Mozo, A., Gómez-Canaval, S.: Data augmentation techniques in time series domain: a survey and taxonomy. arXiv preprint arXiv:2206.13508 (2022)
11. Jalal, A., Kim, Y.H., Kim, Y.J., Kamal, S., Kim, D.: Robust human activity recognition from depth video using spatiotemporal multi-fused features. Pattern Recogn. **61**, 295–308 (2017)
12. Jeong, C.Y., Shin, H.C., Kim, M.: Sensor-data augmentation for human activity recognition with time-warping and data masking. Multimedia Tools Appl. **80**, 20991–21009 (2021)
13. Laput, G., Harrison, C.: Sensing fine-grained hand activity with smartwatches. In: Proceedings of the 2019 CHI Conference on Human Factors in Computing Systems, pp. 1–13 (2019)
14. Li, X., He, Y., Fioranelli, F., Jing, X.: Semisupervised human activity recognition with radar micro-doppler signatures. IEEE Trans. Geosci. Remote Sens. **60**, 1–12 (2021)
15. Mollyn, V., Ahuja, K., Verma, D., Harrison, C., Goel, M.: Samosa: sensing activities with motion and subsampled audio. Proc. ACM Interact. Mob. Wearable Ubiquit. Technol. **6**(3), 1–19 (2022)
16. Park, D.S., et al.: Specaugment: a simple data augmentation method for automatic speech recognition. arXiv preprint arXiv:1904.08779 (2019)
17. Qian, H., Tian, T., Miao, C.: What makes good contrastive learning on small-scale wearable-based tasks? In: Proceedings of the 28th ACM SIGKDD Conference on Knowledge Discovery and Data Mining, pp. 3761–3771 (2022)
18. Quinlan, J.R.: Induction of decision trees. Mach. Learn. **1**, 81–106 (1986)
19. Reiss, A., Stricker, D.: Introducing a new benchmarked dataset for activity monitoring. In: 2012 16th International Symposium on Wearable Computers, pp. 108–109. IEEE (2012)
20. Rish, I., et al.: An empirical study of the Naive Bayes classifier. In: IJCAI 2001 Workshop on Empirical Methods in Artificial Intelligence, vol. 3, pp. 41–46 (2001)

21. Sztyler, T., Stuckenschmidt, H.: On-body localization of wearable devices: an investigation of position-aware activity recognition. In: 2016 IEEE International Conference on Pervasive Computing and Communications (PerCom), pp. 1–9. IEEE (2016)
22. Wright, R.E.: Logistic regression (1995)
23. Xi, R., Hou, M., Fu, M., Qu, H., Liu, D.: Deep dilated convolution on multimodality time series for human activity recognition. In: 2018 International Joint Conference on Neural Networks (IJCNN), pp. 1–8. IEEE (2018)
24. Yang, J., Nguyen, M.N., San, P.P., Li, X., Krishnaswamy, S.: Deep convolutional neural networks on multichannel time series for human activity recognition. In: IJCAI, Buenos Aires, Argentina, vol. 15, pp. 3995–4001 (2015)
25. Yao, S., Hu, S., Zhao, Y., Zhang, A., Abdelzaher, T.: Deepsense: a unified deep learning framework for time-series mobile sensing data processing. In: Proceedings of the 26th International Conference on World Wide Web, pp. 351–360 (2017)

Dfp-Unet: A Biomedical Image Segmentation Method Based on Deformable Convolution and Feature Pyramid

Zengzhi Yang, Yubin Wei, Xiao Yu, and Jinting Guan(✉)

Department of Automation, Xiamen University, Xiamen, China
jtguan@xmu.edu.cn

Abstract. U-net is a classic deep network framework in the field of biomedical image segmentation, which uses a U-shaped encoder and decoder structure to realize the recognition and segmentation of semantic features, but only uses the last layer of the decoder structure for the final prediction, ignoring the feature maps of different levels of semantic strength. In addition, the convolution kernel size used by U-net is fixed, which is poorly adaptable to unknown changes. Therefore, we propose Dfp-Unet based on **d**eformable convolution and **f**eature **p**yramid for biomedical image segmentation. Dfp-Unet uses the idea of feature pyramid to respectively add an additional independent path including convolution and up-sampling operations to each level of the decoder. Then, the output feature maps of all levels are concatenated to obtain the final feature map containing multiple levels of semantic information for final prediction. Besides, Dfp-Unet replaces the convolution in the down-sampling modules with a deformable convolution on the basis of U-net. To verify the performance of Dfp-Unet, four image segmentation data sets including Sunnybrook, ISIC2017, Covid19-ct-scans, and ISBI2012 are used to compare Dfp-Unet with the existing convolutional neural networks (U-net and U-net++), and the experimental results show that Dfp-Unet has high segmentation accuracy and generalization performance.

Keywords: Biomedical Image Segmentation · Dfp-Unet · U-net · Feature Pyramid · Deformable Convolution

1 Introduction

Biomedical imaging refers to the acquisition of internal biological tissue images by researchers through non-invasive ways, providing significant assistance to doctors. The demands of image analysis mainly include the detection and segmentation of target areas such as lesions, organs, cells and fluorescent spots, as well as the classification of pathological states within the images [1]. The development of medical image segmentation algorithms maximizes the presentation of information in images to doctors, reduces human subjective errors, and lowers misdiagnosis rates. What's more, the advancement of image segmentation algorithms has facilitated the development of computer-aided diagnosis systems, driving medical diagnostic systems towards higher efficiency and

D.-N. Yang et al. (Eds.): PAKDD 2024, LNAI 14648, pp. 297–309, 2024.
https://doi.org/10.1007/978-981-97-2238-9_23

accuracy [2]. Currently, biomedical image interpretation heavily relies on professionals, making it difficult to meet the demand for automated diagnosis in the era of big data. Medical imaging is characterized by significant differences in image quality, complexities in image category patterns, difficulties in constructing databases, and great annotation difficulty [3]. Due to these characteristics, the segmentation of medical images often faces considerable challenges.

In recent years, deep learning methods have been successfully applied in the field of image segmentation, particularly convolutional neural networks (CNNs). Compared with traditional segmentation algorithms such as edge-based, region-based, and morphological methods, deep learning methods exhibit significant performance advantages [4]. CNNs were initially designed for image classification tasks. In 2012, Geoffrey Hinton and Alex proposed AlexNet [5], a CNN architecture for image classification. In 2015, Jonathan Long applied fully convolutional neural network (FCNs) to image segmentation tasks by replacing the fully connected layers with convolutional layers and using deconvolution or up-sampling operations in the output layer to obtain results with the same size as the original images [6]. Yuan et al. utilized an end-to-end approach to train a convolutional neural network and designed an FCN to effectively address the segmentation of skin disease lesions [7].

As the development of FCNs progressed, research on encoder-decoder networks reached a climax, particularly in the case of U-net. In 2015, Olaf Ronneberger proposed the U-net, which perfectly solved the problem of segmenting cell edges [8]. Since then, numerous methods based on U-net have emerged. Ariel H. Curiale and Flavio D. Colavecchia proposed RU-Net (Deep Residual U-net Network) model by incorporating ResNet's structure into U-net and used it for myocardial segmentation tasks [9]. Zhou et al. introduced U-net++, which consists of U-net and the skip connections between the encoder and decoder to maintain the same resolution [10]. Subsequently, many networks have been developed as improvements upon U-net, such as R2U-Net [11], Multi Res U-Net [12], H-Dense U-Net [13], among others.

U-net, with its encoder-decoder structure, has gradually become the mainstream CNN in the field of image segmentation. However, U-net also has some shortcomings. Firstly, U-net overlooks the impact of different-scale feature maps, as only the last layer's convolutional result is used for segmentation during the up-sampling process. So, the modules of different scales in the decoder pathway cannot share features, and information between different layers cannot flow within the network, resulting in a considerable number of redundant parameters. Then U-net uses fixed convolutional kernel sizes, leading to poor adaptability to unknown variations [14].

To improve U-net, we propose Dfp-Unet for two-dimensional biomedical image segmentation, which incorporates the ideas of feature pyramid network (FPN) [15] and deformable convolution (DC) [16, 17]. The key idea of FPN is to extract features separately from each level of the decoder pathway and concatenate them, thus enabling the flow of information between modules of different scales in the decoder pathway [18]. Additionally, Dfp-Unet replaces the convolutions used in U-net's down-sampling process with DC, expanding the receptive field of convolutional kernel and allowing more accurate feature extraction from the images. The performance of Dfp-Unet is evaluated using Sunnybrook, ISIC2017, Covid19-ct-scans, and ISBI2012 datasets. It shows that Dfp-Unet shows superior performance relative to U-net and U-net++ to some extent.

2 Dfp-Unet

2.1 Feature Pyramid Network

FPN [15] can obtain feature maps in different scales, effectively combining these feature maps to solve the segmentation task. The input image propagates through two paths: the "bottom-up path" and the "top-down path", as shown in Fig. 1. To construct a pyramid-like structure, in the "bottom-up path", the feature maps are gradually down-sampled from bottom to top to obtain feature maps with different resolutions. The "bottom-up path" includes convolution modules named C1, C2, C3, C4, and C5. As the input image passes through the network, the resolution of the image is reduced by half in each step. They correspond to stride values of 2, 4, 8, 16, and 32. The "top-down path" uses up-sampling operations to increase the resolution of high-level semantic features and generate higher-resolution feature maps. Then, these feature maps are used to recover the detailed information of images. After that, the "bottom-up" feature maps are merged with the "top-down" feature maps using skip connections. This merging is achieved by using 1×1 convolutions to combine information from both paths. Then, the resulting feature maps F2, F3, F4, and F5 are processed using 3×3 convolutions to produce the final predictions P2, P3, P4, and P5. With this approach, the feature map of each layer will give a prediction result. Feature maps with higher resolutions will generate results containing more detailed information, while feature maps with lower resolutions will produce results with more semantic features.

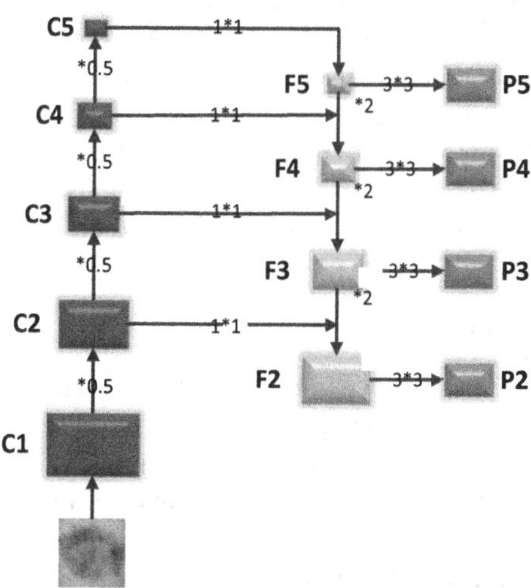

Fig. 1. Structure diagram of FPN.

2.2 Deformable Convolution

The 2D deformable convolution mainly consists of two parts. For the feature map x, it is sampled according to the predefined *Rule*, and then the samples are weighted and summed using the weights w. The rule *Rule* is represented as follows:

$$Rule = \{(-1, -1), (-1, 0), (0, 0) \ldots, (0, 1), (1, 1)\} \tag{1}$$

The deformable convolution introduces an offset $\Delta P_n : \{\Delta P_n | n = 1, \ldots N\}, N = |Rule|$, for each position P_0 on the output feature map y. The output formula is given by:

$$y(P_0) = \Sigma_{P_n \in Rule} w(P_n) x(P_0 + P_n + \Delta P_n) \tag{2}$$

where $x(P_0 + P_n + \Delta P_n)$ represents the output of the sampling point with the added offset from the previous layer, while $y(P_0)$ represents the output at position P_0, and $w(P_n)$ represents the corresponding weights. The sampling of $y(P_0)$ is done at irregular offset positions, and since $P_n + \Delta P_n$ values are usually not integers, bilinear interpolation is used to replace them:

$$x(P) = \Sigma_Q G(Q, P) x(Q) \tag{3}$$

where P represents an arbitrary position $P = P_0 + P_n + \Delta P_n$, Q represents all the integral positions in the feature map, and $G(.,.)$ represents the bilinear interpolation kernel. Since this is a 2D operation, it can be split into two 1D kernels:

$$G(Q, P) = g(Q_x, P_x) g\left(Q_y, P_y\right) \tag{4}$$

where g is a 1D function $g(a, b) = \max(0, 1 - |a - b|)$. Therefore, $G(Q, P)$ is only non-zero for a few Q, making the computation time of Eq. (3) relatively low.

The functionality of DC is illustrated in Fig. 2. An input feature map undergoes a convolution operation to obtain the offset. The scale of this convolution kernel matches the spatial resolution of the current convolution layer. The output offset has the same spatial dimensions as the input feature map, and the number of channels in the offset is $2N$, corresponding to the 2D offsets in the x and y directions. The offset parameters are mainly trained through gradient descent algorithm based on Eqs. (3) and (4).

2.3 Structure of Dfp-Unet

The architecture of Dfp-Unet is shown in Fig. 3. Dfp-Unet retains the complete structure of U-net's encoder and decoder and integrates different-level feature maps in the decoder using additional paths. The network includes four up-sampling and down-sampling operations, resulting in a depth of five layers. In each layer of the decoder, a simple independent path is added, consisting of a 3×3 convolutional layer and an up-sampling layer. The outputs of the encoder are denoted as $up^i (i = 1, 2, \ldots, 5)$. After passing through convolution and up-sampling operations, these feature maps are transformed into five feature maps with the same channel size (64 channels each) and scale. All the feature maps are horizontally concatenated to form a final feature map with 64×5

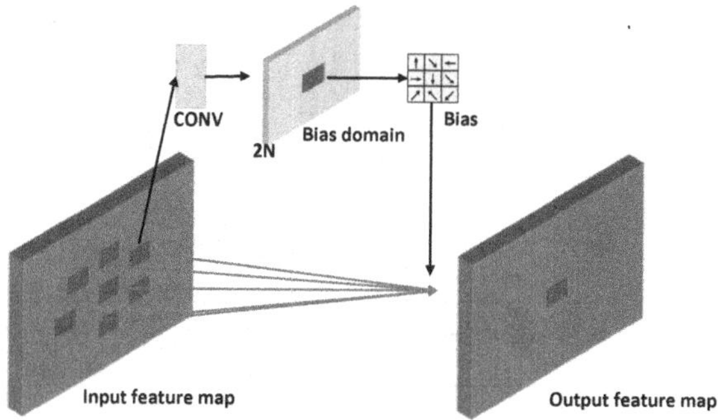

Fig. 2. Diagram of DC.

channels. This final feature map is passed through a 1 × 1 convolutional filter to obtain the predicted output image of the network.

The formula for the convolution layer is:

$$x_{ij}^{l} = \Sigma_{a=0}^{m-1}\Sigma_{b=0}^{m-1}W_{ab}y_{(i+a)(j+b)}^{l-1} \tag{5}$$

The convolutional kernel size within the formula is $m \times m$, and its weight is denoted as W_{ab}. l represents the layer number, y^{l-1} denotes the output from the previous layer, and x^{l} represents the output of the current layer. i and j represent the current position in the feature map. Thus, $y_{(i+a)(j+b)}^{l-1}$ represents the corresponding part of the output from the $(l-1)$-th layer that aligns with the convolutional kernel. By performing a summation with the weight W_{ab}, we obtain the output x_{ij}^{l} for the current position in the l-th layer.

The $ReLU$ activation function uses a simple non-saturating function defined as $ReLU(x) = max(0, x)$. Therefore, the output y_{ij}^{l} for the l-th layer can be expressed as:

$$y_{ij}^{l} = ReLU(x_{ij}^{l}) \tag{6}$$

Since the convolutional kernel size in the final fully connected layer is 1 × 1, the convolutional kernel size m in Eqs. (5) and (6) is 1, and l is equal to L, where L is the index of the last layer. Therefore, the output y_{ij}^{L} of the last layer is given by:

$$y_{ij}^{L} = ReLU(w_{00}y_{ij}^{L-1}) \tag{7}$$

Here, w_{00} represents the weights associated with the 1 × 1 convolutional kernel in the fully connected layer.

Regarding the output y_{ij}^{L-1} of the second-to-last layer, it is obtained by concatenating five additional paths. Therefore, y_{ij}^{L-1} can be expressed as:

$$y_{ij}^{L-1} = concat\left[conv\left(up^{5}, 64\right), \ldots conv\left(up^{1}, 64\right)\right] \tag{8}$$

Here, $concat[\ldots]$ represents the concatenation operation, and $conv\left(up^{i}, 64\right)$ denotes applying a 3 × 3 convolutional operation and up-sampling to the feature maps up^{i}, resulting in feature maps with 64 channels.

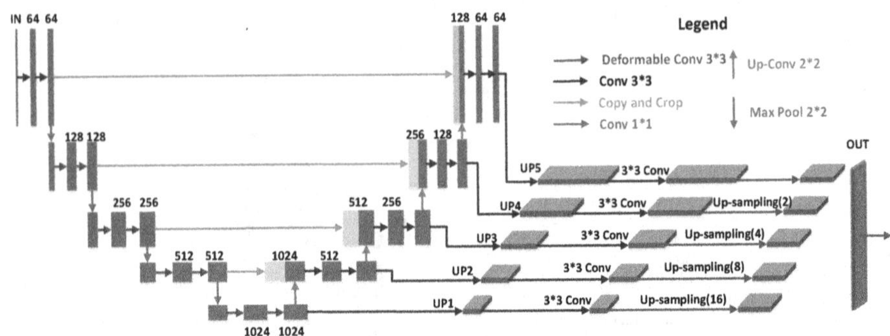

Fig. 3. Diagram of Dfp-Unet

Dfp-Unet fully incorporates all the ideas of U-net while adding additional paths to integrate feature maps of different scales in the decoder path. As a result, the final predictions of Dfp-Unet contain both deep-level features and shallow-level details. Additionally, Dfp-Unet replaces all convolutional operations in the down-sampling process with deformable convolutional operations. By using deformable convolutions, the receptive field of the convolutional kernel is enlarged, allowing the network to capture more fine-grained image details while extracting high-level semantic features. This improvement enhances the accuracy of target recognition and localization tasks. It is worth noting that deformable convolutional operations involve a larger number of trainable parameters, resulting in longer training times. However, during the down-sampling process, skip connections are utilized to copy feature maps to the up-sampling process, facilitating information sharing. By selectively replacing the convolutional modules with DC rather than replacing all convolutional modules, the algorithm achieves improved performance while minimizing the training time.

3 Experimental Data and Process

3.1 Description of Datasets

We validated the performance of the proposed image segmentation model on four datasets, each representing a different type of biomedical tissues and organs: Sunnybrook Left Ventricle Segmentation Task, ISIC2017 Skin Disease Segmentation Dataset, Covid19-ct-scans Lung Segmentation Dataset, and ISBI Cell Segmentation Dataset. The Sunnybrook Left Ventricle Segmentation Task [19] was derived from the Sunnybrook Cardiac Left Ventricle Segmentation, comprising MRI images of 45 different patients' hearts at the ES and ED phases. It contains 10,418 experimental datasets. The ISIC2017 dataset [20] includes 2000 images, covering skin disease types such as 18.7% melanoma, 12.7% seborrheic keratosis, and 68.6% nevi. In this study, all input images were resized to 256 × 256 pixels while preserving the aspect ratio. The Covid19-ct-scans dataset consists of lung CT scan data. After the COVID-19 outbreak in 2019, coronacases.org and radiopaedia.org collected complete chest CT scans of 10 confirmed COVID-19 cases, along with segmentation annotations. The ISBI2012 Cell Tracking Challenge dataset was released by the ISBI organization in 2012 and is available at http://brainiac2.mit.edu/isbi_challenge/. The training data includes 30 images of 512 × 512.

3.2 Experimental Environment and Process

Our experiments relied on the following hardware setup: an NVIDIA Quad P2200 GPU with 128GB of VRAM and an Intel(R) Xeon(R) Silver 4214 CPU. The experiments were conducted in the Python 3.6 environment with PyTorch version 1.10.2 + cu113.

First, we used bilinear interpolation to uniformly scale the input images. The scaled images were then overlaid onto a fixed-size black background image, ensuring uniformity in input sizes to adapt to the network. By the way, the experiments demonstrated that this scaling process did not affect the object contours and edge information.

The data augmentation methods we used included rotation transformation, horizontal and vertical translations, flipping, shearing, and scaling. These transformations were applied to the existing images to generate augmented versions of the data, effectively expanding the dataset.

Then, the training dataset and the test dataset were randomly split in a 70% to 30% ratio, respectively. Within the training dataset, a random subset of 10% of the data was selected as the validation set for performing validation during the training process.

We adopted an end-to-end training approach. For classification problems, standard cross-entropy is used as the loss function [14]:

$$L = -\frac{1}{N}\Sigma_{i=1}^{N}\Sigma_{c=1}^{C} y_{ic} log \hat{y}_{ic} \tag{9}$$

C represents the total number of classes, N is the number of samples, y_{ic} represents the true label value of the i-th input sample for the c-th class, and $\hat{y}_{ic}$ represents the predicted probability of the model that the i-th input sample belongs to the c-th class.

We used the Root Mean Square Propagation (RMSP) algorithm to update the network parameters. The initial value of the learning rate for this experiment was set to 0.0001. Each network was trained for 100 epochs. Additionally, other loss functions such as Tversky loss function or distance-based loss function can be used [21].

After inputting the test set image into the network, the network outputted the prediction probability of each pixel belonging to the target class. The ROC curve was used to determine the segmentation threshold according to the network's output probability. After obtaining the segmentation results, performance metrics were calculated for each image in the test set, including Dice coefficient, Jaccard coefficient, sensitivity and specificity. The average of these evaluation metrics was then taken as the performance measure for the test set in this experiment.

For each network, three independent repeated experiments were conducted. During each of the three independent repetitions, the test set data remained unchanged. The best-performing result among the three independent repetitions was considered as the final performance measure for the algorithm.

4 Experimental Results and Analysis

Sunnybrook Dataset. We compared Dfp-Unet with two existing models (U-net and U-net++). The segmentation results of the Sunnybrook dataset are shown in Fig. 4(a)–(e) which respectively represent the original image, manually annotated image and segmentation results of U-net, U-net++, and Dfp-Unet. It can be found that the segmentation result of Dfp-Unet is closer to the annotated result in shape. Figure 4 (f) and (g) illustrate the Dice coefficient curves on the validation set during the training process and the ROC curve on the test set. Dfp-Unet demonstrates faster convergence to a stable level compared with the other two models, indicating that Dfp-Unet has a faster convergence speed and segmentation accuracy. Table 1 presents the experimental results of Dfp-Unet compared with U-net and U-net++. It can be observed that Dfp-Unet achieves the highest accuracy in terms of the Dice coefficient (91.57%) and Jaccard coefficient (86.35%), ranking first among all three models. In terms of sensitivity and specificity, Dfp-Unet ranks second among all three models.

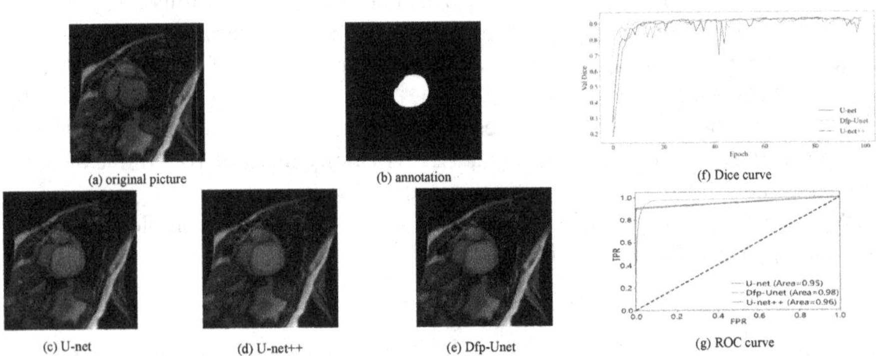

Fig. 4. Image segmentation results of Sunnybrook dataset. (a)–(e) represent the original image, manually annotated image and segmentation results of different models. (f) represents the Dice curve of the validation set and (g) represents the ROC curve of the test set.

Table 1. Results of Sunnybrook dataset.

Method	Indicators of Sunnybrook dataset			
	Dice	Jaccard	Sensitivity	Specificity
U-net	0.8942	0.8332	0.8637	0.9989
U-net++	0.9004	0.8423	0.9157	0.9974
Dfp-Unet	0.9157	0.8635	0.9011	0.9989

ISIC2017 Dataset. The segmentation results of the ISIC2017 dataset are shown in Fig. 5(a)–(e). The segmented contours of Dfp-Unet are more in line with the actual results. Table 2 presents the performance metrics, indicating that the Dfp-Unet achieves the highest accuracy in terms of Dice coefficient and Jaccard coefficient. It slightly underperforms the U-net model in sensitivity and U-net++ model in specificity. In the Dice coefficient curve and ROC curve shown in Fig. 5 (f) and (g), U-net++ achieves the highest Dice coefficient on the validation set, followed by Dfp-Unet. However, on the test set, the performance of U-net++ is lower than Dfp-Unet, indicating that U-net++ suffers from overfitting, while Dfp-Unet exhibits better generalization performance.

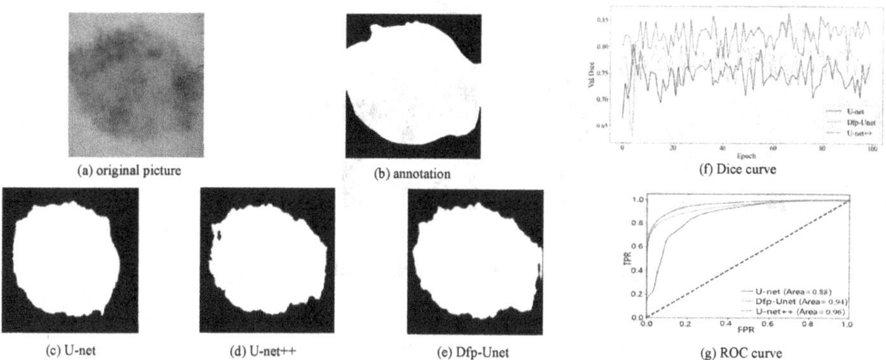

Fig. 5. Image segmentation results of ISIC2017 dataset.

Table 2. Results of ISIC2017 dataset.

Method	Indicators of ISIC2017 dataset			
	Dice	Jaccard	Sensitivity	Specificity
U-net	0.7982	0.6857	0.8424	0.9051
U-net++	0.8293	0.7287	0.7767	0.9826
Dfp-Unet	0.8414	0.7476	0.8389	0.9534

Covid19-ct-Scans Dataset. The segmentation results of the Covid19-ct-scans dataset are presented in Fig. 6(a)–(e). Table 3 shows the evaluation metrics for the Covid19-ct-scans dataset. Dfp-Unet slightly outperforms the other two networks in segmentation performance. Similarly, analyzing the Dice coefficient curve and ROC curve shown in Fig. 6(f) and (g), it can be observed that Dfp-Unet exhibits better generalization performance and is less prone to overfitting compared to the other models.

ISBI2012 Dataset. The results are shown in Fig. 7(a)–(e). Table 4 indicates that Dfp-Unet achieves the highest performance level on the ISBI2012 dataset, outperforming the other networks by 1–5 percentage points. Analyzing the Dice coefficient curve and ROC curve shown in Fig. 7(f) and (g), it can be observed that U-net and U-net++ performed well on the validation set but showed lower performance on the test set. On the other hand, the proposed Dfp-Unet effectively prevents overfitting and demonstrates higher generalization performance.

Overall, Dfp-Unet consistently achieves higher Dice and Jaccard coefficients compared to U-net and U-net++. This indicates that the results are more similar to the truth. In addition, for datasets with limited training data, such as the ISBI2012 dataset, U-net

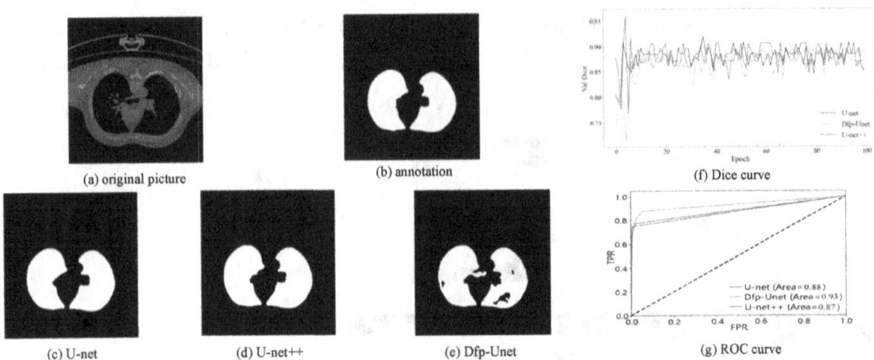

Fig. 6. Image segmentation results of Covid19-ct-scans dataset.

Table 3. Results of Covid19-ct-scans dataset.

Method	Indicators of Covid19-ct-scans dataset			
	Dice	Jaccard	Sensitivity	Specificity
U-net	0.8842	0.8662	0.9680	0.9980
U-net++	0.8856	0.8681	0.9733	0.9979
Dfp-Unet	0.8976	0.8707	0.9407	0.9966

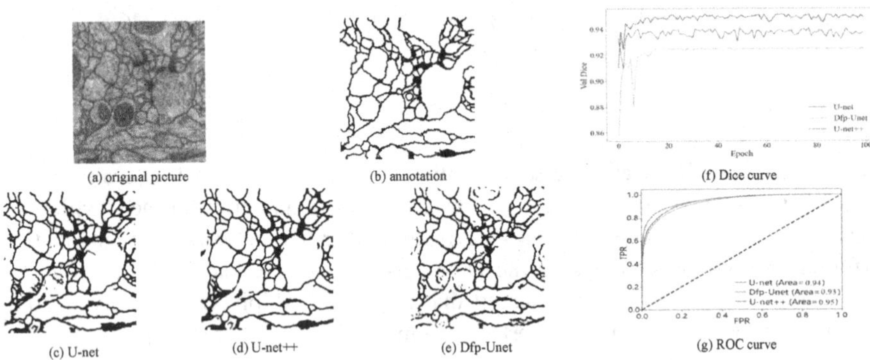

Fig. 7. Image segmentation results of ISBI2012 dataset.

and U-net++ are more prone to overfitting on the training set. However, DC of Dfp-Unet which introduces offsets as trainable parameters significantly improves the network's flexibility, making it less susceptible to overfitting on the training set. Dfp-Unet demonstrates its effectiveness for biomedical image segmentation.

Table 4. Results of ISBI2012 dataset.

Method	Indicators of ISBI2012 dataset			
	Dice	Jaccard	Sensitivity	Specificity
U-net	0.9140	0.8479	0.8850	0.8441
U-net++	0.9232	0.8631	0.9006	0.8454
Dfp-Unet	0.9267	0.8641	0.9344	0.8067

5 Discussion

U-net has significant advantages for biomedical image segmentation. However, it suffers from issues such as feature information blocking leading to parameter redundancy, and fixed convolutional kernel sizes, leading to limited generalization ability. We proposed Dfp-Unet upon U-net, incorporating the idea of FPN by adding extra independent pathways and introducing DC. Dfp-Unet utilizes the concept of feature pyramids to extend the structure of U-net, enabling it to merge all feature information from the up-sampling pathways for the final prediction process. Additionally, the application of DC allows the network to extract more complex image features and varying edge information. As a result, Dfp-Unet achieves higher segmentation accuracy and generalization performance, demonstrating satisfactory segmentation results on different types of biomedical image data.

By using four datasets, this study compares the performance metrics of Dfp-Unet with U-net and U-net++. The results indicate that the proposed Dfp-Unet outperforms U-net and U-net++ and exhibits higher generalization performance, showing less susceptibility to overfitting. Furthermore, the experiments reveal that Dfp-Unet excels at handling low-resolution biomedical images, such as left ventricle and cell contours. It effectively addresses the challenge of low dataset resolution, producing segmentation results closely resembling actual image contours. This capability enables Dfp-Unet to assist medical professionals in making prompt judgments and providing diagnostic support in clinical settings, showcasing practical applications.

Acknowledgments. This work was supported by National Science and Technology Major Project (No. 2021ZD0112600), National Natural Science Foundation of China (61803320) and Natural Science Foundation of Fujian Province of China (2022J05012).

Disclosure of Interests. The authors have no competing interests to declare that are relevant to the content of this article.

References

1. Xu, Y.-Y., Yao, L.-X., Shen, H.-B.: Bioimage-based protein subcellular location prediction: a comprehensive review. Front. Comp. Sci. **12**, 26–39 (2018)

2. Yidong, C., Qin, Z., lan, L., Li, P., Jin, Y.: A review of deep convolutional neural networks in medical image segment. Chin. J. Health Inf. Manag. (2021)
3. Seo, H., et al.: Machine learning techniques for biomedical image segmentation: an overview of technical aspects and introduction to state-of-art applications. Med. Phys. **47**(5), e148–e167 (2020)
4. Pang, S., Du, A., Orgun, M.A., Yu, Z.: A novel fused convolutional neural network for biomedical image classification. Med. Biol. Eng. Comput. **57**(1), 107–121 (2019)
5. Krizhevsky, A., Sutskever, I., Hinton, G.E.: ImageNet classification with deep convolutional neural networks. Commun. ACM **60**(6), 84–90 (2017)
6. Long, J., Shelhamer, E., Darrell, T.: Fully convolutional networks for semantic segmentation. In: Fully Convolutional Networks for Semantic Segmentation, pp. 3431–3440. IEEE (2015)
7. Yuan, Y., Chao, M., Lo, Y.-C.: Automatic skin lesion segmentation using deep fully convolutional networks with jaccard distance. IEEE Trans. Med. Imaging **36**(9), 1876–1886 (2017)
8. Ronneberger, O., Fischer, P., Brox, T.: U-net: convolutional networks for biomedical image segmentation. In: Navab, N., Hornegger, J., Wells, W., Frangi, A. (eds.) U-net: Convolutional Networks for Biomedical Image Segmentation, pp. 234–241. Springer, Cham (2015). https://doi.org/10.1007/978-3-319-24574-4_28
9. Curiale, A.H., Colavecchia, F.D., Kaluza, P., Isoardi, R.A., Mato, G.: Automatic myocardial segmentation by using a deep learning network in cardiac MRI. In: Automatic Myocardial Segmentation by Using a Deep Learning Network in Cardiac MRI, pp. 1–6. IEEE (2017)
10. Zhou, Z., Siddiquee, M.M.R., Tajbakhsh, N., Liang, J.: UNet++: redesigning skip connections to exploit multiscale features in image segmentation. IEEE Trans. Med. Imaging **39**(6), 1856–1867 (2020)
11. Alom, M.Z., Hasan, M., Yakopcic, C., Taha, T.M., Asari, V.K.: Recurrent residual convolutional neural network based on u-net (r2u-net) for medical image segmentation, arXiv preprint (2018)
12. Ibtehaz, N., Rahman, M.S.: MultiResUNet: rethinking the U-Net architecture for multimodal biomedical image segmentation. Neural Netw. **121**, 74–87 (2020)
13. Li, X., Chen, H., Qi, X., Dou, Q., Fu, C.-W., Heng, P.-A.: H-DenseUNet: hybrid densely connected UNet for liver and tumor segmentation from CT volumes. IEEE Trans. Med. Imaging **37**(12), 2663–2674 (2018)
14. Lei, T., Wang, R., Zhang, Y., Wan, Y., Liu, C., Nandi, A.K.: DefED-Net: deformable encoder-decoder network for liver and liver tumor segmentation. IEEE Trans. Radiat. Plasma Med. Sci. **6**(1), 68–78 (2022)
15. Lin, T.-Y., Dollár, P., Girshick, R., He, K., Hariharan, B., Belongie, S.: Feature pyramid networks for object detection. In: Feature Pyramid Networks for Object Detection, pp. 2117–2125. IEEE (2017)
16. Dai, J., et al.: Deformable convolutional networks. In: Deformable Convolutional Networks, pp. 764–773. IEEE (2017)
17. Zhu, X., Hu, H., Lin, S., Dai, J.: Deformable convnets v2: more deformable, better results. In: Deformable Convnets v2: More Deformable, Better Results, pp. 9308–9316. IEEE (2019)
18. Moradi, S., et al.: MFP-Unet: a novel deep learning based approach for left ventricle segmentation in echocardiography. Phys. Med. **67**, 58–69 (2019)
19. Fonseca, C.G., et al.: The cardiac atlas project–an imaging database for computational modeling and statistical atlases of the heart. Bioinformatics **27**(16), 2288–2295 (2011)

20. Codella, N.C., et al.: Skin lesion analysis toward melanoma detection: a challenge at the 2017 international symposium on biomedical imaging (ISBI), hosted by the international skin imaging collaboration (ISIC). In: Skin Lesion Analysis Toward Melanoma Detection: A Challenge at the 2017 International Symposium on Biomedical Imaging (ISBI), hosted by the International Skin Imaging Collaboration (ISIC), pp. 168–172. IEEE (2018)
21. Karimi, D., Salcudean, S.E.: Reducing the hausdorff distance in medical image segmentation with convolutional neural networks. IEEE Trans. Med. Imaging **39**(2), 499–513 (2020)

ACHIM: Adaptive Clinical Latent Hierarchy Construction and Information Fusion Model for Healthcare Knowledge Representation

Gaohong Liu[1,2,3], Jian Ye[1,2,3](✉), and Borong Wang[1,2,3]

[1] Institute of Computing Technology, Chinese Academy of Sciences, Beijing 100190, China
{liugaohong21s,wangborong23s}@ict.ac.cn

[2] University of Chinese Academy of Sciences, Beijing 100049, China

[3] Beijing Key Laboratory of Mobile Computing and Pervasive Device, Beijing, China
jye@ict.ac.cn

Abstract. Utilize electronic health records (EHR) to forecast the likelihood of a patient succumbing under the current clinical condition. This assists healthcare professionals in identifying clinical emergencies promptly, enabling timely intervention to alter the patient's critical state. Existing healthcare prediction models are typically based on clinical features of EHR data to learn a patient's clinical representation, but they frequently disregard structural information in features. To address this issue, we propose **A**daptive **C**linical latent **H**ierarchy construction and **I**nformation fusion **M**odel (ACHIM), which adaptively constructs a clinical potential level without prior knowledge and aggregates the structural information from the learned into the original data to obtain a compact and informative representation of the human state. Our experimental results on real-world datasets demonstrate that our model can extract fine-grained representations of patient characteristics from sparse data and significantly improve the performance of death prediction tasks performed on EHR datasets.

Keywords: representation learning · healthcare prediction · adaptive graph generation · information fusion

1 Introduction

In recent years, as a result of the geriatric population and pandemics, people are more concerned about their own health issues in order to receive prompt treatment. Unlike the previous paradigm of "waiting for patients to develop illnesses before seeking medical care", timely healthcare prediction and treatment before the condition becomes severe can prevent irreversible illnesses. The swift adoption of EHR in modern hospitals has gathered a significant volume of "golden data", consequently facilitating various healthcare applications, including mortality prediction.

D.-N. Yang et al. (Eds.): PAKDD 2024, LNAI 14648, pp. 310–321, 2024.
https://doi.org/10.1007/978-981-97-2238-9_24

The objective of mortality prediction is to forecast whether a patient in the ICU will succumb after undergoing various sequences of physiological states. This prediction is based on the characteristics of multiple time series physiological indicators, such as Capillary Refill Rate, Diastolic Blood Pressure, Fraction Inspired Oxygen, and others. This constitutes a comprehensive process within the realm of smart medical treatment. By issuing early warnings for specific patient states in the initial stages and promptly administering appropriate medications in response to these states, the objective of reducing mortality is realized.

The crux of this task lies in extracting a representation of the human body state through clinical sequence indicators, followed by implementing predictions based on this derived representation. [12] proposed a novel health status representation model termed ConCare. This model independently embedded the time series of each feature to learn the representation of health status and performed health prediction by conducting a more in-depth analysis of the individual's health context. [23] expanded upon ConCare by incorporating patient similarities and attained state-of-the-art performance. In addition, previous research such as GRAM [2] and MIME [4] aimed to transform the inherent multilevel structure of EHR data into multilevel embeddings when the quantity of data was limited. [5] made use of transformers [20] to learn underlying structural information.

These works have achieved a good performance, but also ignored some hidden problems. ConCare ignored external information that could have aided the learning process and often existed implicitly in the real world. The patient correlations considered by GRASP [23] are based on raw physiological indicators and, in general, also do not include the underlying hierarchical structure between features. GRAM [2] and MIME [4] require prior labeling, which is impractical for large datasets. [5] still relied on external prior knowledge and rule-based initial conditional probabilities.

To address this problem, We introduce ACHIM, a model designed to automatically discover the latent hierarchical structure among medical features from EHR data and aggregate it with the original data to obtain a compact and informative representation of patient health status. ACHIM aims to establish a personal health representation and undertake death prediction tasks, thereby contributing to the comprehensive process of smart medical care. It can determine the likelihood of subsequent mortality based on existing physiological indicators, promptly adjust medication as needed, and enhance overall physiological health conditions.

Specifically, our model considers every feature to be a node in a graph and discovers the potential intrinsic structure in an adaptive manner, without the need for additional prior knowledge. We also use KL divergence to constrain the learning process of the potential structure. The original representation of the human body state derived from medical feature data lacks essential compact structural information. Through adaptive information aggregation, the problem of insufficient information can be effectively solved.

We demonstrate that our method yields comparable performance on large-scale datasets. In summary, the key contributions are as follows:

- We present ACHIM, an adaptive and precise end-to-end learning model for health state representations. The experimental results on three real-world datasets demonstrate that our model outperforms all baseline methods in a significant and consistent manner.
- ACHIM is a novel end-to-end model that takes the inherent relationships between features and hierarchical information into consideration. By adaptively reconstructing graph networks, it addresses the problem of effectively integrating hierarchical information and intrinsic associations in features, which previous models struggled with due to a lack of prior knowledge. Our method is applicable to the vast majority of existing datasets, thereby reducing model complexity and facilitating practical applications.
- Our model offers insight into analogous multi-feature convergent representation learning issues. In contrast to existing methods that rely on prior knowledge, our approach generates graph networks adaptively based on existing discrete features and aggregates a large quantity of hierarchical information through graph convolution to minimize the entropy between predicted and target information.

2 Related Work

Knowledge extraction from informative EHR data has become a prominent research topic in healthcare prediction tasks [10,14–17,22,24] in recent years. Deep learning models have shown superior performance in mortality prediction [12,18,19], patient subtyping [1], and diagnosis prediction [6,8,9,13]. Despite their distinctions, the essence of these healthcare tasks is to extract a representation of the patient's health state from the EHR data and then perform the prediction task using this representation. Currently, extraction of patient representation is predominantly divided into two situations:

(1) The status of the patient is determined exclusively by the inherent characteristics of the data. [3], for instance, proposed a two-level neural attention model that detects influential past visits and significant clinical variables within those visits to achieve high accuracy while remaining clinically interpretable. [1] learned a subspace decomposition of the cell memory, allowing time decay to disregard the memory content in accordance with the passage of time. [12] embeds feature sequences independently by modeling time-aware distribution and extracting feature interrelationships. [23] identifies similar patients with relevant information to supplement the EHR data's lack of information. These methods model individual state representations based on their inherent basic features to produce accurate predictions. In contrast, our model simultaneously employs natural data dependencies and inherent hierarchical information to discover the patient's status representation.

(2) The driving force behind advances in healthcare duties also includes external information. Due to the data collection procedure's accidental nature, there

are many missing values in medical data [21]. Some researchers have incorporated ontology structure information that does not exist in actual data in order to improve performance. For instance, [2] considered the frequency information of medical concepts in the EHR data and their ontology-related progenitors. Then, it optimized the medical concept information by adaptively combining them using an attention mechanism. Later, [11] improved on this approach by utilizing medical ontology information not just in the representation learning phase but also in the final prediction phase for residual fusion. Not all data, however, contain hierarchies. When structure information is unavailable, [5] suggests using graph convolutional networks to discover the concealed structure of each encounter. Inspired by this, our model employs fine-grained mappings of characteristics to learn structural information without prior probability, resulting in the best performance for representation learning.

3 Methods

Figure 1 provides an overview of the ACHIM model. We will give a brief introduction to the output at the end of each submodule. Each patient encounter (i.e., the t-th visit $\nu^{(t)}$) consists of three types of metrics: vital signs, laboratory values, and demographics. The real EHR data is shown in Table 1.

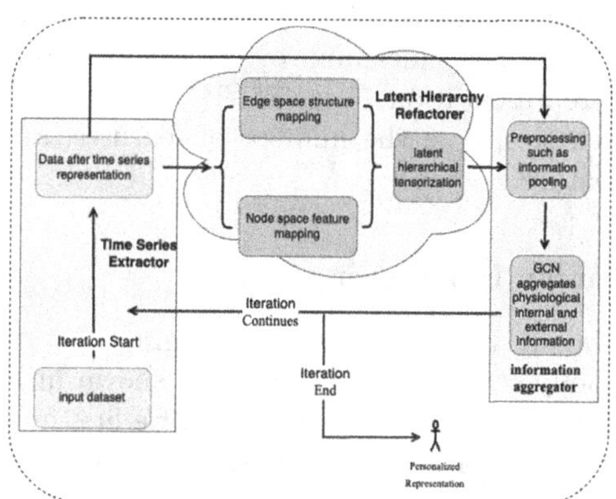

Fig. 1. The ACHIM model overview.

3.1 Time Series Extractor

Within EHR data, the varying time intervals necessitate accounting for the state influence in each preceding time interval while calculating the physiological representation pertaining to the current timestamp. Our observation indicated that timestamps further removed from the present have a reduced impact. Hence, we devised a negatively correlated time series model to address this issue.

Table 1. The basic format of the cleaned EHR dataset. (the empty space indicates that the physiological indicator features are missing at this time step)

Hours	Capillary refill rate	Diastolic blood pressure	...	Fraction inspired oxygen	Glascow coma scale eye opening
12.0			...		Spontaneously
12.06			...		
⋮	⋮	⋮	⋱	⋮	⋮
23.5		55.0	...		
24.0		56.0	...		

$$q_{n,T}^{(emb)} = W_q^{(emb)} \cdot C_{n,T}, \quad k_{n,t}^{(emb)} = W_k^{(emb)} \cdot C_{n,t}, \tag{1}$$

$$\xi = \beta \cdot \tan(\frac{q_{n,T}^{emb} \cdot k_{n,t}^{emb}}{\Delta t \cdot (\left\| q_{n,T}^{emb} \right\|_2 \cdot \left\| k_{n,t}^{emb} \right\|_2)}) \tag{2}$$

$q_{n,T}^{(emb)}$ represents the basic solution vector in the time series feature modeling process, and $k_{n,t}^{(emb)}$ is used to represent the original value vector that will be used to calculate the time interval correlation with the basic vector. Where β is a learnable parameter used to control the learning rate. The farther the current time is from the last timestamp, the larger the generated Δt value, and the smaller the resulting weight ξ. This is consistent with medical knowledge and effectively solves the time series problem. Finally, we multiply the physiological characteristics of the previous timestamp by the weight ξ and accumulate them to obtain a feature representation $C_i \in \mathbb{R}^{b \times n \times m}$ after time series extraction. Where b is the batch size, n is the number of features, and m is the vector dimension of each feature.

3.2 Latent Hierarchy Refactorer

We used latent hierarchy to supplement the original features to improve the learning of human body state representation. As shown in Fig. 2. Before delving into learning the latent hierarchical structure, we first map the physiological indicators into a high-dimensional semantic space, aiming to glean initial structural insights through their correlation. This correlation serves a dual purpose: initially integrating similar feature relationships and laying the groundwork for subsequent application of KL divergence to limit the shift in the latent structure. as shown below:

$$P_{i,k}^{(j)} = sim((C_i^{(j-1)} \cdot W_p^{(j)}), C_k^{(j-1)}) \tag{3}$$

$$C_i^{(j)*} = \sum_{k=1}^{n} P_{i,k}^{(j)} \cdot C_{i,k}^{(j-1)}. \tag{4}$$

where $W_p^{(j)}$ is a learnable parameter, $P_{i,k}^{(j)}$ is a correlation with the dual purpose, We further optimize the nodes based on the intrinsic relationship to yield

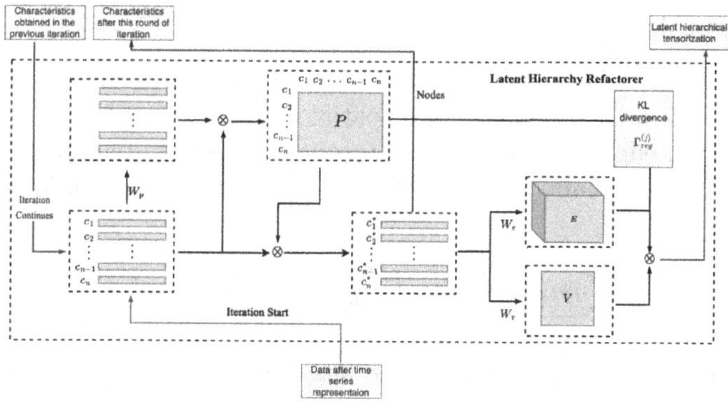

Fig. 2. Latent Hierarchy Refactorer.

$C_i^{(j)*}$. We map the optimized feature vectors into "edge space" and "node space" respectively for subsequent graph structure simulation tensor edge learning.

$$E^{(j)} = Relu(C^{(j)*} \otimes W_e^{(j)}), \quad V^{(j)} = Relu(C^{(j)*} \otimes W_v^{(j)}). \tag{5}$$

where $W_e^{(j)}$ and $W_v^{(j)}$ respectively mean the trainable parameters of the two different high-level spatial mappings in j-th iteration. Note that $W_e^{(j)} \in \mathbb{R}^{b \times m \times m}$, where b is the dimension of the edge vector. While edges are commonly represented as scalars, we use multidimensional tensors to discover high-level semantic structures between features.

In the process of learning the edges of the tensor structure, we regard the node feature $V_n^{(j)}$ in the node space as the source of the problem vector, which is used to calculate the intrinsic correlation of different edges in the edge space based on the correlation of node features. Then the value representation of the edge space is weighted according to the edge correlation, and the final tensor edge $L^{(j)} \in \mathbb{R}^{b \times n \times n \times m}$. Where b is the batch size, n is the number of features, and $n \times n \times m$ represents the dimension of the tensor potential hierarchy of n pairs of features, which is m.

We utilize KL divergence to penalize edges that stray from reality, thereby ensuring the accuracy of the information.

$$\begin{aligned} \Gamma_{reg}^{(j)} &= KL(P \| L^{(j)}) \quad, j = all, \\ \Gamma_{reg}^{(j)} &= KL(L^{(j-1)} \| L^{(j)}) \quad, j > 1. \end{aligned} \tag{6}$$

$\Gamma_{reg}^{(j)}$ is referred to as the "Regularization loss". To accomplish a stable learning process, we not only replace the error distribution of the edges during the initial learning process, but also encode the variability of the new edges relative to the old ones in subsequent iterations. The previous feature correlation $P_{i,k}^{(j)}$ is regarded as the initial edge distribution, which is used to prevent deviation from the real situation during the tensor edge learning process of the graph structure.

3.3 Information Aggregator

We employ channel attention to enrich the comprehension of channel information within latent hierarchical tensors and spatial attention to enhance insights into information-rich edges in space. Simultaneously, we apply maximum pooling to eliminate redundancy and utilize average pooling to refine high-level semantic information.

$$A^{(j)}_{n,n} = \frac{maxpooling_b(L^{(j)}_{n,n,b}) + avgpooling_b(L^{(j)}_{n,n,b})}{2} \tag{7}$$

The pooling operation scalarizes the latent hierarchical structure to obtain the adjacency matrix edges of the $A \in \mathbb{R}^{n \times n}$ used to aggregate the graph convolution information.

Through graph convolution [7], we integrate the captured latent hierarchical structure with the original physiological information to enhance information. This process yields a concise and information-rich representation of human physiological states.

$$C^{j+1} = Relu(\hat{D}^{-\frac{1}{2}} \hat{A} \hat{D}^{-\frac{1}{2}}\ C^j\ W) \tag{8}$$

where C^{j+1} denotes the representation of the aggregation node after j-th round of iterations. $W \in \mathbb{R}^{N_{hid} \times N_{hid}}$ is the shared projection matrix for all iteration loops. $\hat{D}$ is the degree matrix with respect to $\hat{A}$.

3.4 Healthcare Prediction

The structurally condensed and informative physiological characteristics that we derived can be directly used as representations of human states for subsequent prediction tasks. It only requires adding a multi-class fully connected layer after the human body representation.

$$\hat{y} = \sigma(W^1_{fin} \cdot Relu(W^0_{fin} \cdot S)) \tag{9}$$

Finally, the total loss can be presented as a combination of cross-entropy loss and regularization loss in the Sec. 3.2.

$$\Gamma = \Gamma_{cross-entropy} + \sum_j \Gamma^{(j)}_{reg} \tag{10}$$

4 Experiment

We perform a binary classification disease prediction task on the MIMIC-III dataset, Cardiology dataset and Tongji Hospital Dataset. We evaluate the contribution of the modules proposed by our model to the whole human state modeling process. Overall details are shown below.

We estimate the performance of the ACHIM using the area under the receiver operating characteristic curve (AUROC), area under the precision-recall curve

(AUPRC), and the minimum of precision and sensitivity Min(Se, P+). When dealing with highly skewed data, it is not enough to use AUROC to present binary classification. AUPRC gives a more informative picture of a model's performance.

4.1 Experimental Results

In this subsection, we will demonstrate the efficacy of the various models on the three datasets of electronic health records and provide additional analysis. Table 2 depicts each model's performance. The number in "()" represents the 5-fold cross-validation standard deviation. The results demonstrate that our model outperforms other baseline approaches on all datasets with significantly and consistently superior metrics.

Table 2. Experimental results on three open-source EHR datasets: MIMIC-III, Cardiology and TJH.

	MIMIC-III Dataset			Cardiology Dataset			TJH		
Methods	AUROC	AUPRC	min(Se,P+)	AUROC	AUPRC	min(Se,P+)	AUROC	AUPRC	min(Se,P+)
GRU*	.7473(.0121)	.3429(.0286)	.3759(.0189)	.7599(.0075)	.2942(.0293)	.3143(.0327)	.8956(.0347)	.8864(.0057)	.8793(.0179)
RETAIN	.8254(.0091)	.4466(.0326)	.4560(.0245)	.8456(.0099)	.3814(.0154)	.3939(.0105)	.9832(.0034)	.9816(.0045)	.9400(.0192)
T-LSTM	.7856(.0107)	.3748(.0236)	.4017(.0269)	.7927(.0060)	.2629(.0129)	.3131(.0162)	.9816(.0120)	.9767(.0141)	.9250(.0293)
ConCare	.8480(.0056)	.4668(.0130)	.4909(.0098)	.9255(.0143)	.6626(.0926)	.6496(.0489)	.9847(.0096)	.9639(.0385)	.9415(.0272)
GRASP	.8475(.0078)	.4915(.0240)	.4866(.0092)	.9391(.0045)	.7276(.0083)	.6768(.0176)	.9901(.0035)	.9871(.0064)	.9566(.0098)
SAFARI	.8443(.0092)	.4949(.0265)	.4863(.0170)	.9316(.0054)	.6546(.0408)	.6358(.0428)	.9480(.0484)	.9252(.0638)	.8437(.0961)
ACHIM$_{-LHR}$	.8175(.0074)	.4019(.0188)	.4590(.0081)	.9356(.0031)	.7421(.0149)	.6831(.0211)	.9598(.0059)	.9579(.0068)	.9185(.0130)
ACHIM$_{-IA}$	.8321(.0114)	.421(.0169)	.4690(.0102)	.9359(.0073)	.7472(.0017)	.6911(.0281)	.9798(.0053)	.9791(.0068)	.9205(.0190)
ACHIM$_{-RE}$	.8498(.0076)	**.4969(.0179)**	.4844(.0184)	.9372(.0041)	.7197(.0285)	.6763(.0219)	.9866(.0071)	.9830(.0100)	.9380(.0290)
ACHIM	**.8501(.0067)**	.4954(.0182)	**.4913(.0118)**	**.9457(.0030)**	**.7627(.0170)**	**.7091(.0184)**	**.9908(.0033)**	**.9876(0.0056)**	**.9595(.0165)**

ConCare, GRASP, and SAFARI are the baselines with the most competitive methods. ConCare extracts individual features independently and models their intrinsic relationship using MLP and self-attentive mechanisms to produce an effective representation of individual states, which shows that modeling correlations between features is effective. With GRASP and SAFARI, we have observed that information enhancement can be achieved based on similar human body representations of patients, which is essentially the utilization of external information. ACHIM is based on state representation, which not only concentrates on the original features but also constructs a semantic hierarchical structure in an adaptive manner without requiring prior knowledge. Our model integrates the two portions of information from the latent hierarchical and original features to produce a concise representation of the patient's condition. Experiments demonstrate that integrating latent structure information into raw features has a substantial impact on performance improvement.

We observe that the cardiology dataset has a larger number of cases than MIMIC-III, resulting in an increase in the performance of our model. This finding demonstrates that at equal data densities, employing big data not only enables the discovery of the inherent hierarchical structure among original features but also indicates that greater data volume leads to an improved effect. In addition, we found that all models performed substantially better on the TJH dataset than on the first two datasets. This is attributed to the denser nature of the TJH data, resulting in fewer missing values. It underscores that for enhancing the performance of human body representation, data quality holds greater significance than sample size. It can be seen that our model can better encode the patient state representation whether it is faced with dense data or sparse data. On three datasets with varying sample sizes, our model outperforms extant state-of-the-art models, demonstrating its superior generalization ability.

4.2 Ablation Experiments

In Table 2, we observed that ACHIM performs better than three of its variants. In particular, $ACHIM_{-LHR}$ lacks Latent Hierarchy Refactorer(LHR) of raw data and instead represents the human state solely based on the feature itself. This is an important content of ACHIM, which can realize the information integration of human body state in high-level semantics. Through the reciprocal feedback of internal and external information, we achieve our objective of information enhancement. Experiments demonstrate that $ACHIM_{-LHR}$ performs substantially worse than variant $ACHIM_{-IA}$.

The variant $ACHIM_{-IA}$ only contains the latent structure information learned by the original features, it and $ACHIM_{-LHR}$ are two extremes, but $ACHIM_{-IA}$ performance is significantly better than that of $ACHIM_{-LHR}$, which shows that learning latent semantic structure information is effective for the learning of personal state representation.

Augmenting the understanding of latent structures can substantially enhance model performance, albeit accompanied by a proportional increase in training time, as indicated in Table 3. Nonetheless, our primary objective is to optimize performance, making this trade-off acceptable.

The variant $ACHIM_{-RE}$ does not incorporate regularization loss. The variant $ACHIM_{-RE}$ performance drop supports our contention that restricting the edges of the graph structure based on the correlations between nodes is an effective early-stage training strategy. In addition, $ACHIM_{-RE}$ demonstrates the importance of using regularization constraints to restrict the size of the hypothesis space and reduce empirical risk during each iteration.

4.3 Qualitative Visualization of Patient Representation

To qualitatively prove the interpretability of our model with the best performance during representation learning, we use individual representations obtained from test data in the 4,228 cross-validation subsets on the MIMIC-III dataset for PCA dimensionality reduction and Plot 2-D points. As shown in Fig. 3, each

Table 3. The differences in training duration and prediction performance between variants $ACHIM_{-LHR}$ and $ACHIM$. Variant $ACHIM_{-LHR}$ lacks the incorporation of the underlying hierarchical structure, a feature present in the comprehensive model $ACHIM$. Moreover, variant $ACHIM_{-LHR}$ excludes the information fusion process, resulting in significantly shorter training times compared to $ACHIM$

	MIMIC-III Dataset			Cardiology Dataset			TJH Dataset		
	$ACHIM_{-LHR}$	$ACHIM$	Variation	$ACHIM_{-LHR}$	$ACHIM$	Variation	$ACHIM_{-LHR}$	$ACHIM$	Variation
Training time / epoch	265 s	317 s	19.62%↑	148s	190 s	28.37%↑	92 s	105 s	14.13%↑
AUROC	.8175(.0074)	.8501(.0067)	3.99%↑	.9356(.0031)	.9457(.0030)	1.08%↑	.9598(.0059)	.9908(.0033)	3.23%↑

point represents the two-dimensional information of the patient's personal representation after dimensionality reduction. Two different colors represent whether a patient died of a disease or not. Ideally, points with the same color should be in the same cluster, and there should be a large distance between the margins between different clusters. However, in practice, this is often difficult to achieve.

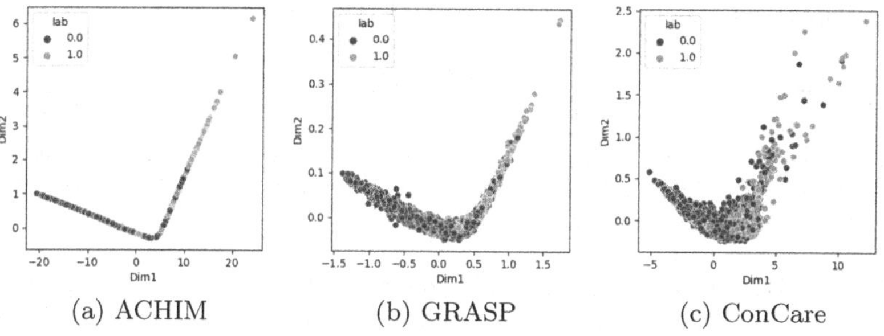

(a) ACHIM (b) GRASP (c) ConCare

Fig. 3. PCA Scatterplots of Patients Representation Learned by Predictive Models on the MIMIC-III dataset.

From the Fig. 3a, we observe that the scatterplots obtained by aggregating raw medical features with their learned hierarchies are strongly clustered. This shows that the learning process of the ACHIM model is perfectly matched to the consultation process, and there is a matching problem between similar diseases.

On the contrary, the scatter plot representations of GRASP and ConCare are a little worse. The results of Fig. 3b and Fig. 3c have certain bloated information because the model cannot fully learn the correct patient representation information. But GRASP takes into account the similarity between patients, and the degree of dispersion is slightly better compared to ConCare. This indicates that the patient representation learned by our model by aggregating scattered raw features and latent hierarchical information is correct. This provides a solid basis for health task prediction.

5 Conclusion

In this paper, we propose ACHIM, a novel and efficient model for health prediction based on personal state representation. By aggregating the raw data and latent hierarchical structure of the clinical patients, ACHIM realizes the semantically compact representation of the human body state and accomplishes information enhancement. The most crucial point compared to all existing models is that our model does not require any additional prior knowledge, it is applicable to various EHR datasets, and it has excellent generalization. We conducted experiments using three datasets derived from the actual world. ACHIM demonstrates remarkable performance in human state representation. Our current focus aims at enhancing performance rather than addressing optimization concerns. Our model employs potential edges between feature pairs, leading to a notable increase in computational complexity. Future efforts will concentrate on optimizing the model's performance.

Acknowledgments. This work is supported by the National Key Research and Development Program of China (No. 2022YFB3904702), Key Research and Development Program of Jiangsu Province (No.BE2018084), Opening Project of Beijing Key Laboratory of Mobile Computing and Pervasive Device, and Industrial Internet Innovation and Development Project of 2021 (TC210A02M, TC210804D).

References

1. Baytas, I.M., Xiao, C., Zhang, X., Wang, F., Jain, A.K., Zhou, J.: Patient subtyping via time-aware LSTM networks. In: Proceedings of the 23rd ACM SIGKDD International Conference on Knowledge Discovery and Data Mining, pp. 65–74 (2017)
2. Choi, E., Bahadori, M.T., Song, L., Stewart, W.F., Sun, J.: Gram: graph-based attention model for healthcare representation learning. In: Proceedings of the 23rd ACM SIGKDD International Conference on Knowledge Discovery and Data Mining, pp. 787–795 (2017)
3. Choi, E., Bahadori, M.T., Sun, J., Kulas, J., Schuetz, A., Stewart, W.: Retain: an interpretable predictive model for healthcare using reverse time attention mechanism. In: Advances in Neural Information Processing Systems **29** (2016)
4. Choi, E., Xiao, C., Stewart, W., Sun, J.: Mime: Multilevel medical embedding of electronic health records for predictive healthcare. In: Advances in Neural Information Processing Systems **31** (2018)
5. Choi, E., et al.: Learning the graphical structure of electronic health records with graph convolutional transformer. In: Proceedings of the AAAI Conference on Artificial Intelligence. vol. 34, pp. 606–613 (2020)
6. Gao, J., et al.: Camp: co-attention memory networks for diagnosis prediction in healthcare. In: 2019 IEEE International Conference on Data Mining (ICDM), pp. 1036–1041. IEEE (2019)
7. Kipf, T.N., Welling, M.: Semi-supervised classification with graph convolutional networks. arXiv preprint arXiv:1609.02907 (2016)
8. Lee, W., Park, S., Joo, W., Moon, I.C.: Diagnosis prediction via medical context attention networks using deep generative modeling. In: 2018 IEEE International Conference on Data Mining (ICDM), pp. 1104–1109. IEEE (2018)

9. Ma, F., Chitta, R., Zhou, J., You, Q., Sun, T., Gao, J.: Dipole: Diagnosis prediction in healthcare via attention-based bidirectional recurrent neural networks. In: Proceedings of the 23rd ACM SIGKDD International Conference on Knowledge Discovery and Data Mining, pp. 1903–1911 (2017)
10. Ma, F., et al.: Unsupervised discovery of drug side-effects from heterogeneous data sources. In: Proceedings of the 23rd ACM SIGKDD International Conference on Knowledge Discovery and Data Mining, pp. 967–976 (2017)
11. Ma, F., You, Q., Xiao, H., Chitta, R., Zhou, J., Gao, J.: Kame: knowledge-based attention model for diagnosis prediction in healthcare. In: Proceedings of the 27th ACM International Conference on Information and Knowledge Management, pp. 743–752 (2018)
12. Ma, L., et al.: Concare: Personalized clinical feature embedding via capturing the healthcare context. In: Proceedings of the AAAI Conference on Artificial Intelligence. vol. 34, pp. 833–840 (2020)
13. Ma, T., Xiao, C., Wang, F.: Health-atm: a deep architecture for multifaceted patient health record representation and risk prediction. In: Proceedings of the 2018 SIAM International Conference on Data Mining, pp. 261–269. SIAM (2018)
14. Miotto, R., Wang, F., Wang, S., Jiang, X., Dudley, J.T.: Deep learning for healthcare: review, opportunities and challenges. Brief. Bioinform. **19**(6), 1236–1246 (2018)
15. Rajkomar, A., et al.: Scalable and accurate deep learning with electronic health records. NPJ Digital Med. **1**(1), 1–10 (2018)
16. Suo, Q., et al.: Deep patient similarity learning for personalized healthcare. IEEE Trans. Nanobiosci. **17**(3), 219–227 (2018)
17. Suo, Q., et al.: Personalized disease prediction using a CNN-based similarity learning method. In: 2017 IEEE International Conference on Bioinformatics and Biomedicine (BIBM), pp. 811–816. IEEE (2017)
18. Suresh, H., Gong, J.J., Guttag, J.V.: Learning tasks for multitask learning: heterogenous patient populations in the ICU. In: Proceedings of the 24th ACM SIGKDD International Conference on Knowledge Discovery & Data Mining, pp. 802–810 (2018)
19. Tan, Q., et al.: Data-gru: Dual-attention time-aware gated recurrent unit for irregular multivariate time series. In: Proceedings of the AAAI Conference on Artificial Intelligence. vol. 34, pp. 930–937 (2020)
20. Vaswani, A., et al.: Attention is all you need. In: Advances in Neural Information Processing Systems **30** (2017)
21. Xu, Y., Biswal, S., Deshpande, S.R., Maher, K.O., Sun, J.: Raim: recurrent attentive and intensive model of multimodal patient monitoring data. In: Proceedings of the 24th ACM SIGKDD International Conference on Knowledge Discovery & Data Mining, pp. 2565–2573 (2018)
22. Yuan, Y., et al.: A novel channel-aware attention framework for multi-channel EEG seizure detection via multi-view deep learning. In: 2018 IEEE EMBS International Conference on Biomedical & Health Informatics (BHI), pp. 206–209. IEEE (2018)
23. Zhang, C., Gao, X., Ma, L., Wang, Y., Wang, J., Tang, W.: Grasp: generic framework for health status representation learning based on incorporating knowledge from similar patients. In: Proceedings of the AAAI Conference on Artificial Intelligence. vol. 35, pp. 715–723 (2021)
24. Zhang, S., Xie, P., Wang, D., Xing, E.P.: Medical diagnosis from laboratory tests by combining generative and discriminative learning. arXiv preprint arXiv:1711.04329 (2017)

Using Multimodal Data to Improve Precision of Inpatient Event Timelines

Gabriel Frattallone-Llado[1], Juyong Kim[2(✉)], Cheng Cheng[2], Diego Salazar[3], Smitha Edakalavan[4], and Jeremy C. Weiss[3]

[1] Universidad de Puerto Rico, PR 00926-1117 San Juan, Puerto Rico
gabriel.frattallone2@upr.edu

[2] Carnegie Mellon University, Pittsburgh, PA 15206, USA
{juyongk,ccheng2}@andrew.cmu.edu

[3] National Library of Medicine, Bethesda, MD 20894, USA
{diego.salazar,jeremy.weiss}@nih.gov

[4] University of Pittsburgh, Pittsburgh, PA 15260, USA
SME57@pitt.edu

Abstract. Textual data often describe events in time but frequently contain little information about their specific timing, whereas complementary structured data streams may have precise timestamps but may omit important contextual information. We investigate the problem in healthcare, where we produce clinician annotations of discharge summaries, with access to either unimodal (text) or multimodal (text and tabular) data, (i) to determine event interval timings and (ii) to train multimodal language models to locate those events in time. We find our annotation procedures, dashboard tools, and annotations result in high-quality timestamps. Specifically, the multimodal approach produces more precise timestamping, with uncertainties of the lower bound, upper bounds, and duration reduced by 42% (95% CI 34–51%), 36% (95% CI 28–44%), and 13% (95% CI 10–17%), respectively. In the classification version of our task, we find that, trained on our annotations, our multimodal BERT model outperforms unimodal BERT model and Llama-2 encoder-decoder models with improvements in F1 scores for upper (10% and 61%, respectively) and lower bounds (8% and 56%, respectively). The code for the annotation tool and the BERT model is available (link).

Keywords: Temporal Information · Timeline Construction · Multimodal Data · Absolute Timeline Prediction

1 Introduction

Temporal data mining involves the extraction of temporal information from different sources and modalities of data, and it has broad application in fields such as law, finance, and healthcare. For instance, in criminal recidivism prediction, the event timeline for a defendant could be extracted from both texts

G. Frattalone-Llado and J. Kim-Equal contribution.

D.-N. Yang et al. (Eds.): PAKDD 2024, LNAI 14648, pp. 322–334, 2024.
https://doi.org/10.1007/978-981-97-2238-9_25

in probation office documents and tables in psychiatric health records [2]; In stock price movement and volatility prediction, financial time series could be extracted from financial news, daily stock market price tables, and verbal and vocal cues in earning calls [1]. In clinical risk prediction, patient timeline could be extracted from electronic health records with both unstructured clinical notes and structured tabular data [10]. This paper focuses on providing a multimodal extraction system and a benchmark dataset for clinical timelines.

Precise clinical event timelines are crucial for prognosis and prediction tasks. These forecasting tasks have been studied with varied prediction times and unstructured data sources [7]. Discharge summaries provide the most complete information. In the 2012 Informatics for Integrating Biology and Bedside (i2b2) Challenge [9], 310 discharge summaries were annotated with temporal information, including clinical events, temporal expressions, and their temporal relations. This approach yields a relative timeline of clinical events, rather than absolute, leaving many events without the precise timing needed for forecasting tasks. To achieve a more complete event timeline, i2b2 events were annotated with absolute time values [3], by bounding the events with closed intervals in calendar times and temporal uncertainties on the bounds. Their annotation procedure was unimodal, as annotators had access only to the discharge summary text.

Meanwhile, there is a consensus that the integration of structured and unstructured data has a significant impact on constructing models and predicting target variables [7]. In this project, we take advantage of the combination of unstructured and structured data in a multimodal approach, which has proven beneficial in other applications [4,5]. Our work adopts a version of the probabilistic bounds described and applies them to discharge summaries from the i2b2 dataset in order to generate absolute inpatient event timelines. We introduce the following: a visualization and annotation tool, an annotation process with a three-pass system, two types of annotations to better represent the nature of clinical events, and the multimodal annotation approach. By combining the information from unstructured and structured data, this multimodal approach should yield a more precise (*i.e.*, less uncertain) timeline for inpatient events.

Our multimodal approach contributes the following: (i) we introduce absolute timeline intervals without assumption of independence of endpoint uncertainties, (ii) we find that multimodal annotations lead to more precise timelines than the unimodal annotations. (iii) we verify the annotation quality by mapping our annotations to temporal relations, where our relations compare favorably against benchmark annotations (i2b2), and (iv) we demonstrate that a fine-tuned multimodal encoder (BERT) architecture outperforms fine-tuned unimodal encoder and off-the-shelf generative encoder-decoder architectures (Llama-2). Overall, we show the importance of annotation from multimodal data sources, both in the annotation process and for machine learning predictive performance.

2 Methods

We use the i2b2 dataset [12], a compendium of de-identified electronic health records from Partner Healthcare and Beth Israel Deaconess Medical Center, con-

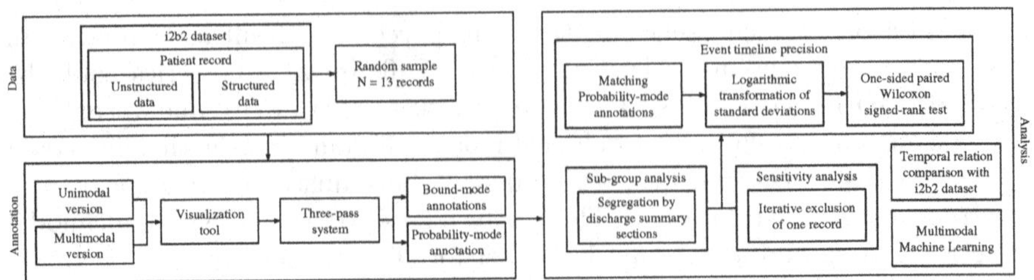

Fig. 1. Flowchart of our method.

taining discharge summaries with annotated clinical time/events. We located i2b2 patient's records in MIMIC-III with matching discharge summaries, allowing us to collect both structured and unstructured records. Thirteen records were randomly selected from the training set of the 2012 i2b2 temporal relations challenge data [9]. For each of these records, a human annotator with medical experience identified clinical events in the discharge summary and timestamped their endpoints on a timeline.

In our study, an annotation comprises a contiguous, highlighted text, representing a clinical event, and a time interval. Based on the use of the structured and unstructured data, two annotation versions were generated: multimodal and unimodal. In the unimodal version, the annotator only had access to the discharge summary and the admission and discharge times. In the multimodal version, the annotator also had access to the full structured data. Out of the total thirteen records, five were annotated using both unimodal and multimodal versions, while the remaining records were annotated from only multimodal. Upon acceptance, the annotation tool, the annotation files , and the analyses will be made public. A flowchart representation of the process describes the steps of our annotation and analysis (Fig. 1).

2.1 Annotation Tool

The R Shiny annotation tool displays all the unstructured and structured data for any given record (Fig. 2). The structured data is displayed on a graph, where the x-axis contains absolute time values and the y-axis contains event identifiers. The user annotates time intervals by clicking and dragging on the graph (blue overlay). Structured data events contained within this overlay are displayed in a table that the user can search and select to be relevant. On the unstructured data, or discharge summary, section, the user selects the relevant span for annotation by highlighting the text. The polarity of negative events is specified with a checkbox. Both types of annotation (Bounds- and Probability- mode) require the user to generate an overlay on the graphical timeline. In Bounds-mode annotation, either the lower or upper bound can be omitted by unchecking a box, which represents *indefiniteness*. In Probability-mode annotation, users also select standard deviations to represent their uncertainty about the lower bound, upper bound, and duration of the event. The choices are pre-defined as follows: 1, 3,

Fig. 2. The web-based annotation tool.

or 10 s; 1, 3, 10, or 30 min; 1, 2, 4, 6, 12, or 24 h; 2 d; 1 or 2 weeks; 1 month; 1, 10, or 100 years.

2.2 Annotation Process

The annotation process consisted of a three-pass system defined as follows: (1) the annotator peruses the document in its entirety to become familiarized with its content, (2) the annotator makes annotations at the paragraph level, setting lower and upper bounds that apply to all the events in that paragraph, and (3) the annotator makes annotations on particular events, which could represent words or phrases. Events are selected according to the i2b2 annotation guidelines [8]. For simplicity, the one event attribute recorded is the polarity. As stated previously, two types of annotations are available: Bounds-mode and Probability-mode. In Bounds-mode annotation, lower and upper bounds are defined, with the expectation that the event(s) occurs at some point between them, and no information is specified about event duration or timing uncertainty. One can omit either bound when the value is unknowable, as in the lower bound for certain conditions from the past medical history. In this case, the bound defaults to negative or positive infinity, as appropriate. Bounds-mode annotation is predominantly used in the second pass, because each paragraph may contain various events with different timing attributes. In Probability-mode annotation, the selected lower and upper bounds represent the mean values of two distributions. The mean value of the event's duration is calculated as their difference. All three distributions are assumed to be normal and the annotator must select a standard deviation for each one. These standard deviations represent the annotator's level of uncertainty about the event's timing and will serve as a surrogate of timeline precision for data analysis. Probability-mode annotation is used predominantly in the third pass, where the duration and approximate bounds of particular events may be reasonably determined.

2.3 Statistical Methods

The annotation process above was performed on the thirteen randomly-selected records from the i2b2 training dataset [9]. Five records are annotated from both unimodal and multimodal data by performing the process twice for each of these records, while the rest are annotated solely from the multimodal version of the data. In total, eighteen annotation files were generated, comprising five unimodal and thirteen multimodal versions. These files contain 4884 annotations in total, of which 1156 were in Bounds-mode and 3728 were in Probability-mode.

We performed two descriptive analyses. For the first analysis, we focus on the ten annotation files originated from both unimodal and multimodal data to study and compare the annotation practices across both versions. We defined the main effect as the precision of the event timelines (multimodal vs unimodal), *i.e.*, the difference in uncertainty of events is obtained by comparing the standard deviation for the lower bound, upper bound, and duration for multimodal and unimodal annotations. To compare these event timelines, we selected exact matches from the Probability-mode annotations, and we performed a one-sided paired Wilcoxon signed-rank test on the logarithmic transformation of the standard deviations. The resulting estimated differences between the two versions correspond to a scaling factor, which reflects the degree of increase or decrease in uncertainty, which we call *precision factor*. Confidence intervals were calculated using the bootstrap. To verify the robustness of the results, a sensitivity analysis was performed by re-running the test while iteratively excluding one record pair. Sub-analyses were also performed on different sections of the discharge summaries, termed Medical History (history of present illness, past medical history, home medications, allergies, social and family history), Hospital Stay (hospital course, procedures, service), Examinations and Findings (physical exam, laboratory tests, images, and studies), and Discharge (condition, disposition, instructions, medications, diagnosis).

For the second analyses, we sought to assess the compatibility between our annotations and the i2b2 dataset. We aligned the events in the thirteen multimodal annotation files to the events in the corresponding i2b2 dataset. Since the i2b2 dataset provides only temporal relations between two events - "BEFORE", "AFTER", or "OVERLAP" - while ours provides the absolute time of events, we extracted those i2b2 temporal relations where both text endpoints could be matched with our annotated events, and compared to the temporal relations computed from our annotations. We used character-level intersection over union to match the text span of the i2b2 events and ours. We restricted our events to be aligned to be Probability-mode and calculated a pair of z-scores: one indicating the likelihood that one event precedes the other and another for the opposite order. If exactly one z-score exceeds a predetermined threshold, we determined that one event precedes the other, otherwise they overlap. After matching the temporal relations between our dataset and the i2b2 dataset, we report the F1 score, inter-annotator agreement, and accuracy of the types of relations.

2.4 Multimodal Learning

To test the utility of multimodal data over single-modal data, we study a reduced version of our absolute timeline prediction. The task involves binary classifications of whether the lower bound (LB) and upper bound (UB) are finite, identifying the type of annotation (Bounds- or Probability-mode), and multi-class classifications of the bounds and the standard deviations of LB, UB, and duration. Each of the bound and standard deviation classifications consists of three classes defined as follows. Bounds are classified relative to admission time, with thresholds at admission and 24 h post-admission. Standard deviations for LB and UB are grouped into three classes: under 2 h, 4 h to 1 day, and over 2 d. Duration standard deviations follow a similar pattern, classified as under 1 h, 2 h to 1 day, and over 2 d.

The experiment on the classification version of our dataset employed two BERT-based models. The first model, named Unimodal BERT, is a span classification model [13] where the contextualized BERT embeddings of the annotated text span from clinical notes along with its context are fed to a feedforward network for classification. The second model, named Multimodal BERT, also incorporates the structured event data by applying multi-head attention. In this setup, the BERT embedding of the text span serves as the query, and the keys and values are the contextualized embeddings of the names and the one-hot encodings of the timestamps in the table. The resulting weighted sum of the one-hot encodings is taken logarithm and then added to the logits of the mean predictions. The base BERT of both models is initialized with the BlueBERT-Base [6].

To fit the input length shorter than 128, the maximum sequence length of BlueBERT, we filtered out paragraph-level events that were annotated in the second pass. The fourteen annotation files[1] are split into 5 groups to perform 5-fold cross-validation, and we report the average of test performances across three different random seeds. For all the experiments, we trained the models for 20 epochs with a learning rate of 5e-5. For comparison, we also report the performance of the majority selection baseline, the baseline using Llama-2 (llama-2–13b-chat-hf) on the annotated clinical notes (same as Unimodal BERT) to generate the LB and UB through few shot prompting [11], and the baseline selecting the LB and UB based on the ground truth selection of the structured events.

[1] One additional annotation file that 2012 i2b2 dataset does not include is added.

3 Results

3.1 Comparison Across Different Modalities

An exploratory data analysis of the 10 paired annotation files (5 multimodal and 5 unimodal) revealed a total of 2718 annotations, of which 693 were in Bounds-mode and 2025 were in Probability-mode. For each record, the number of annotations and their distribution between Bounds and Probability modes were very similar between the unimodal and multimodal versions. Furthermore, 95.7% of the events were annotated as exact matches across versions, meaning that the selected text had the same start and end positions, *i.e.*, span, regardless of annotation mode (Table 1A).

Across discharge summary sections, there were relatively more Bounds-mode annotations in the Medical History (67% bounds) and Discharge (35%) sections, and fewer in the larger Hospital Stay (7%) and Examinations and Findings (17%), indicating increased ability to annotate intraencounter events in Probability-mode. Across sections, there was high agreement between unimodal and multimodal annotation, shown by the proportion of exact matches (Table 1B).

All matching events annotated in Probability-mode were used to perform a Wilcoxon signed-rank test. We compared the standard deviations annotated for lower bound, upper bound, and duration. The standard deviations reported in the multimodal version were significantly smaller than those reported in the unimodal version. This trend was consistent across the lower bound (p-value <0.001), upper bound (p-value <0.001), and duration (p-value <0.001). In general, we increased the precision on average by a factor of 1.42, 1.36, and 1.13 for the lower bound, upper bound, and duration, respectively. These values suggest that the uncertainty in multimodal annotations was reduced by 42%, 36%, and 13%, compared to unimodal annotations for the corresponding types of bounds and duration (Fig. 3).

Similarly, we conducted a comparison of the standard deviations pertaining to bounds and duration types across various sections present in clinical notes (Fig. 3). Our findings indicate that, in the case of bounds, annotations exhibited an improvement in precision in all the sections. However, for the duration, there was no observed change in uncertainty, except for the Medical History, Exams and Findings, and Discharge sections.

In the sensitivity analysis, the standard deviation effect estimates (lower, upper, and duration) remained close to sample estimate, *i.e.*, the effect of increased precision in the multimodal annotations was maintained. The estimated precision factors were for lower bound, upper bound, and duration: 1.42 (95% CI [1.34–1.51]), 1.36 (95% CI [1.28–1.44]), and 1.13 (95% CI [1.10–1.17]) respectively. The Wilcoxon sign-rank test again showed differences between both versions of annotation (p-value < 0.001).

Table 1. Comparison of annotation practices across unimodal and multimodal versions (A) and across discharge summary sections (B), with number of annotations per record/per section, their breakdown into Bounds-mode and Probability-mode, and the exact match between the versions.

Comparison of Annotation Practices Across Versions

Record[1]	Version	Annotations	Bounds-Mode # (%)	Probability-Mode # (%)	Exact Matches # (%)
A	Unimodal	161	45 (28.0%)	116 (72.0%)	156 (96.9%)
A	Multimodal	165	46 (27.8%)	119 (72.1%)	156 (94.5%)
B	Unimodal	379	53 (14.0%)	326 (86.0%)	354 (93.4%)
B	Multimodal	380	60 (15.8%)	320 (84.2%)	354 (93.2%)
C	Unimodal	182	58 (31.9%)	124 (68.1%)	167 (91.8%)
C	Multimodal	180	58 (32.2%)	122 (67.8%)	167 (92.8%)
D	Unimodal	352	125 (35.5%)	227 (64.5%)	347 (98.6%)
D	Multimodal	361	115 (31.9%)	246 (68.1%)	347 (96.1%)
E	Unimodal	278	64 (23.0%)	214 (77.0%)	276 (99.3%)
E	Multimodal	280	69 (24.6%)	211 (75.4%)	276 (98.6%)
Total	-	**2718**	**693 (25.5%)**	**2025 (74.5%)**	**2600 (95.7%)**

[1]Record numbers have been redacted

(A) Across Versions

Comparison of Annotation Practices Across Sections

Section	Version	Annotations	Bounds-Mode # (%)	Probability-Mode # (%)	Exact Matches # (%)
Medical History	Unimodal	244	161 (66.0%)	83 (34.0%)	234 (95.9%)
Medical History	Multimodal	249	167 (67.1%)	82 (32.9%)	234 (94.0%)
Exams and Findings	Unimodal	484	29 (6.0%)	455 (94.0%)	467 (96.5%)
Exams and Findings	Multimodal	489	32 (6.5%)	457 (93.5%)	467 (95.5%)
Hospital Stay	Unimodal	366	59 (16.1%)	307 (83.9%)	354 (96.7%)
Hospital Stay	Multimodal	369	62 (16.8%)	307 (83.2%)	354 (95.9%)
Discharge	Unimodal	248	96 (38.7%)	152 (61.3%)	234 (94.4%)
Discharge	Multimodal	248	87 (35.1%)	161 (64.9%)	234 (94.4%)

(B) Across Sections

3.2 Comparison with I2b2 Dataset (2012)

The total number of events in the thirteen annotation files derived from the multimodal data is 3532, with 2721 in Probability-mode annotations. The corresponding i2b2 training data files have 1024 events ("EVENT" tag) and 2140 temporal relations ("TLINK" tag), with 901 comparing events only. Note these i2b2 files do not cover the entire discharge summary. By aligning the text spans with the character-level IOU threshold of 0.5, 761 of the Probability-mode events are matched with the i2b2 events, and 490 of the i2b2 temporal relations can be mapped to our dataset. We use -1.0 for the threshold of z-scores to determine the temporal relations between two Probability-mode events. The confusion matrix between the i2b2 temporal relations and our annotations' temporal relations is shown in Fig. 4. The accuracy, the macro F1-score, and the Cohen's kappa score between the i2b2 and our temporal relations are 0.698, 0.599, and 0.383, respectively. In other words, 70% of the temporal links between the two events agree

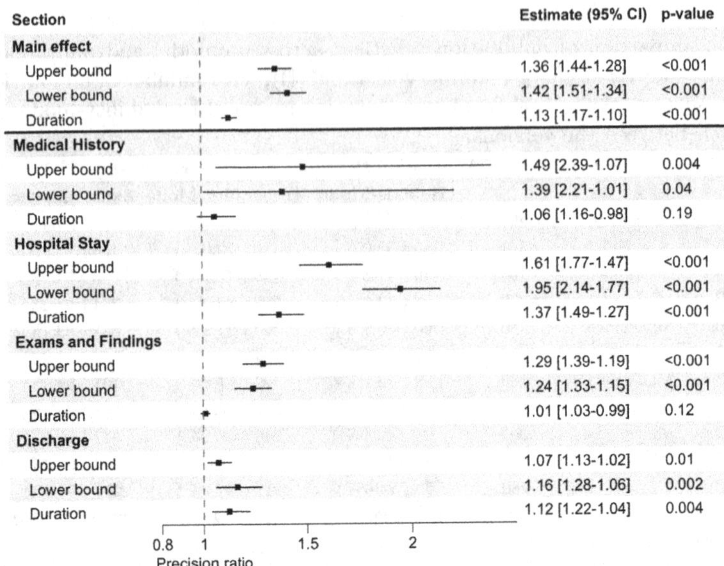

Section
Estimate (95% CI) p-value
Main effect
Upper bound 1.36 [1.44-1.28] <0.001
Lower bound 1.42 [1.51-1.34] <0.001
Duration 1.13 [1.17-1.10] <0.001
Medical History
Upper bound 1.49 [2.39-1.07] 0.004
Lower bound 1.39 [2.21-1.01] 0.04
Duration 1.06 [1.16-0.98] 0.19
Hospital Stay
Upper bound 1.61 [1.77-1.47] <0.001
Lower bound 1.95 [2.14-1.77] <0.001
Duration 1.37 [1.49-1.27] <0.001
Exams and Findings
Upper bound 1.29 [1.39-1.19] <0.001
Lower bound 1.24 [1.33-1.15] <0.001
Duration 1.01 [1.03-0.99] 0.12
Discharge
Upper bound 1.07 [1.13-1.02] 0.01
Lower bound 1.16 [1.28-1.06] 0.002
Duration 1.12 [1.22-1.04] 0.004
0.8
1
1.5
2
Precision ratio

Fig. 3. Precision factor between the two types of annotations. A value greater than 1 indicates a reduction of uncertainty in multimodal annotations (p-values obtained from the Wilcoxon sign-rank test).

between our annotations and the i2b2 dataset. The reason for the low kappa score is imbalanced distribution of the categories, leading to a high expected agreement of random assignments.

To more thoroughly compare the temporal relations between the i2b2 dataset and ours, we conducted human assessment. For each type of mismatched relation, we randomly selected three examples, resulting in a total of 18 temporal relations. Then, a human with medical experience, who was not involved in the annotation process, evaluated the selected examples by examining the input note. This individual was asked to judge which annotation represented temporal relations more accurately. The results of the assessment are shown in Table 2. Out of the 18 examples, 9 were found to be better annotated in our dataset, while 8 were more accurately represented in the i2b2 dataset. This result suggests that the temporal relations solely from textual information are not complete, emphasizing the importance of incorporating structured data, but also suggests potential improvement in our multimodal based annotation process.

3.3 Multimodal Learning

Table 3 shows the results of the classification version of absolute timeline prediction. The F1 score of the lower bound prediction (Mean-LB) and the upper bound prediction (Mean-UB) of Multimodal BERT improved 10% (0.604 vs. 0.551) and 8% (0.680 vs. 0.632) from Unimodal BERT, respectively. This improvements,

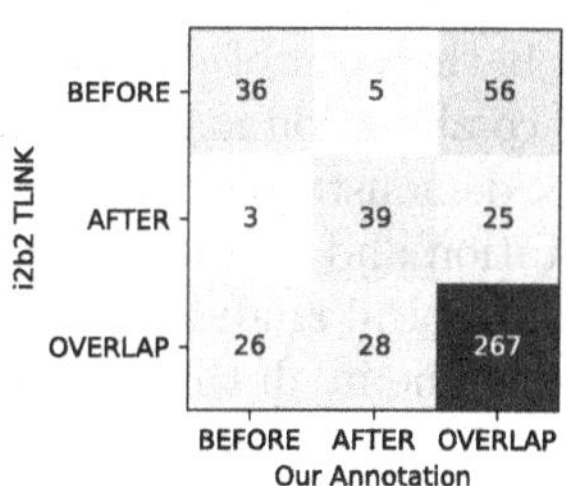

Fig. 4. Confusion matrix between the temporal relations of the 2012 i2b2 dataset and our dataset

Table 2. Human assessment on the 18 sampled mismatched temporal relations

Temporal relation (i2b2 / Ours)	Judgment		
	i2b2	Ours	Both
BEFORE / AFTER	2	1	0
BEFORE / OVERLAP	2	1	0
AFTER / BEFORE	1	1	1
AFTER / OVERLAP	0	3	0
OVERLAP / BEFORE	2	1	0
OVERLAP / AFTER	1	2	0
Total	**8**	**9**	**1**

Table 3. Results of the classification version of absolute timeline prediction on our dataset. The best results among the non-oracle methods are highlighted in bold.

Classification Version of Absolute Timeline Prediction

	Type			Mean		Std Dev		
Method	LB inf$^{\dagger 1}$	UB inf$^{\dagger 1}$	Anno$^{\dagger 1}$	LB^2	UB^2	LB^2	UB^2	Dur^2
Tabular (Oracle)	-	-	-	0.594	0.789	-	-	-
Majority	0.859	1.000	0.840	0.224	0.238	0.220	0.207	0.267
Llama-2	-	-	-	0.374	0.441	-	-	-
Unimodal BERT	0.935	1.000	0.908	0.551	0.632	**0.446**	**0.447**	0.573
Multimodal BERT	**0.936**	1.000	**0.912**	**0.604**	**0.680**	0.433	0.436	**0.579**

†LB/UB inf: definiteness of LB/UB, Anno: Bounds- or Probability-mode
^{1}Accuracy, ^{2}Macro-averaged F1 score

stemming from the integration of multi-head attention into the final mean prediction logits, demonstrate the benefits of integrating unstructured text with structured patient data. Comparing with Llama-2, our Multimodal BERT shows 61% (0.604 vs. 0.374) and 54% (0.680 vs. 0.441) higher F1 score in bound predictions. While the structured data baseline (Tabular (Oracle)) showed the best results in the upper bound mean prediction, this was under the idealized assumption of perfect attention on the structured data. The inferior lower bound mean prediction of the structured data baseline, compared to its upper bound, stems from the mismatch in labels between events anchored to the admission and the admission time itself.

4 Discussion and Conclusions

Clinical events from unstructured data provide information about the patient's progression, but placing them on an absolute timeline can be challenging. Yet while the timing of events in structured data is more certain, the structured data may miss events or other predictive insights. When both types of data are in alignment following the patient's clinical course, their combination may generate more complete and precise event timelines. Our work demonstrates the benefit of the multimodal approach, both in quality of annotation and prediction.

When comparing Probability-mode annotations, statistical analysis revealed superior precision for the multimodal version of the timeline in all three temporal entities (lower bound, upper bound, and duration). The precision factor for the bounds was similar at 1.40 and 1.34 (lower and upper, respectively), whereas for the duration it was at 1.13. This follows the intuition that the duration of many clinical events can be estimated based on clinical knowledge. In particular, the specific position of events on a timeline is more uncertain and depends on many factors that cannot be predicted with clinical knowledge alone, and having relevant structured data with precise timestamps greatly reduced timing uncertainty for the lower and upper bounds of events.

A subgroup analysis compared the precision factor of the timelines within different sections of the discharge summary. In the case of the bounds, multimodal annotation increased precision in all sections, but most prominently in the Hospital Stay and Medical History. The improvement in the Hospital Stay section is expected since the vast majority of the structured data parallels the patient's clinical course from admission to discharge. The Medical History section usually describes events that occurred prior to admission. Likewise, some of the structured data is generated during the patient's time in the emergency department, prior to admission. The Exam and Findings section also showed a moderate improvement in precision with the multimodal version. It contains laboratory tests and imaging studies, which could frequently be referenced to structured data with precise timestamps. Thus, the uncertainty of event timing is significantly reduced when the structured data is aligned with the unstructured data,

In the case of event duration, the multimodal version yielded statistically significant improvement in precision only in the Discharge and Hospital Stay section. Events in the Discharge section are very likely to have their upper bound anchored to the time of discharge, a event that was also available in the unimodal annotation. Thus, little improvement is seen in the precision of the upper bound when the rest of the structured data is made available. This aligns with the result that the upper bound experiences less improvement than the lower bound in the Hospital Stay since events here also extend until discharge.

This study has several limitations. First, one annotator was used, which precludes measurement of inter-annotator agreement. Additionally, the uncertainties defined for the bounds and duration could violate the positive semi-definite condition of a multivariate normal distribution, which may be oversimplified for the annotator's belief about the interval. Since a small sample size was used, the selected discharge summaries may not be representative, *e.g.*, due to differences in chief complaints, institutional policies, or note templates. Finally, there was no functionality for adding additional meta-data to events or recording temporal relations without established timelines.

The reported findings support the use of multimodal data to generate more precise event timelines when compared to unstructured data alone. The benefit is especially prominent when a large quantity of structured data aligns with the unstructured data. Further areas of study could include working with a larger sample size and analyzing differences when subjects are measured more frequently, *e.g.*, in critical care units versus the hospital floor.

The compatibility analysis with the i2b2 dataset validated that our dataset provides a complement to the existing text-based annotations. In the multimodal learning experiments, a BERT model leveraging structured events through multi-head attention improved F1 scores for predicting lower and upper bounds over the unimodal BERT. This demonstrates how temporal localization of clinical events benefits from jointly modeling text and structured data sources. Overall, these strongly support the enhanced utility of our multimodal annotation approach for generating more precise absolute timelines of inpatient events.

Acknowledgments. This research was supported in part by the Intramural Research Program of the National Library of Medicine (NLM), National Institutes of Health.

References

1. Ang, G., Lim, E.P.: Guided attention multimodal multitask financial forecasting with inter-company relationships and global and local news. In: Proceedings of the 60th Annual Meeting of the Association for Computational Linguistics (May 2022)
2. Cheng, T.T., Cua, J.L., Tan, M.D., Yao, K.G., Roxas, R.E.: Information extraction from legal documents. In: 2009 Eighth International Symposium on Natural Language Processing, pp. 157–162. IEEE (2009)
3. Leeuwenberg, A., Moens, M.F.: Towards extracting absolute event timelines from english clinical reports. IEEE/ACM Trans. Audio, Speech, and Language Process. **28**, 2710–2719 (2020)
4. Liu, S., Wang, X., et al.: Multimodal data matters: language model pre-training over structured and unstructured electronic health records. IEEE J. Biomed. Health Inform. **27**(1), 504–514 (2022)
5. Moldwin, A., Demner-Fushman, D., Goodwin, T.R.: Empirical findings on the role of structured data, unstructured data, and their combination for automatic clinical phenotyping. AMIA Summits on Translational Science Proceedings (2021)
6. Peng, Y., Chen, Q., Lu, Z.: An empirical study of multi-task learning on BERT for biomedical text mining. In: Proceedings of the 19th SIGBioMed Workshop on Biomedical Language Processing, pp. 205–214 (2020)
7. Seinen, T.M., Fridgeirsson, E.A., et al.: Use of unstructured text in prognostic clinical prediction models: a systematic review. J. Am. Med. Inform. Assoc. **29**(7), 1292–1302 (2022)
8. Sun, W., Rumshisky, A., Uzuner, O.: Annotating temporal information in clinical narratives. J. Biomed. Inform. **46**, S5–S12 (2013)
9. Sun, W., Rumshisky, A., Uzuner, O.: Evaluating temporal relations in clinical text: 2012 i2b2 challenge. J. Am. Med. Inform. Assoc. **20**(5), 806–813 (2013)
10. Tayefi, M., et al.: Challenges and opportunities beyond structured data in analysis of electronic health records. Computational Statistics (2021)

11. Touvron, H., Martin, L., et al.: Llama 2: Open foundation and fine-tuned chat models. arXiv preprint (2023)
12. Uzuner, Ö., South, B.R., Shen, S., DuVall, S.L.: 2010 i2b2/va challenge on concepts, assertions, and relations in clinical text. J. Am. Med. Inform. Assoc. **18**(5), 552–556 (2011)
13. Zhong, Z., Chen, D.: A frustratingly easy approach for entity and relation extraction. In: Proceedings of the 2021 Conference of the North American Chapter of the Association for Computational Linguistics, pp. 50–61 (2021)

Adversarial-Robust Transfer Learning for Medical Imaging via Domain Assimilation

Xiaohui Chen and Tie Luo(✉)

Computer Science Department, Missouri University of Science and Technology, Rolla, MO 65409, USA
{xcqmk,tluo}@mst.edu

Abstract. Extensive research in Medical Imaging aims to uncover critical diagnostic features in patients, with AI-driven medical diagnosis relying on sophisticated machine learning and deep learning models to analyze, detect, and identify diseases from medical images. Despite the remarkable accuracy of these models under normal conditions, they grapple with trustworthiness issues, where their output could be manipulated by adversaries who introduce strategic perturbations to the input images. Furthermore, the scarcity of publicly available medical images, constituting a bottleneck for reliable training, has led contemporary algorithms to depend on pretrained models grounded on a large set of natural images—a practice referred to as transfer learning. However, a significant *domain discrepancy* exists between natural and medical images, which causes AI models resulting from transfer learning to exhibit heightened *vulnerability* to adversarial attacks. This paper proposes a *domain assimilation* approach that introduces texture and color adaptation into transfer learning, followed by a texture preservation component to suppress undesired distortion. We systematically analyze the performance of transfer learning in the face of various adversarial attacks under different data modalities, with the overarching goal of fortifying the model's robustness and security in medical imaging tasks. The results demonstrate high effectiveness in reducing attack efficacy, contributing toward more trustworthy transfer learning in biomedical applications.

Keywords: Medical images · natural images · transfer learning · colorization · texture adaptation · adversarial attacks · robustness · trustworthy AI

1 Introduction

Since its inception, Artificial Intelligence (AI) has evolved into a powerful tool across various domains. Particularly in the realm of medical diagnosis and treatment, AI has demonstrated impressive performance in predicting a range of diseases such as cancer, often treated as a classification problem. Beyond diagnosis,

D.-N. Yang et al. (Eds.): PAKDD 2024, LNAI 14648, pp. 335–349, 2024.
https://doi.org/10.1007/978-981-97-2238-9_26

applying AI to medical problems for object detection and segmentation has also become focal points for researchers. The year 2018 marked a milestone toward real-world integration of AI in clinical diagnosis, where IDX-DR became the first FDA-approved AI algorithm designed for automatic screening of Diabetic Retinopathy. On the research arena, numerous algorithms have been proposed that showcase remarkable performance in disease detection and diagnosis across diverse patient data modalities, such as MRI, CT, and Ultrasound. However, the lack of publicly available medical data, often attributed to privacy and the high cost of human expert annotation, remains a critical challenge for advancing medical AI research. In the meantime, model performance heavily depends on the size, quality, and diversity of training data to extract meaningful patterns. To address this challenge, researchers have turned to *transfer learning*, which takes models pretrained on large, extensive datasets of natural images and then fine-tunes the models' weights on the (smaller) medical datasets.

However, natural and medical images have inherent differences from each other, forming a gap that is largely overlooked when applying transfer learning. This gap, which we refer to as "domain discrepancy", encompasses two particular aspects that eventually lead to heightened *vulnerability* of medical AI models to adversarial attacks. First, the *monotonic biological textures* in medical images tend to mislead deep neural networks to paying extra attention to (larger) areas *irrelevant* to diagnosis. Second, medical images typically have simpler features than natural images, yet applying overparameterized deep networks to learn such simple patterns can result in a sharp loss landscape, as empirically observed in [19]. Both large attention regions and sharp losses render medical models trained via transfer learning susceptible to *adversarial examples*, the most common attack by adding small, imperceptible perturbations to original input images to induce significant changes in model output, resulting in mispredictions.

In this study, we introduce a novel approach called *Domain Assimilation* to bridge the gap between medical images and natural images. The aim is to align the characteristics of medical images more closely with those of natural images, thereby enhancing the robustness of models against adversarial attacks without compromising the accuracy of unaltered transfer learning. Our contributions are as follows:

- We propose a novel domain assimilation approach embodied by a texture-color-adaption module integrated into transfer learning. This module transforms medical images to resemble natural images more closely, thereby reducing domain discrepancy and enhancing model robustness against attacks.
- To prevent over-adaptation, which can lead to the loss of essential information in medical images and result in misdiagnoses, we introduce a novel Gray-Level Co-occurrence Matrix (GLCM) loss into the training process. This new loss incorporates texture preservation into the optimization process, which is instrumental to ensuring integrity of medical data and reliable diagnoses.
- We conduct extensive experiments across multiple modalities, including MRI, CT, X-ray, and Ultrasound, and evaluate the performance under various

adversarial attacks, such as FGSM, BIM, PGD, and MIFGSM. Our results demonstrate that the proposed texture-color adaption with GLCM loss effectively enhances the robustness of transfer learning while maintaining competitive model accuracy.

2 Related Work

The inherent differences between medical and natural images pose challenges and some efforts were invested to narrow their disparities [16,22]. A typical technique is grayscale image colorization, which is a prominent area in computer vision aiming to transform single-hue images into vibrant, colorful representations to enhance details and information for various applications. This field has garnered significant interest in the research community, finding applications in historical image restoration, architectural visualization, and the conversion of black-and-white movies into colorful ones, among others. Numerous existing works address colorization, either leveraging human-provided input such as scribbles or text, or utilizing color information from reference images [7,13]. Existing methods aim to predict the values of channels within chosen color spaces (e.g., RGB, YUV) by approaching colorization as a regression problem. In these colorization tasks, evaluation typically involves two steps: (1) convert color images into grayscale and generate colored output using designed algorithms; (2) use the original color images as ground truth for comparison with the generated output. However, tasks that need fine-grained colorization entail extensive data training, and one-on-one comparison to ground truth result in high computational overhead.

Furthermore, colorization of medical images poses additional challenges compared to natural images due to the absence of color space information in the original medical dataset. Prior research has underscored the importance of learning from natural images to address the domain gap with medical images. In a pioneering work [16], a bridging technique was employed, enabling the trained projection function from source natural images to transfer to bridge images derived from the same medical imaging modality. Subsequently, this transferred projection function was utilized to map target medical images to their respective feature space. In a subsequent study [22], a three-stage colorization-enhanced transfer learning pipeline was proposed. This involved allowing the colorization module to learn from a frozen pretrained backbone, followed by comprehensive training of the entire network on the provided dataset. The final step included training the network on its ultimate classification layer using a distinct dataset, aiming to enhance transferability. Recent work by Wang et al. [28] introduced a hybrid method incorporating both exemplar and automatic colorization for lung CT images. This approach utilized referenced natural images of meat, drawing inspiration from their similar color appearance to human lungs. Most recently, a self-supervised GAN-based colorization framework was proposed by [6]. This framework aims to colorize medical images in a semantic-aware manner and addresses the challenge of lacking paired data, a common requirement in supervised learning.

While these existing approaches offer various advantages, many necessitate human intervention and rely on large sets of source or referenced images for learning color information to transfer to target medical images, resulting in a lengthy process. Additionally, texture information has been overlooked but is a crucial aspect in medical imaging tasks.

3 Preliminaries

Medical Images vs. Natural Images. Medical images typically manifest as grayscale, single-channel images, showcasing distinctive features from those found in natural images. The pixel-level values and spatial relationships within medical images convey inherent anatomical diversities among individual patients, often unveiling critical diagnostic features, for example, lesions. In contrast, natural images predominantly adopt the RGB color model, influenced by diverse illumination effects. Unlike the relatively standardized nature of medical images, natural images exhibit a broader spectrum of randomness and variability, encompassing a wide array of colors, objects, scenes, and textures. Figure 1 shows histogram comparisons between different imaging modalities [2,3,5,27] and natural images [23]. A histogram is widely used in image analysis to understand pixel intensity value distribution. Here, we are plotting both frequency and probability density curve for a better demonstration. As we can see from Fig. 1a–d, within the same imaging modality, e.g., Brain MRI, the value distribution pattern is quite similar, however, the difference between medical images(a-d) and natural images(e-h) is rather apparent. We also compared natural images from different synsets in the well-known ImageNet2012 [23] dataset. As shown in the histograms, natural images intuitively contain more variations across different synsets and even within the same synset.

While first-order histogram-based statistics, such as mean, variance, skewness, and kurtosis, offer critical information on gray-level distribution, they lack the ability to depict spatial relationships at different intensity levels [1]. To further enhance the learning process, the importance of texture information within a medical image becomes apparent. Texture analysis involves scrutinizing the visual attributes, configuration, and distribution of elements constituting an object within an image. It delves into the spatial organization and recurrent patterning of pixel intensities, dispersed across the entirety of the image or specific regions. This interplay of intensities forms the fundamental essence defining the overall visual construct of the image [4,26].

The Co-occurrence Matrix, also known as the Gray Level Co-occurrence Matrix (GLCM) [10], has been widely employed for texture analysis, especially on medical images [17,21,25]. The dimension of GLCM is defined by the number of intensity levels in an image; for example, a 4-value grayscale image would result in a $4 \times 4 = 16$ matrix. It serves as a texture descriptor, extracting second-order statistics and understanding the distribution of pairwise pixel graylevel values at a given distance and orientation, resulting in a specific offset. In the equation (1), given an image I, (i, j) represents the grayscale values, (x, y) represents the

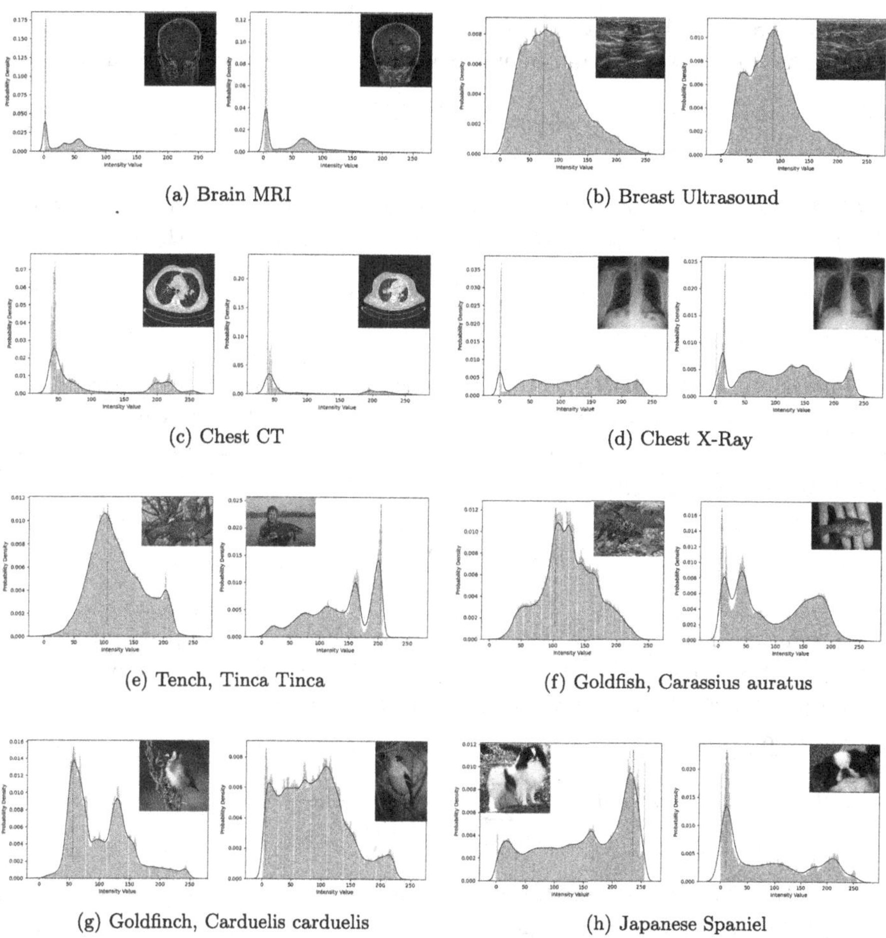

(a) Brain MRI

(b) Breast Ultrasound

(c) Chest CT

(d) Chest X-Ray

(e) Tench, Tinca Tinca

(f) Goldfish, Carassius auratus

(g) Goldfinch, Carduelis carduelis

(h) Japanese Spaniel

Fig. 1. Histograms showing pixel intensity value distribution of medical images and natural images. Conversion from RGB to grayscale was done for natural images for comparison purposes.

spatial locations of pixels, and $(\nabla x, \nabla y)$ is the defined offset. In GLCM, common orientations include 0°(horizontal), 45°(front diagonal), 90°(vertical), 135°(back diagonal), and so on.

$$C_{\nabla x, \nabla y}(i,j) = \sum_{x=1}^{n}\sum_{y=1}^{m} \begin{cases} 1, & \text{if } I(x,y) = i \text{ and } I(x+\nabla x, y+\nabla y) = j \\ 0, & \text{otherwise} \end{cases} \quad (1)$$

FROM GLCM we can further extract second-order statistics as follows.

Angular Second Moment (ASM) calculates the sum of the square of each value in the GLCM matrix and depicts the level of smoothness in an image.

$$ASM = \sum_{i,j} P(i,j)^2 \tag{2}$$

Contrast quantifies how distinct the pixel intensity pairs are in terms of their differences. It reflects the amount of local intensity variation in the image.

$$Contrast = \sum_{i,j} (i-j)^2 P(i,j) \tag{3}$$

Homogeneity reflects the degree to which the pixel intensities in the image tend to be close to each other. A higher homogeneity value indicates that the pixel pairs in the image have similar intensity values, resulting in a more homogeneous and uniform appearance.

$$Homogeneity = \sum_{i,j} \frac{P(i,j)}{1+|i-j|^2} \tag{4}$$

Correlation quantifies how correlated or linearly related the pixel intensities are in terms of their spatial arrangement. It indicates the degree to which the intensities at one pixel location can be predicted based on the intensities at another location.

$$Correlation = \sum_{i,j} \frac{(i-\mu_i)(j-\mu_j)P(i,j)}{\sqrt{\sigma_i^2 \sigma_j^2}} \tag{5}$$

Dissimilarity reflects how dissimilar or different the pixel intensities are in terms of their spatial arrangement.

$$Dissimilarity = \sum_{i,j} P(i,j)|i-j| \tag{6}$$

Adversarial Attacks. In the domain of neural networks, particularly for a classification problem, we are presented with a set of input samples $X \in \mathbb{R}^d$, each associated with ground truth labels y drawn from the distribution D. Guided by the choice of the loss function $L(\theta, x, y)$, where θ denotes the model parameters subject to training, our objective is to identify the parameters that minimize the risk $\mathbb{E}_{(x,y)\sim D}[L(\theta, x, y)]$. The challenge of adversarial robustness is cast as a saddle-point problem, as articulated in the work of Madry et al. [20] (see Equation (7)), wherein the crux lies in solving both the inner maximization and outer minimization problem. In this study, we employ four gradient-based attacks as described in the following to assess the performance of our models.

$$\min_\theta \rho(\theta), \text{ where } \rho(\theta) = \mathbb{E}_{(x,y)\sim D}[\max_{\delta \in S} L(\theta, x+\delta, y)] \tag{7}$$

Fast Gradient Sign Method (FGSM) [9] introduced a single-step adversarial perturbation towards the input data with some magnitude along the input gradient direction. See (8).

$$X_{adv} = X + \epsilon * sign(\nabla_x L(\theta, X, y)) \tag{8}$$

Basic Iterative Method (BIM) [18] proposed an iterative version of FGSM, also known as IFGSM, which perturbs the input data iteratively with smaller step size. See (9)–(10).

$$X^0_{adv} = X, \tag{9}$$

$$X^{t+1}_{adv} = Clip_{x,\epsilon}\left\{X^t_{adv} + \alpha * sign(\nabla_x L(\theta, X^t_{adv}, y))\right\} \tag{10}$$

Momentum Iterative Fast Gradient Sign Method (MIFGSM) [8] proposed an extension of IFGSM that adds a momentum term to the optimization process to help attacks escape from local maxima. See (11)–(13).

$$g_0 = 0, X^0_{adv} = X,\ \alpha = \epsilon/T,\ T \text{ is the number of iterations} \tag{11}$$

$$g_{t+1} = \mu \times g_t + \frac{\nabla_x L(\theta, X^t_{adv}, y)}{\|\nabla_x L(\theta, X^t_{adv}, y)\|_1},\ \mu \text{ is the decay factor} \tag{12}$$

$$X^{t+1}_{adv} = X^t_{adv} + \alpha \times sign(g_{t+1}) \tag{13}$$

Projected Gradient Descent (PGD) [20] proposed an extension to fast gradient sign methods which projects adversarial examples back to the ϵ-ball of x and is considered one of the strongest first-order attack. See (14)–(15).

$$X^0_{adv} = X + \big(U^d(-\epsilon, \epsilon) \text{ if random start}\big), \tag{14}$$

$$X^t_{adv} = \Pi_\epsilon\Big(X^{t-1}_{adv} + \alpha \times sign(\nabla_x L(\theta, X^{t-1}_{adv}, y))\Big) \tag{15}$$

4 Method

Medical images are monochromatic, single-channel grayscale representations. In the process of transforming them into RGB images, one can either generate values for distinct color channels or directly produce three-channel images by leveraging deep neural networks. Analogously to the colorization task, Convolutional Neural Networks (CNNs) can be employed to extract low-level information, such as textures, from images through operations like convolution and pooling.

In the realm of transfer learning, we leverage a pretrained backbone obtained from an extensive dataset, utilizing it as a feature extractor to address specific tasks. The objective of this research is to facilitate the adaptive and voluntary learning of texture and color information by two distinct modules during the training process. This is achieved by employing a pretrained network and optimizing the three-channel output to enhance the classification performance.

Inspired by previous work [22], we introduce lightweight texture and colorization modules preceding the pretrained backbone. The modules generate a three-channel image, subsequently fed into the pretrained models to produce a final classification outcome. See Fig. 2 for a demonstration of the input-output-flow of the two modules.

Problem Formulation. In the context of a generic classification problem, where the input $X \in R^d$, corresponding target labels y, and parameters θ are considered, our objective is to address the following challenge where our primary objective is to optimize the model parameters in order to minimize the loss between predicted and target labels.

$$\theta = \underset{\theta}{\operatorname{argmin}} \, L(H(X, \theta), y) \tag{16}$$

To effectively capture both texture and color information, our approach involves the integration of two distinct modules, dividing the overall network into four components: a texture module T, a color module C, a pretrained backbone B, and a final classifier F. Before the colorization process, it is imperative to preserve crucial texture features inherent in the input image. This is achieved by initially generating a single-channel image through the texture module, which is subsequently input into the color module. The color module's role is to learn three-channel information and output a colored image. Finally, the output is fed into the pretrained model for classification. This formulation of the problem leads to the following.

$$< \theta_T, \theta_C, \theta_B, \theta_F > = \underset{\theta_T, \theta_C, \theta_B, \theta_F}{\operatorname{argmin}} \; L\left(F\left(B\Big(C\big(T(X, \theta_T), \theta_C\big), \theta_B\Big), \theta_F\right), y\right) \tag{17}$$

Observing the proven efficacy of texture as significant features in medical image tasks such as classification and segmentation [4,15,24], it is noted that colorization, while enhancing the 'natural' appearance of medical images, may potentially distort the original texture information due to its fine-grained labeling at the pixel level. Despite its ability to improve alignment with pretrained models, we seek to mitigate the impact of colorization during the training process. To achieve this, we introduce a normalized Gray-Level Co-occurrence Matrix (GLCM) loss, in addition to the cross-entropy loss, thereby producing optimized results. This extends the problem into the following.

$$\begin{aligned} < \theta_T, \theta_C, \theta_B, \theta_F > = \underset{\theta_T, \theta_C, \theta_B, \theta_F}{\operatorname{argmin}} & \Bigg(\alpha \times \text{CrossEntropyLoss}\Big(F(B(C(T(X, \theta_T), \theta_C), \theta_B), \theta_F), y\Big) + \\ & (1 - \alpha) \times \text{GLCMLoss}\Big(C(X, \theta_C), X\Big)\Bigg) \end{aligned} \tag{18}$$

where α is a predefined weight. For GLCM loss, we first compute a GLCM matrix for the input before and after the colorization procedure, respectively. Then, the GLCM loss is expressed as the feature distance between the two matrices. Our objective is to minimize this distance since it represents the distortion introduced by colorization as compared to the original texture. This distance is formulated as a *L-infinite norm* which is equivalent to the equation below:

$$\mathrm{GLCMLoss} = \max_{1 \leqslant i \leqslant m} \sum_{j=1}^{n} \left| SOT\Big(grayscale\big(C(X)\big)\Big)_{i,j} - SOT(X)_{i,j} \right| \tag{19}$$

where SOT denotes *second-order texture* features of GLCM, m is batch size and n is the total number of SOT features. We consider the five SOT features described by (2)–(6) with a distance of 3 for 8 orientations (0/45/90/135/180/225/270 degrees). Hence, $n = 5 \times 8 = 40$. All the 40 features are normalized in the same range when constructing the $m \times n$ SOT matrix.

Our texture module follows a simplified autoencoder architecture, wherein the encoder comprises convolutional, batch normalization, ReLU, and max-pooling layers, ultimately generating a single-channel output. The decoder, on the other hand, utilizes transpose convolution operations to reconstruct the original input. The color module mirrors the structure of the texture module but adopts a shallower configuration. It outputs a three-channel image, subsequently fed into the pretrained backbone. For a visual representation of the architecture, refer to Fig. 3.

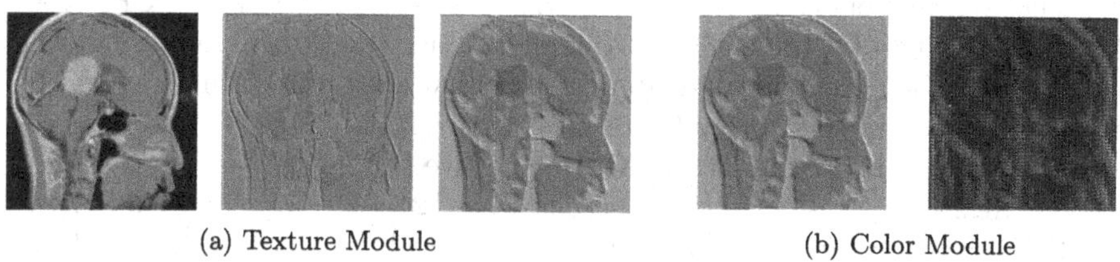

Fig. 2. Example of the input and output of texture and color modules. (a) from left to right: original brain MRI image, encoder output, decoder output. (b) from left to right: input from texture module, three-channel image output.

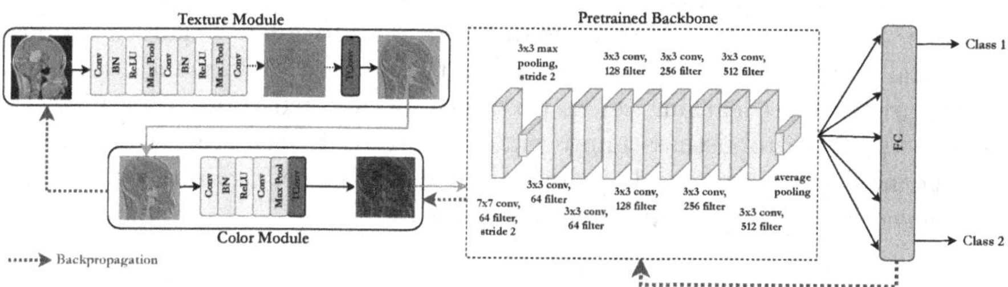

Fig. 3. Architecture of our proposed texture-color adaption alongside the backbone (ResNet18 as an example) and final classifier.

5 Experiments

5.1 Dataset

Our experiments encompassed four datasets representing distinct imaging modalities - Brain MRI [5], Chest CT [3], Chest XRay [14] and Breast Ultrasound [2]. Given the substantial class size imbalance inherent in medical images, we employed downsampling with a random sampling strategy to balance the datasets. Additionally, to address data scarcity concerns for specific classes and formulate a binary classification problem, we grouped various classes together. For example, within the Brain MRI dataset, we consolidate glioma, meningioma, and pituitary into a class designated as *tumor*. See Table 1 for an overview of the dataset.

5.2 Setup

Our dataset uniformity is maintained through the resizing and center-cropping of images, resulting in dimensions of (224, 224, 1), effectively reducing training costs. All experiments adopt a standardized hyperparameter set, including 300 epochs, a batch size of 32, a learning rate of 0.0001, and early stopping with a patience parameter set to 30. The experiments are conducted on GCP V100-SXM2-16GB with 4 GPUs to achieve a significant acceleration in processing. The experiment was conducted by performing the follow steps 1 and 2 independently, and then subjecting their output models to attacks as in step 3 for robustness comparison: **1) Fine-tune Base Models:** Three commonly-used pretrained models—ResNet18, ResNet50 [11], and DenseNet121 [12]—are adopted. Following [19], we replace the last layer of these three models by a new sequential layer consisting of a dense layer, a dropout layer, and a final dense layer. We then fine-tune these 3 models on our medical datasets. **2) Train Base Models with our Texture and Color Modules:** The network architecture now consists of four components—Texture Module, Color Module, Pretrained Backbone, and Final Classifier—which are trained in an end-to-end fashion. GLCM and cross-entropy losses (with a predefined $\alpha = 0.98$) are computed for each batch, and the overall loss is propagated backward through the entire network for updating parameters. **3) Adversarial Attacks on Trained Models:** We launch various adversarial attacks on all the above models to compare performance degradation of basic transfer learning versus texture-color-adapted transfer learning, to assess the possible robustness improvement. Note that both steps 1 and 2 use the same medical datasets, and all the models are fine-tuned/trained until nearly convergence, in order to ensure a fair comparison.

5.3 Evaluation

We assess the performance of various models based on their testing accuracy. The evaluation is conducted incrementally on a fine-tuned base model, the model with texture and color adaptation, and the model with texture and color adaptation combined with GLCM loss. To examine how different models respond to gradient-based adversarial attacks-specifically, FGSM, BIM, MIFGSM, and PGD-in terms of performance degradation, we subject our trained models to different attack perturbation sizes denoted as ϵ (1/255, 2/255, 3/255, 4/255, 5/255, 6/255, 7/255, 8/255). The goal is to evaluate model performance under attacks with and without our proposed approach.

Table 1. Dataset Overview

Dataset Name	Classes	Class Size
Brain MRI	no-tumor, tumor (glioma, meningioma, pituitary)	1595, 1595
Chest XRay	normal, pneumonia	1583, 1583
Chest CT	no-cancer (normal and benign), cancer	536, 536
Breast Ultrasound	no-cancer (normal and benign), cancer	210, 210

5.4 Results

As depicted in Table 2, our comprehensive experiments span four distinct medical imaging modalities-MRI, Ultrasound, CT, and X-Ray-employing three selected pretrained models: ResNet18, ResNet50, and DenseNet121. We compare the performance across different approaches. Notably, for all imaging modalities except Breast Ultrasound images, all three adapted approaches demonstrated comparable results compared to the fine-tuned base models. The incorporation of GLCM loss notably contribute to performance improvements or maintenance across most models. However, both the table and Fig. 4 highlight that the adaptation of color and texture introduces distortion, and in some cases, undesirable noise to the original image. The limited size of the ultrasound dataset might as well contribute to the challenges in learning and the model's struggle to converge effectively. This effect was more pronounced in modalities such as Ultrasound, which contains more distinguishable and complex textures compared to other imaging modalities, resulting in a performance degradation. Our proposed GLCM loss has effectively mitigated this undesirable distortion, as evidenced by the results presented in Table 2. Its efficacy is particularly evident on ultrasound compared to other imaging modalities, underscoring the criticality of preserving texture information.

Further in evaluating model robustness against gradient-based adversarial attacks-FGSM, BIM, MIFGSM, and PGD-we observe that our texture-color-adapted models with GLCM loss enhanced robustness by increasing the difficulty of generating successful attacks. Figure 6a illustrates that the base model achieve <20% accuracy under any attack (except for BIM) even with a very small perturbation magnitude (e.g., $\epsilon = 1/255$); BIM reduces model accuracy to around 30% at $\epsilon = 1/255$ but soon collapses the model with an accuracy of zero at $\epsilon = 2/255$. In contrast, armed with our domain assimilation strategy, the model's robustness increases notably: under all the attacks with $\epsilon = 1/255$, the model maintains a reasonably good accuracy of about 90%, 80%, 80%, and 60% under the attacks BIM, PGD, FGSM, and MIFGSM, respectively. At a stronger attack strength of $\epsilon = 3/255$, where the base model collapses to zero accuracy for almost all the attacks, our strategy helps maintain an accuracy of about 40% under BIM and FGSM attacks. On the other hand, Fig. 6b shows that our approach was not effective for ultrasound. This can be attributed to the characteristic captured by Table 2, which reveals that ultrasound images contain more complex texture and hence were too sensitive to our adaptation. This calls for more advanced methods to be developed in future research. The vanilla FGSM, represented by the dashed gray line in Fig. 6a and 6b, exhibits outlier behavior due to its characteristic of advancing gradients for a *single* step only. As such, augmenting the epsilon value would not enhance its efficacy but can potentially precipitate its significant deviation towards suboptimal solutions, as opposed to *iterative* gradient ascending as in other attack methods. Hence for FGSM, emphasis should be placed on very low epsilon values such as 1/255.

As part of our analysis, we also explored the transferability of learned parameters across different imaging modalities. Figure 5 illustrates the transferred colorization from a model trained on Breast Ultrasound images to colorize Chest CT and Brain MRI images. The results suggest that our adapted models can be effectively transferred to different modalities. With further fine-tuning, we believe these models can be even better adapted to diverse datasets.

Table 2. Model Accuracy before and after TC adaptation (w/ or w/o correction by GLCM loss). No Attack.

			Brain MRI	Breast Ultrasound	Chest CT	Chest X-Ray
ResNet18	Base w/ Fine Tuning		97.9%	74.6%	97.5%	92.2%
	Adapted	Texture-Color (TC)	96.4%	71.4%	95.6%	90.9%
		TC+GLCMLoss	96.9%	71.4%	95.6%	91.8%
ResNet50	Base w/ Fine Tuning		97.5%	79.4%	97.5%	92.0%
	Adapted	Texture-Color (TC)	97.3%	76.2%	96.3%	91.6%
		TC+GLCMLoss	97.3%	77.8%	96.9%	91.8%
DenseNet121	Base w/ Fine Tuning		98.1%	73.0%	98.8%	92.6%
	Adapted	Texture-Color (TC)	97.5%	66.7%	98.1%	92.2%
		TC+GLCMLoss	97.3%	68.3%	98.1%	92.2%

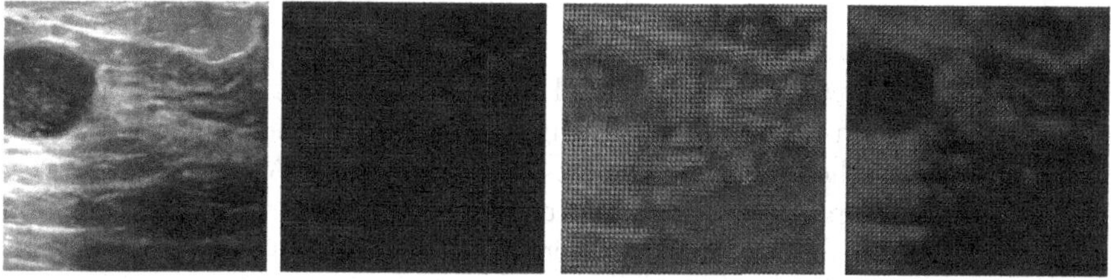

Fig. 4. Demonstration of our incremental workflow. From left to right: original Breast Ultrasound image, adaptation with only colorization, with texture adaption added, with GLCM loss added.

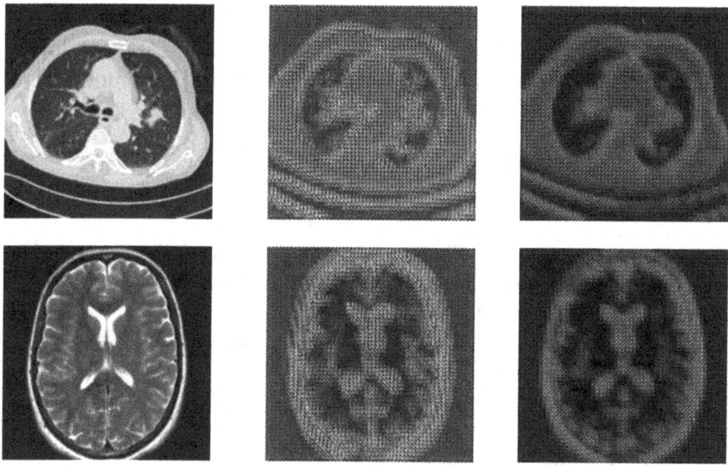

Fig. 5. Demonstration of transferability of our learned parameters across different modalities, using our colorization module as an example. Here colorization was learned from Breast Ultrasound color images and applied to Chest CT (upper row) and Brain MRI (lower row). From left to right: original image, after colorization, and colorization with GLCM loss.

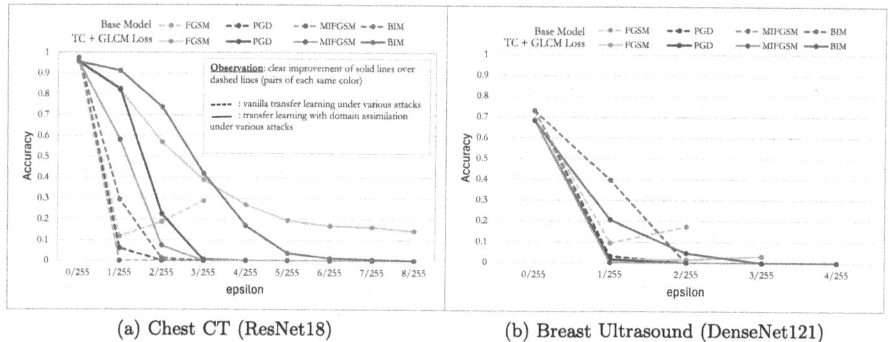

(a) Chest CT (ResNet18) (b) Breast Ultrasound (DenseNet121)

Fig. 6. Robustness comparison under various adversarial attacks.

6 Conclusion

In this study, we introduce texture and color adaption into transfer learning to bridge the domain discrepancy existing in AI-based medical imaging, consequently addressing the heightened vulnerability of medical AI models to adversarial attacks. Our proposed approach consists of a texture-color adaptation module that dynamically learns parameters in conjunction with pre-trained models, and a GLCM loss that retains essential texture information to restrict distortion, fostering a resilient model over various imaging modalities. Our evaluation demonstrate enhanced model robustness against adversarial attacks, specifically gradient-based adversarial examples created by BIM, PGD, FGSM, and MIFGSM. Our analysis also discover the challenges posed by imaging modalities with intricate textures, exemplified by Ultrasound images. Our findings validate the domain assimilation idea and the effectiveness of the tamed adaption approach, yet also pointing out potential future work on improvement for different imaging modalities.

References

1. Aggarwal, N., Agrawal, R.K.: First and second order statistics features for classification of magnetic resonance brain images. J. Signal Inf. Process. **03**(02), 146–153 (2012). https://doi.org/10.4236/jsip.2012.32019
2. Al-Dhabyani, W., Gomaa, M., Khaled, H., Fahmy, A.: Dataset of breast ultrasound images. Data Brief **28**, 104863 (2020). https://doi.org/10.1016/j.dib.2019.104863. https://www.sciencedirect.com/science/article/pii/S2352340919312181
3. Alyasriy, H., AL-Huseiny, M.: The iq-othnccd lung cancer dataset (2021). https://doi.org/10.17632/bhmdr45bh2.2
4. Castellano, G., Bonilha, L., Li, L., Cendes, F.: Texture analysis of medical images. Clin. Radiol. **59**(12), 1061–1069 (2004). https://doi.org/10.1016/j.crad.2004.07.008
5. Chaki, J., Wozniak, M.: Brain tumor MRI dataset (2023). https://doi.org/10.21227/1jny-g144. https://dx.doi.org/10.21227/1jny-g144
6. Chen, S., Xiao, N., Shi, X., Yang, Y., Tan, H., Tian, J., Quan, Y.: Colormedgan: a semantic colorization framework for medical images. Appl. Sci. **13**(5) (2023). https://doi.org/10.3390/app13053168. https://www.mdpi.com/2076-3417/13/5/3168
7. Cheng, Z., Yang, Q., Sheng, B.: Deep colorization. In: 2015 IEEE International Conference on Computer Vision (ICCV), pp. 415–423 (2015). https://doi.org/10.1109/ICCV.2015.55
8. Dong, Y., Liao, F., Pang, T., Su, H., Zhu, J., Hu, X., Li, J.: Boosting adversarial attacks with momentum (2018)
9. Goodfellow, I.J., Shlens, J., Szegedy, C.: Explaining and harnessing adversarial examples (2015)
10. Haralick, R.M., Shanmugam, K., Dinstein, I.: Textural features for image classification. IEEE Trans. Syst. Man Cybern. **SMC-3**(6), 610–621 (1973). https://doi.org/10.1109/TSMC.1973.4309314
11. He, K., Zhang, X., Ren, S., Sun, J.: Deep residual learning for image recognition (2015)

12. Huang, G., Liu, Z., van der Maaten, L., Weinberger, K.Q.: Densely connected convolutional networks (2018)
13. Huang, S., Jin, X., Jiang, Q., Liu, L.: Deep learning for image colorization: current and future prospects. Eng. Appl. Artif. Intell. **114**(C) (2022). https://doi.org/10.1016/j.engappai.2022.105006
14. Kermany, D., Zhang, K., Goldbaum, M.: Labeled optical coherence tomography (OCT) and chest X-Ray images for classification (2018). https://doi.org/10.17632/rscbjbr9sj.2
15. Kiani, F.: Texture features in medical image analysis: a survey (2022)
16. Kim, H.G., Choi, Y., Ro, Y.M.: Modality-bridge transfer learning for medical image classification. In: 2017 10th International Congress on Image and Signal Processing, BioMedical Engineering and Informatics (CISP-BMEI), pp. 1–5 (2017). https://doi.org/10.1109/CISP-BMEI.2017.8302286
17. Kumar, D., et al.: Feature extraction and selection of kidney ultrasound images using GLCM and PCA. Procedia Comput. Sci. **167**, 1722–1731 (2020)
18. Kurakin, A., Goodfellow, I., Bengio, S.: Adversarial examples in the physical world (2017)
19. Ma, X., et al.: Understanding adversarial attacks on deep learning based medical image analysis systems. Pattern Recognit. **110**, 107332 (2021). https://doi.org/10.1016/j.patcog.2020.107332
20. Madry, A., Makelov, A., Schmidt, L., Tsipras, D., Vladu, A.: Towards deep learning models resistant to adversarial attacks. In: International Conference on Learning Representations (2018). https://openreview.net/forum?id=rJzIBfZAb
21. Mall, P.K., Singh, P.K., Yadav, D.: GLCM based feature extraction and medical x-ray image classification using machine learning techniques. In: 2019 IEEE Conference on Information and Communication Technology, pp. 1–6 (2019). https://doi.org/10.1109/CICT48419.2019.9066263
22. Morra, L., Piano, L., Lamberti, F., Tommasi, T.: Bridging the gap between natural and medical images through deep colorization (2020)
23. Russakovsky, O., et al.: ImageNet large scale visual recognition challenge. Int. J. Comput. Vision (IJCV) **115**(3), 211–252 (2015). https://doi.org/10.1007/s11263-015-0816-y
24. Sharma, N., Ray, A., Sharma, S., Shukla, K., Pradhan, S., Aggarwal, L.: Segmentation and classification of medical images using texture-primitive features: application of bam-type artificial neural network. J. Med. Phys. **33**(3), 119 (2008). https://doi.org/10.4103/0971-6203.42763
25. Singh, D., Kaur, K.: Classification of abnormalities in brain MRI images using GLCM, PCA and SVM. Int. J. Eng. Adv. Technol. (IJEAT) **1**(6), 243–248 (2012)
26. Varghese, B.A., Cen, S.Y., Hwang, D.H., Duddalwar, V.A.: Texture analysis of imaging: what radiologists need to know. Am. J. Roentgenol. **212**(3), 520–528 (2019). https://doi.org/10.2214/AJR.18.20624
27. Wang, X., Peng, Y., Lu, L., Lu, Z., Bagheri, M., Summers, R.M.: ChestX-ray8: hospital-scale chest x-ray database and benchmarks on weakly-supervised classification and localization of common thorax diseases. In: 2017 IEEE Conference on Computer Vision and Pattern Recognition (CVPR). IEEE (2017). https://doi.org/10.1109/cvpr.2017.369
28. Wang, Y., Yan, W.Q.: Colorizing grayscale CT images of human lungs using deep learning methods. Multimedia Tools Appl. **81**(26), 37805–37819 (2022). https://doi.org/10.1007/s11042-022-13062-0

Author Index

D.-N. Yang et al. (Eds.): PAKDD 2024, LNAI 14648, pp. 351–352, 2024.
https://doi.org/10.1007/978-981-97-2238-9

Printed in the United States
by Baker & Taylor Publisher Services